ISNM 80:
International Series of Numerical Mathematics
Internationale Schriftenreihe zur Numerischen Mathematik
Série internationale d'Analyse numérique
Vol. 80

Edited by
Ch. Blanc, Lausanne; R. Glowinski, Paris;
H. O. Kreiss, Pasadena; J. Todd, Pasadena

Birkhäuser Verlag
Basel · Boston

General Inequalities 5

5th International Conference on General Inequalities,
Oberwolfach, May 4–10, 1986

Edited by
W. Walter

1987

Birkhäuser Verlag
Basel · Boston

Editor

Prof. Dr. Wolfgang Walter
Universität Karlsruhe
Mathematisches Institut I
Kaiserstraße 12
D–7500 Karlsruhe 1

Library of Congress Cataloging in Publication Data

International Conference on General Inequalities (5th :
1987 : Oberwolfach, Germany)
General inequalities 5.

(International series of numerical mathematics ;
vol. 80 = Internationale Schriftenreihe zur numerischen
Mathematik)
English, French, German.
Bibliography: p.
Includes index.
1. Inequalities (Mathematics)--Congresses.
I. Walter, Wolfgang, 1927- . II. Title.
III. Title: General inequalities five. IV. Series:
International series of numerical mathematics ; v. 80.
QA295.I57 1987 512.9'7 87-18393

CIP-Kurztitelaufnahme der Deutschen Bibliothek

General inequalities : ... Internat. Conference on
General Inequalities. – Basel ; Boston : Birkhäuser
1 mit Parallelt.: Allgemeine Ungleichungen
NE: Internationale Tagung über Allgemeine Ungleichungen; PT
5. Oberwolfach, May 4–10, 1986. – 1987.
(International series of numerical mathematics ;
Vol. 80)
ISBN 3-7643-1799-X (Basel)
ISBN 0-8176-1799-X (Boston)

Printed in Germany
ISBN 3-7643-1799-X
ISBN 0-8176-1799-X

FOREWORD

The Fifth International Conference on General Inequalities was held from May 4 to May 10, 1986, at the Mathematisches Forschungsinstitut Oberwolfach (Black Forest, Germany). The organizing committee consisted of W.N. Everitt (Birmingham), L. Losonczi (Debrecen) and W. Walter (Karlsruhe). Dr. A. Kovačec served efficiently and enthusiastically as secretary to the conference. The meeting was attended by 50 participants from 16 countries.

In his opening address, W. Walter had to report on the death of five colleagues who had been active in the area of inequalities and who had served the mathematical community: P.R. Beesack, G. Pólya, D.K. Ross, R. Bellman, G. Szegö. He made special mention of G. Pólya, who had been the last surviving author of the book *Inequalities* (Cambridge University Press, 1934), who died at the age of 97 years and whose many and manifold contributions to mathematics will be recorded elsewhere, in due course.

Inequalities continue to play an important and significant role in nearly all areas of mathematics. The interests of the participants to this conference reflected the many different fields in which both classical and modern inequalities continue to influence developments in mathematics. In addition to the established fields, the lectures clearly indicated the importance of inequalities in functional analysis, eigenvalue theory, convexity, number theory, approximation theory, probability theory, mathematical programming and economics.

On the occasion of this conference, special attention was payed to the recent solution of the Bieberbach conjecture. In two prepared lectures, given on the invitation of the committee, Dr. N. Steinmetz (Karlsruhe) reviewed the history and the proof of the correctness of the conjecture. His excellent presentation showed the importance of a number of inequalities required for the proof, and how an inequality for a solution of a linear system with constant coefficients could significantly simplify part of the proof as a whole.

The problems and remarks sessions yielded many new ideas and intriguing conjectures.

All the participants came under the influence of the remarkable atmosphere now such an established feature of the Institute.

The conference was closed by W.N. Everitt who, in paying tribute to all those who had contributed to the progress of the meeting, asked that the best thanks of all the participants be presented to the staff of the Institute for their unique contribution in the form of excellent hospitality and quiet and effective service.

W.N. Everitt

L. Losonczi

W. Walter

PARTICIPANTS

J. ACZÉL, University of Waterloo, Waterloo, Ontario, Canada

R.P. AGARWAL, National University of Singapore, Kent Ridge, Singapore

C. ALSINA, Universitat Politècnica de Barcelona, Barcelona, Spain

C. BANDLE, Universität Basel, Basel, Switzerland

A. BEN-ISRAEL, University of Delaware, Newark, Delaware, USA

C. BENNEWITZ, University of Uppsala, Uppsala, Sweden

B. CHOCZEWSKI, University of Mining and Metallurgy, Kraków, Poland

A. CLAUSING, Westfälische Wilhelms-Universität, Münster, West Germany

W. EICHHORN, Universität Karlsruhe, Karlsruhe, West Germany

M. ESSÉN, University of Uppsala, Uppsala, Sweden

W.N. EVERITT, University of Birmingham, Birmingham, England

F. FEHÉR, Rheinisch-Westfälische Technische Hochschule Aachen, Aachen, West Germany

I. FENYÖ, Polytechnical University of Budapest, Budapest, Hungary

R. GER, Silesian University, Katowice, Poland

M. GOLDBERG, Technion-Israel Institute of Technology, Haifa, Israel

W. HAUSSMANN, Gesamthochschule Duisburg, Duisburg, West Germany

P. HEYWOOD, University of Edinburgh, Edinburgh, Scotland, U.K.

H.-H. KAIRIES, Technische Universität Clausthal, Clausthal-Zellerfeld, West Germany

H. KÖNIG, Universität des Saarlandes, Saarbrücken, West Germany

H. KÖNIG, Universität Kiel, Kiel, West Germany

A. KOVAĆEC, An der Niederhaid 21, Wien, Austria

N. KUHN, Universität des Saarlandes, Saarbrücken, West Germany

M.K. KWONG, Northern Illinois University, DeKalb, Illinois, USA

L. LOSONCZI, L. Kossuth University, Debrecen, Hungary

E.R. LOVE, University of Melbourne, Parkville, Victoria 3052, Australia

G. LUMER, Université de l'Etat, Mons, Belgium

A.W. MARSHALL, University of British Columbia, Vancouver, British Columbia, Canada

H.W. McLAUGHLIN, Rensselaer Polytechnical Institute, Troy, N.Y., USA

PARTICIPANTS (Continued)

R.J. NESSEL, Rheinisch-Westfälische Technische Hochschule Aachen, Aachen, West Germany

C.T. NG, University of Waterloo, Waterloo, Ontario, Canada

Z. PÁLES, Kossuth Lajos University, Debrecen, Hungary

J. RÄTZ, Universität Bern, Bern, Switzerland

Q.I. RAHMAN, Université de Montréal, Montréal, Canada

D.C. RUSSELL, York University, Downsview, Ontario, Canada

B. SAFFARI, Université de Paris-Orsay, Orsay, France

F.J. SCHNITZER, Montanuniversität Leoben, Leoben, Austria

P. SCHÖPF, Universität Graz, Graz, Austria

J. SCHRÖDER, Universität Köln, Köln, West Germany

B. SCHWEIZER, University of Massachusetts, Amherst, Massachusetts, USA

B. SMITH, California Institute of Technology, Pasadena, California, USA

R.P. SPERB, Eidgenössische Technische Hochschule Zürich, Zürich, Switzerland

N. STEINMETZ, Universität Karlsruhe, Karlsruhe, West Germany

P.M. VASIĆ, Faculty of Electrotechnics, Belgrade, Yugoslavia

P. VOLKMANN, Universität Karlsruhe, Karlsruhe, West Germany

B.J. WALLACE, La Trobe University, Bundoora, Victoria 3083, Australia

W. WALTER, Universität Karlsruhe, Karlsruhe, West Germany

CH.-L. WANG, University of Regina, Regina, Saskatchewan, Canada

K. ZELLER, Universität Tübingen, Tübingen, West Germany

A. ZETTL, Northern Illinois University, DeKalb, Illinois, USA

SCIENTIFIC PROGRAM OF THE CONFERENCE

Monday, May 5

Opening of the conference — W. WALTER

Early morning session — Chairman: W. WALTER

E.R. LOVE: An inequality for geometric means

C. BENNEWITZ: The HELP inequality in a regular case

Late morning session — Chairman: M. GOLDBERG

W.N. EVERITT: An example of the Hardy-Littlewood type of integral inequalities

I. FENYÖ: Inequalities concerning convolutions of kernels of integral equations

Problems and remarks

Early afternoon session — Chairman: HEINZ KÖNIG

R.J. NESSEL: Approximation theory in the space of Riemann integrable functions

C. ALSINA: On the stability of a functional equation arising in probabilistic normed spaces

R. GER: Subadditive multifunctions and Hyers-Ulam stability

Late afternoon session — Chairman: D.C. RUSSELL

J. RÄTZ: Some remarks around the Cauchy-Schwarz inequality

C.T. NG: A functional inequality

Problems and remarks

* * * * *

Tuesday, May 6

Early morning session — Chairman: M. ESSÉN

N. STEINMETZ: The Bieberbach conjecture I

P. VOLKMANN: Ein Existenzsatz für gewöhnliche Differentialgleichungen in geordneten Banachräumen

Late morning session Chairman: A. ZETTL

A. CLAUSING: Experimenting with operator inequalities

M. ESSÉN: Rearrangements and optimization problems for certain linear second order differential equations

R. SPERB: Optimal inequalities in a semilinear boundary value problem on a two dimensional Riemannian manifold

Problems and remarks

Early afternoon session Chairman: P. HEYWOOD

W. HAUSSMANN: Uniqueness inequality and best harmonic L^1-approximation

A. ZETTL: Norm inequalities for derivatives and differences

M.K. KWONG: Inequalities between norms of functions and their derivatives

Late afternoon session Chairman: B. CHOCZEWSKI

R.J. WALLACE: Sequential search for zeroes of $2(2^n-1)$-st derivative

R.P. AGARWAL: Linear and nonlinear discrete inequalities in n independent variables

Problems and remarks

* * * * *

Wednesday, May 7

Early morning session Chairman: E.R. LOVE

N. STEINMETZ: The Bieberbach conjecture II

H.W. McLAUGHLIN: Inequalities arising from discrete curves

Late morning session Chairman: J. ACZÉL

A.W. MARSHALL: Extensions of Markov's inequality for random variables taking values in a linear topological space

L. LOSONCZI: Nonnegative trigonometric polynomials and related quadratic inequalities

B. CHOCZEWSKI: A linear iterative functional inequality of third order

Problems and remarks

Afternoon excursion and discussion

* * * * *

Thursday, May 8

Early morning session Chairman: R.J. NESSEL

HERMANN KÖNIG: Strict inequalities for projective constants

K. ZELLER: Positivity in absolute summability

A. KOVAČEC: On an extension of the Bruhat order of the symmetric group

H.-H. KAIRIES: Inequalities for the q-factorial functions

Late morning session Chairman: C. ALSINA

J. ACZÉL: Entropies, generalized entropies, inequalities and the maximum entropy principle

Z. PÁLES: How to make fair decisions?

B. SAFFARI: Refinements of norm inequalities on functions of mean value zero

Problems and remarks

Early afternoon session Chairman: K. ZELLER

M. GOLDBERG: Multiplicativity and mixed-multiplicativity for operator-norms and matrix-norms

N. KUHN: Almost t-convex functions

A. BEN-ISRAEL: F-convexity

Late afternoon session Chairman: A.W. MARSHALL

P. SCHÖPF: Zwei Ungleichungen für konvexe bzw. sternförmige Funktionen

P.M. VASIĆ: Sur quelques interpolations

Problems and remarks

* * * * *

Friday, May 9

Early morning session Chairman: P. VOLKMANN

G. LUMER: Parabolic maximum principles, diffusion equations and population dynamics

F. FEHÉR: P-estimates for ultra products of Banach lattices

B. SMITH: Convolution orthogonality inequalities

Late morning session Chairman: I. FENYÖ

W. EICHHORN: Ungleichungen in der Theorie des Messens

CH.-L. WANG: Inequalities and Mathematical Programming III

Problems and remarks

Closing of the conference W.N. EVERITT

PREFEACE

The Fifth Conference on General Inequalities, henceforth referred to as "GI 5", was held at the Mathematisches Forschungsinstitut Oberwolfach/Germany from May 4 to May 10, 1986. The present proceedings contain research articles which were presented at GI 5 and also some contributions by authors who were unable to attend the conference. While most Oberwolfach conferences deal with a special field of research, the GI meetings have always reflected the fact that inequalities occur in almost all fields of mathematics and play a significant role in many of them. The articles, which come from very different areas, have been divided in several chapters. It is already a tradition of the GI conferences that the morning and afternoon sessions terminate with a problems and remarks session. In the last chapter some of these activities are recorded.

In the time between GI 4 and GI 5 a famous inequality problem had been resolved, the Bieberbach conjecture. During the conference, N. Steinmetz gave two one-hour lectures on the conjecture, its history and, of greater importance, its proof. The first article gives an account on this beautiful presentation.

The editor is delighted that this volume is again enhanced with beautiful drawings. As in GI 4, they are the work of Mrs. Joy Russell. The editor expresses his sincere gratitude to Mrs. Russell for this enrichment of the present volume which gives it a very special character.

The editor expresses his sincere thanks to all those who have devoted their efforts and knowledge to make this volume a book worthy of its predecessors. Several colleagues have acted as referees. Most of the proofreading was done by Ms. Sabina Schmidt and Dr. Reinhard Redlinger, and, to be sure, by the authors. The index has been compiled by Ms. Schmidt. The editorial secretary, Ms. Irene Jendrasik, has typed about half of the articles, sustained the correspondence with the authors and performed all technical preparations with unusual care and expertise. The editor is grateful to them all and to Birkhäuser Verlag for a concordant and effective collaboration.

Karlsruhe, March 1987

Wolfgang Walter, Editor
Universität Karlsruhe

CONTENTS

THE BIEBERBACH CONJECTURE

INEQUALITIES FOR SUMS, SERIES AND INTEGRALS

INEQUALITIES IN ANALYSIS AND APPROXIMATION

INEQUALITIES OF FUNCTIONAL ANALYSIS

FUNCTIONAL EQUATIONS AND INEQUALITIES

INEQUALITIES FOR DIFFERENTIAL OPERATORS

INEQUALITIES IN ECONOMICS, OPTIMIZATION AND APPLICATIONS

PROBLEMS AND REMARKS

SKETCHES

by

Joy Russell

The Bieberbach Conjecture

Chess men at the Institute

International Series of
Numerical Mathematics, Vol. 80

DE BRANGES' PROOF OF THE BIEBERBACH CONJECTURE

Norbert Steinmetz

Abstract. This is a report on de Branges' proof of the Bieberbach conjecture based on two one-hour-lectures given at the Oberwolfach Conference on General Inequalities (May 5 - 9, 1986). The main objective was to present to the audience an idea of the function-theoretic background, especially of Löwner's beautiful theory.

1. THE CLASS S

The class S consists of all power series

(1) $$f(z) = z + \sum_{n=2}^{\infty} a_n z^n$$

representing univalent ('schlicht') functions in the unit disk $\mathbb{D} = \{|z| < 1\}$. As an application of the so-called 'Area Theorem' Bieberbach [2] proved in 1916 the inequality

(2) $$|a_2| \le 2 ,$$

where equality holds only for rotations of the Koebe function

(3) $$k(z) = \frac{z}{(1-z)^2} = z + 2z^2 + 3z^3 + \dots ,$$

i.e., for $f(z) = \bar{\omega}k(\omega z)$, $|\omega| = 1$. Since S is a fairly large family, every global result of course has important consequences. We mention some statements equivalent to (2):

(2a) $f(\mathbb{D})$ contains the disk $|w| < \frac{1}{4}$

(Koebe's *One-Quarter-Theorem*),

This paper is in final form and no version of it will be submitted for publication elsewhere.

(2b) $$|f(z)| \le \frac{|z|}{(1-|z|)^2}$$

(Growth Theorem) and

(2c) $$|f'(z)| \le \frac{1+|z|}{(1-|z|)^3}$$

(Koebe's *Verzerrungssatz* in its quantitative version).

The Koebe function is extremal for many problems within the class S, and Bieberbach conjectured in a footnote of [2] that it is also extremal for the coefficient problem:

BIEBERBACH CONJECTURE (1916):

(B) $$\begin{cases} |a_n| \le n \quad (n = 2,3,4,\dots) \quad \text{for} \quad f \in S \,; \\ |a_n| = n \quad \text{for some } n \quad \text{implies } f(z) = \bar{\omega}k(\omega z)\,. \end{cases}$$

Every $f \in S$ generates an odd function

(4) $$g(z) = z(f(z^2)/z^2)^{1/2} = z + \sum_{n=1}^{\infty} b_n z^{2n+1}$$

in S, and conversely every odd function in S may be obtained in this manner. If the Cauchy-Schwarz inequality is applied to

(5) $$a_{n+1} = \sum_{k=0}^{n} b_k b_{n-k} \quad (b_0 = 1)\,,$$

it is easily seen that the

ROBERTSON CONJECTURE ([22], 1936):

(R) $$\begin{cases} 1 + \sum_{k=1}^{n} |b_k|^2 \le n+1 \quad \text{for} \quad n = 1,2,3,\dots \\ \text{and every odd } g \in S \end{cases}$$

is stronger than Bieberbach's conjecture.

The more general Paley-Littlewood conjecture [14] $|b_n| \le 1$ (which is 'quite obvious': $(k(z^2))^{1/2} = \sum_{n=0}^{\infty} z^{2n+1}$) has been disproved by Fekete and Szegö [8], who showed that max $|b_2| = \frac{1}{2} + e^{-2/3} > 1$.

Since $f \in S$ has no zero except $z = 0$,

$$(6) \qquad \log \frac{f(z)}{z} = \sum_{n=1}^{\infty} c_n z^n , \qquad c_1 = a_2 ,$$

is analytic in $\mathbb{D}$. We remark that the logarithmic coefficients of the Koebe function are $c_n = 2/n$. The inequality $|c_n| \le 2/n$, however, is false even in magnitude (Pommerenke [21] has given an example with $\sup n|c_n| = +\infty$).

Milin [16] conjectured that the logarithmic coefficients of the Koebe function are maximal in an average sense. The precise statement is

MILIN'S CONJECTURE (1971):

$$(M) \qquad \sum_{k=1}^{n} (n+1-k)\left(k|c_k|^2 - \frac{4}{k}\right) \le 0 \quad \text{for} \quad n = 1,2,3,\dots,$$

and actually de Branges [5] proved this (and even more).

2. SOME HISTORICAL REMARKS

In 1923 Löwner [15] derived the inequality $|a_3| \le 3$ from his parametric representation of the so-called slit-mappings, which is now called Löwner theory and plays an important role in de Branges' proof. Some aspects will be presented in Section 5.

For individual $n = 4,6,5$ the Bieberbach conjecture has been proved between 1955 and 1972 by various authors using various methods. We mention the papers of Garabedian and Schiffer [11] $(n = 4)$, Pederson [19] $(n = 6)$, Ozawa [18] $(n = 6)$ and Pederson and Schiffer [20] $(n = 5)$.

The first uniform estimate $|a_n| < en$ is due to Littlewood [13], while FitzGerald [9] proved $|a_n| < \sqrt{7/6}\, n$. Another result in this direction is Hayman's [12] $\lim_{n\to\infty} |a_n|/n \le 1$.

Some other remarkable results should be mentioned (for more detailed information the reader is referred to Duren's book [7]) concerning special subclasses of S. The Bieberbach conjecture is known to be true for the class of starlike functions (functions in S whose image domains are starlike with respect to the origin, Nevanlinna [17]) and for typical real functions (functions in S having real coefficients, Dieudonné [6]). The Koebe function belongs to both classes.

3. FROM MILIN'S TO BIEBERBACH'S CONJECTURE

The link between both conjectures is an inequality due to Lebedev and Milin [16], which has nothing to do with univalent functions.

SECOND LEBEDEV-MILIN INEQUALITY. *Let*

$$\phi(z) = 1 + \sum_{n=1}^{\infty} \beta_n z^n \quad \text{and} \quad \psi(z) = \log \phi(z) = \sum_{n=1}^{\infty} \gamma_n z^n, \ \gamma_1 = \beta_1,$$

be analytic in some disk $|z| < r$. *Then*

$$(7) \qquad 1 + \sum_{k=1}^{n} |\beta_k|^2 \le (n+1) \exp\left[\frac{1}{n+1} \sum_{k=1}^{n} (n+1-k)\left(k|\gamma_k|^2 - \frac{1}{k}\right)\right]$$

holds for $n = 1,2,\dots$, *and equality occurs only if* $|\gamma_1| = 1$ (*actually it occurs if and only if* $\gamma_k = \gamma_1^k/k$ $(1 \le k \le n)$ *and* $|\gamma_1| = 1$).

Proof. The proof starts from $\phi' = \psi'\phi$ which implies $n\beta_n = \sum_{k=0}^{n-1} \beta_k (n-k)\gamma_{n-k}$ and so

$$(8) \qquad n^2 |\beta_n|^2 \le \sum_{k=0}^{n-1} |\beta_k|^2 \sum_{k=1}^{n} k^2 |\gamma_k|^2 = B_{n-1} \cdot C_n .$$

This gives

$$(9) \quad B_n \le B_{n-1}\left(1 + \frac{C_n}{n^2}\right) = \frac{n+1}{n} B_{n-1}\left(1 + \frac{C_n - n}{n(n+1)}\right) \le \frac{n+1}{n} B_{n-1} \exp\left(\frac{C_n - n}{n(n+1)}\right).$$

Hence

$$(10) \qquad B_n \le (n+1) \exp\left(\sum_{k=1}^{n} \frac{C_k - k}{k(k+1)}\right)$$

follows, where equality may only occur if $C_k = k$ $(1 \le k \le n)$ and so $|\gamma_1| = 1$. If the sum on the right hand side is evaluated by Abel's summation, it is seen that (10) is equivalent with (7).

To prove that (M) implies (B), let $f \in S$ be arbitrary and apply the Lebedev-Milin inequality to the functions

$$\phi(z) = \frac{g(\sqrt{z})}{\sqrt{z}} = 1 + \sum_{n=1}^{\infty} b_n z^n$$

and

$$\psi(z) = \log \phi(z) = \frac{1}{2} \log \frac{f(z)}{z} = \sum_{n=1}^{\infty} \frac{c_n}{2} z^n ,$$

where g is given by (6). This gives

$$(11) \qquad |a_{n+1}| \le (n+1) \exp \left[\frac{1}{n+1} \sum_{k=1}^{n} (n+1-k)\left(k\left|\frac{c_k}{2}\right|^2 - \frac{1}{k}\right)\right] ,$$

and (B) is true for n+1 if (M) is true for n. Equality is possible only if $|c_1/2| = |a_2/2| = 1$, i.e. if $|a_2| = 2$, and so f is a rotation of the Koebe function by Bieberbach's result.[1)] This is an important step since the method of de Branges does not immediately allow to identify the extremals.

4. SOME PROPERTIES OF CONFORMAL MAPPINGS

4.1 Boundary behaviour. Let the domain Ω be bounded by a Jordan curve and let f be a conformal mapping of $\mathbb{D}$ onto Ω. Then, by a famous theorem of Carathéodory [4], f has a continuous extension to $\overline{\mathbb{D}}$ which provides a topological mapping of $\overline{\mathbb{D}}$ onto $\bar{\Omega}$.

If Ω is a domain bounded by a single Jordan arc J tending to infinity, we may open the plane by a square-root-transform. After a suitable Möbiustransform Ω is mapped conformally onto a Jordan domain Ω', and every point on J except the end points has two images on the boundary of Ω'. Combining this elementary result with Carathéodory's theorem we conclude that, if f maps $\mathbb{D}$ conformally onto Ω, then f extends continuously onto $\overline{\mathbb{D}}$, and $\partial\mathbb{D}$ is divided into two arcs which are separately mapped onto J, while their common end points are mapped onto the end points of J (in other words: every point on J except the end points defines exactly two prime ends). The inverse function is continuous at the end points of J and has (different) 'one-sided' limits at the other points.

4.2 Kernel convergence. Let Ω_n, n = 0,1,2,..., be simply connected domains, which are different from the whole plane and contain a fixed disk $|w| < \rho$. By the Riemann mapping theorem there is a

1) This was kindly pointed out to me by Jochen Becker.

unique conformal mapping of $\mathbb{D}$ onto Ω_n such that $f_n(0) = 0$ and $f_n'(0) > 0$.

The kernel Ω_o of the sequence $(\Omega_n)_{n\geq 1}$ is defined to be the union of all domains H such that $0 \in H$ and $H \subseteq \Omega_n$ for $n \geq n(H)$. The sequence $(\Omega_n)_{n\geq 1}$ is said to converge to its kernel Ω_o if Ω_o is also the kernel of every subsequence (Ω_{n_k}).

We mention two examples: Every increasing sequence converges to its kernel $\Omega_o = \bigcup_{n\geq 1}\Omega_n$, and every decreasing sequence converges to its kernel which is the component of $\left(\bigcap_{n\geq 1}\Omega_n\right)^o$ containing the origin.

Another famous theorem of Carathéodory [3] states that $(\Omega_n)_{n\geq 1}$ converges to its kernel Ω_o if and only if the sequence of associated mapping functions (f_n) tends to f_o, locally uniformly in $\mathbb{D}$.

4.3 Slit mappings. It is clear that the set of functions in S which are analytic in $\overline{\mathbb{D}}$ form a dense subset of S (always with respect to the topology of locally uniform convergence). If $f \in S$ is analytic in $\overline{\mathbb{D}}$, then its image domain $f(\mathbb{D})$ is bounded by an analytic Jordan curve C: $w = w(\tau)$, $0 \leq \tau \leq 1$. If $|w(0)| \geq |w(\tau)|$, as we may assume, then C_n: $w = w(\tau)$ for $\frac{1}{n} \leq \tau \leq 1$ and $w = w(0)+(\tau-1)w'(0)$ for $\tau > 1$ is a smooth Jordan arc which defines a slit domain $\Omega_n = \mathbb{C} \setminus C_n$. If f_n maps $\mathbb{D}$ onto Ω_n ($f_n(0) = 0$, $f_n'(0) > 0$), then by Carathéodory's theorem on kernel convergence f_n tends to f, and so does $f_n/f_n'(0)$ which belongs to S.

Thus it is enough to prove Milin's conjecture for slit mappings in S.

5. THE LÖWNER DIFFERENTIAL EQUATION

In 1923 Löwner [15] proved that every slit mapping f may be embedded into a continuous family $e^{-t}f(z,t)$ of slit mappings, where $f(z,t)$ is a solution of a linear first order partial differential equation. The precise statement is

LÖWNER'S THEOREM. _Let_ f _be a slit mapping. Then there exist slit mappings_ $e^{-t}f(z,t)$, $0 \le t < \infty$, _and a continuous function_ κ: $[0,\infty) \to \partial\mathbb{D}$ _such that_

(12) $$f_t = \frac{1+\kappa(t)z}{1-\kappa(t)z}\, zf_z ,$$

(13) $$f(z,0) = f(z) , \quad f(0,t) = 0 , \quad f_z(0,t) = e^t$$

and

(14) $$f,\ f_z \text{ and } f_t \text{ are continuous in } \mathbb{D}\times[0,\infty) .$$

The function $f(z,t)$ is sometimes called the Löwner chain generated by f. For the Koebe function we have $k(z,t) = e^t k(z)$ and, unfortunately $\kappa(t) = -1$ (if κ were +1, Löwner would already have solved the Bieberbach conjecture).

Proof. Let J_o: $w = w(\tau)$, $0 \le \tau < \infty$, be the complement of $f(\mathbb{D})$ and denote the subarc starting at $w(t)$ by J_t: $w = w(\tau)$, $t \le \tau < \infty$. By the Riemann mapping theorem there exists a unique conformal mapping $f(z,t)$ of $\mathbb{D}$ onto $\Omega_t = \mathbb{C} \setminus J_t$ normalized by $f(0,t) = 0$ and $f_z(0,t) > 0$. We will first show that f is continuous in $\mathbb{D}\times[0,\infty)$ and that $f_z(0,t)$ is a continuous, strictly increasing and unbounded function of t in the range $0 \le t < \infty$.

The first part is a consequence of Carathéodory's theorem on kernel convergence: If $t_n \to t$, then (Ω_{t_n}) tends to its kernel Ω_t, which implies $f(z,t_n) \to f(z,t)$ locally uniformly in $\mathbb{D}$. Thus, $f(z,s) \to f(z,t)$ as $s \to t$, uniformly in every fixed disk $|z| \le r < 1$, and so $f(z,t)$ is continuous in $\{|z| \le r\} \times [0,\infty)$. By Cauchy's formula,

(15) $$f_z(0,t) = \frac{1}{2\pi i} \int_{|z|=\frac{1}{2}} \frac{f(z,t)}{z^2}\, dz$$

is a continuous function of t. For $0 \le s < t < \infty$, $h(z) = f^{-1}(f(z,s),t)$ maps $\mathbb{D}$ into $\mathbb{D}$ letting the origin fixed. Schwarz's lemma thus gives $h'(0) = \dfrac{f_z(0,s)}{f_z(0,t)} < 1$ (note that h is not the identity map), i.e., $f_z(0,t)$ is strictly increasing. Another application of Schwarz's lemma shows that $f_z(0,t)$ is unbounded: Given $R > 0$, there exists a number t_R such that $|w(\tau)| > R$ for $\tau > t_R$, thus $\phi(w) = f^{-1}(Rw,t)$ maps $\mathbb{D}$ into itself, $\phi(0) = 0$, and so $\phi'(0) = R/f_z(0,t) \le 1$, i.e., $f_z(0,t) \ge R$ for $t \ge t_R$.

Since the function $\log f_z(0,t)$ maps $[0,\infty)$ onto itself it may as well serve as a parameter for the slit J_o, and with this new parametrization

$$f(z,t) = e^t z + \dots \tag{16}$$

maps $\mathbb{D}$ conformally onto $\Omega_t = \mathbb{C} \setminus J_t$.

Now consider the conformal mapping

$$h(z) = h(z,s,t) = f^{-1}(f(z,s),t)$$

for fixed s and t, $0 \le s < t < \infty$ (see figure).

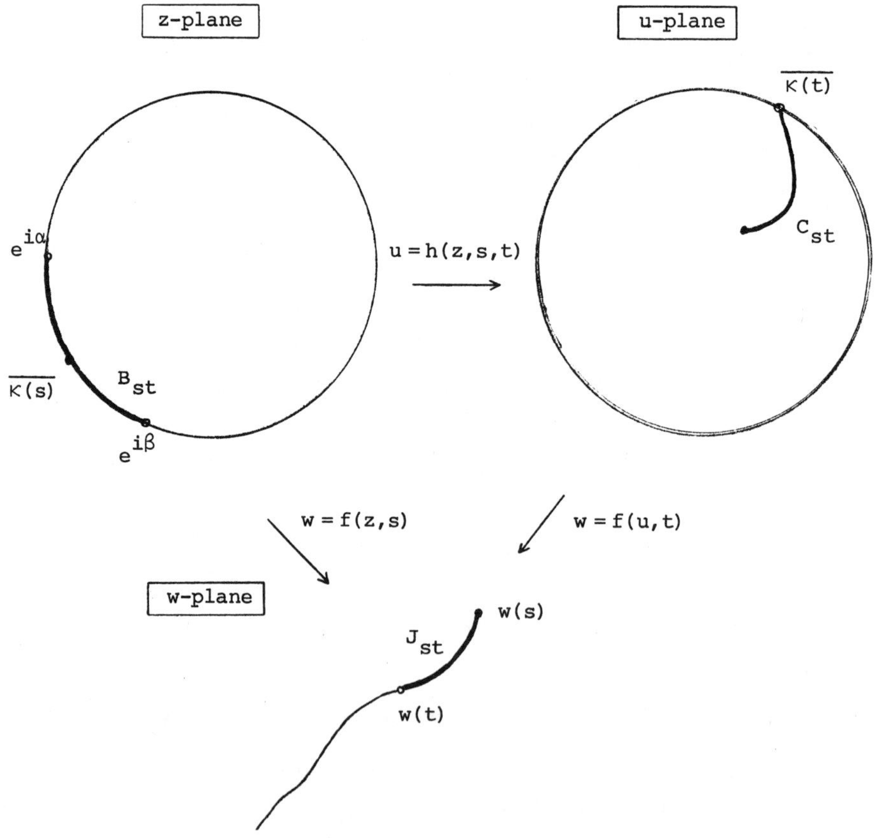

The arc B_{st} with end points $e^{i\alpha}$ and $e^{i\beta}$ is mapped onto C_{st} by $u = h(z,s,t)$ and onto J_{st}: $w = w(\tau)$, $s \le \tau \le t$ by $w = f(z,s)$. The point $\overline{\kappa(s)}$ corresponds to $w(s)$, similarly $u = \overline{\kappa(t)}$ is mapped onto $w = w(t)$ by $w = f(u,t)$. Note that, for fixed s and t, h is continuous on $\overline{\mathbb{D}}$. Since $f^{-1}(w,s)$ is continuous at $w = w(s)$ we have $B_{st} \to \overline{\kappa(s)}$ as $t \searrow s$, in particular $e^{i\alpha}, e^{i\beta} \to \overline{\kappa(s)}$ as $t \searrow s$.

The function

(17) $$H(z) = \log \frac{h(z,s,t)}{z} = u(z) + iv(z) , \quad H(0) = u(0) = s-t$$

is analytic in $\mathbb{D}$ and continuous in $\overline{\mathbb{D}}$, its real part is negative on B_{st} and vanishes on the rest of $\partial\mathbb{D}$. Thus, the Poisson-Schwarz formula gives

(18) $$H(z) = \frac{1}{2\pi} \int_{\alpha}^{\beta} u(e^{i\theta}) \frac{e^{i\theta}+z}{e^{i\theta}-z} \, d\theta ,$$

and this shows that H is analytic in $\mathbb{C} \setminus B_{st}$.

Applying first the mean value theorem of integral calculus and then the Gauß mean value theorem give

(19) $$H(z) = (s-t)\left[\frac{\overline{\kappa(s)}+z}{\overline{\kappa(s)}-z} + o(1)\right] \quad \text{as} \quad t \searrow s ,$$

locally uniformly in $\mathbb{C} \setminus \{\overline{\kappa(s)}\}$, and so $h(z,s,t) = ze^{H(z)}$ converges to the identity map. Given ε, $0 < \varepsilon < 1$, the circle γ_ε: $|z-\overline{\kappa(s)}| = \varepsilon$ separates 0 from B_{st} and so its image $\Gamma_\varepsilon = h(\gamma_\varepsilon)$ separates 0 from C_{st}. If $t-s$ is small enough we have $|h(z,s,t)-z| < \varepsilon$ on γ_ε and so Γ_ε and C_{st} are contained in the disk $|u-\overline{\kappa(s)}| < 2\varepsilon$. This proves $C_{st} \to \overline{\kappa(s)}$ and $\kappa(t) \to \kappa(s)$ as $t \searrow s$. Similarly, we have $C_{st} \to \overline{\kappa(t)}$, $B_{st} \to \overline{\kappa(t)}$ and $\kappa(s) \to \kappa(t)$ as $s \nearrow t$, in particular $\kappa(t)$ is a continuous function.

Now set $z = f^{-1}(u,s) = g(u,s)$ in (19) to deduce

$$\frac{\log g(u,t) - \log g(u,s)}{t-s} = - \frac{\overline{\kappa(s)}+g(u,s)}{\overline{\kappa(s)}-g(u,s)} + o(1)$$

as $t-s \to 0$, and so

(20) $$\frac{\partial}{\partial t} g(u,t) = - \frac{1+\kappa(t)g(u,t)}{1-\kappa(t)g(u,t)} ,$$

which is equivalent to (12) (it requires some care to prove that $f(z,t)$ is differentiable with respect to t, since g is the inverse of f with respect to the first variable).

The continuity statements follow easily. First of all,

(21) $$f_z(z,t) = \frac{1}{2\pi i} \int_{|u|=r} \frac{f(u,t)}{(u-z)^2}\,du \quad (|z| < r)$$

is continuous in $\{|z| < r\} \times [0,\infty)$, thus in $\mathbb{D} \times [0,\infty)$, and the continuity of f_t follows from the differential equation (12).

6. DE BRANGES' PROOF

Let f be a slit mapping embedded in a Löwner chain $f(z,t)$. Then $e^{-t}f(z,t)$ belongs to S for every $t \geq 0$ and has logarithmic coefficients $c_k(t)$ of class $C^1[0,\infty)$. Milin's conjecture then follows from

(22) $$\sum_{k=1}^{n} (n+1-k)\left(k|c_k(0)|^2 - \frac{4}{k}\right) \leq 0 .$$

The main idea of de Branges [5] was to replace $c_k = c_k(0)$ by $c_k(t)$ and the weights $n+1-k$ by ingeneously chosen weight functions $\sigma_k(t)$ with initial values

A) $$\sigma_k(0) = n+1-k .$$

Then (22) follows if it is possible to show that

(23) $$S(t) = \sum_{k=1}^{n} \left(k|c_k(t)|^2 - \frac{4}{k}\right)\sigma_k(t)$$

increases to zero as t tends to infinity.

We shall follow the arguments of FitzGerald and Pommerenke [10], which seem to be more transparent, while de Branges proved even more than Milin's conjecture.

From

(24) $$\log \frac{e^{-t}f(z,t)}{z} = \sum_{n=1}^{\infty} c_n(t)z^n$$

follows

(25a) $$\frac{f_t}{f} = 1 + \sum_{n=1}^{\infty} c_n' z^n ,$$

(25b) $$z\frac{f_z}{f} = 1 + \sum_{n=1}^{\infty} nc_n z^n ,$$

and so Löwner's differential equation (12) gives

(26) $$c_n' = nc_n + 2\kappa^n + 2\sum_{j=1}^{n-1} jc_j\kappa^{n-j} .$$

Set

(27) $$p_n = \sum_{j=1}^{n} jc_j\bar{\kappa}^{j}$$

such that

(28) $$c_n' = \kappa^n(p_{n-1}+p_n+2) .$$

Thus, if (23) is differentiated, we may replace the derivatives c_n' by (28) to derive

(29) $$S' = 2\sum_{k=1}^{n} (|p_k|^2 + 2\,\mathrm{Re}\,p_k)(\sigma_k-\sigma_{k+1}) + \sum_{k=1}^{n} (|p_k-p_{k-1}|^2-4)\frac{\sigma_k'}{k} .$$

Now comes a courageous step, namely, to assume that the weight functions satisfy

B) $$\frac{\sigma_k'}{k} + \frac{\sigma_{k+1}'}{k+1} = \sigma_{k+1} - \sigma_k .$$

Then S' turns out to be a negative sum of squares multiplied by σ_k'/k,

(30) $$S' = -\sum_{k=1}^{n} |p_{k-1}+p_k+2|^2 \frac{\sigma_k'}{k} .$$

Obviously, S is nondecreasing if

C) $$\sigma_k' \le 0 \qquad (1 \le k \le n) ,$$

and S tends to zero if

D) $$\sigma_k \to 0 \qquad (t \to \infty ,\ 1 \le k \le n)$$

and

E) $$|c_k(t)| \le C_k \quad (0 \le t < \infty ,\ 1 \le k \le n) .$$

The last assertion is well known, since S is a normal family in the sense of Montel.

What is left is to prove that A) - D) can be satisfied simultaneously. However, by A) and B) the weights σ_k are uniquely determined as the solutions of the initial value problem

$$(31)\quad \begin{aligned} \sigma_1' &= -\sigma_1 + 2\sigma_2 - 2\sigma_3 + - \dots & , \quad \sigma_1(0) &= n\ , \\ \sigma_2' &= \quad\ - 2\sigma_2 + 4\sigma_3 - + \dots & , \quad \sigma_2(0) &= n-1\ , \\ &\vdots \\ \sigma_{n-1}' &= \qquad\qquad -(n-1)\sigma_{n-1} + 2(n-1)\sigma_n & , \quad \sigma_{n-1}(0) &= 2\ , \\ \sigma_n' &= \qquad\qquad\qquad\qquad -n\sigma_n & , \quad \sigma_n(0) &= 1\ . \end{aligned}$$

It is easy to deduce D), since the eigenvalues of (31) are -1,-2, ...,-n, so that $\sigma_k(t) = O(e^{-kt})$ as $t \to \infty$.

The most difficult part is to prove C). Explicit integration gives

$$(32)\qquad -\frac{\sigma_k'}{k} = e^{-kt} \sum_{j=0}^{n-k} P_j^{(2k,0)}(1-2e^{-t})\ ,$$

where

$$(33)\qquad P_m^{(a,b)} = \frac{(a+1)_m}{m!} \sum_{j=0}^{m} \frac{(-m)_j\,(m+a+b+1)_j}{j!\,(a+1)_j} \left(\frac{1-x}{2}\right)^j$$

in the m-th Jacobi polynomial with parameters (a,b). In 1976 Askey and Gasper [1] considered sums of Jacobi polynomials and proved

$$(34)\qquad \sum_{j=0}^{m} P_j^{(a,0)}(x) \ge 0$$

for $-1 \le x \le 1$ and $a \ge -2$, which shows that C) is also fulfilled.

It is an open problem whether C) may be deduced directly from (31) without integration, using only differential inequality techniques. It seems to the author that $n = 4$ is the most general case.

REFERENCES

1. R. Askey and G. Gasper, Positive Jacobi sums II. Amer. J. Math. 98 (1976), 709-737.

2. L. Bieberbach, Über die Koeffizienten derjenigen Potenzreihen, welche eine schlichte Abbildung des Einheitskreises vermitteln. Sitz.-Ber. Preuss. Akad. d. Wiss. (1916), 940-955.

3. C. Carathéodory, Untersuchungen über die konformen Abbildungen von festen und veränderlichen Gebieten. Math. Ann. 72 (1912), 107-144.

4. C. Carathéodory, Zur Ränderzuordnung bei konformer Abbildung. Nachr. Königl. Ges. Wiss. Göttingen, Math. Phys. Kl. (1913), 509-518.

5. L. de Branges, A proof of the Bieberbach conjecture. Acta Math. 154 (1985), 137-152.

6. J. Dieudonné, Sur les fonctions univalentes. C.R. Acad. Sci. Paris 192 (1931), 1148-1150.

7. P.L. Duren, Univalent functions. Springer-Verlag, New York - Berlin - Heidelberg - Tokyo, 1983.

8. M. Fekete und G. Szegö, Eine Bemerkung über ungerade schlichte Funktionen. J. London Math. Soc. 8 (1933), 85-89.

9. C.H. FitzGerald, Quadratic inequalities and coefficient estimates for schlicht functions. J. Analyse Math. 29 (1976), 203-231.

10. C.H. FitzGerald and Ch. Pommerenke, The De Branges theorem on univalent functions. Trans. Amer. Math. Soc. 290 (1985), 683-690.

11. P.R. Garabedian and M. Schiffer, A proof of the Bieberbach conjecture for the fourth coefficient. J. Rat. Mech. Anal. 4 (1955), 427-465.

12. W.K. Hayman, The asymptotic behaviour of p-valent functions. Proc. London Math. Soc. 5 (1955), 257-284.

13. J.E. Littlewood, On inequalities in the theory of functions. Proc. London Math. Soc. 23 (1925), 481-519.

14. J.E. Littlewood and R.E.A.C. Paley, A proof that an odd schlicht function has bounded coefficients. J. London Math. Soc. 7 (1932), 167-169.

15. K. Löwner, Untersuchungen über schlichte konforme Abbildungen des Einheitskreises, I. Math. Ann. 89 (1923), 103-121.

16. I.M. Milin, Univalent functions and orthonormal systems. Izdat. "Nauka": Moscow 1971 (in Russian). Engl. Transl.: Amer. Math. Soc., Providence, R.I., 1977.

17. R. Nevanlinna, Über die konforme Abbildung von Sterngebieten. Översikt av Finska Vertenskaps-Soc. Förh. 63 (A) (1920-21), 1-21.

18. M. Ozawa, On the Bieberbach conjecture for the sixth coefficient. Kodai Math. Sem. Rep. 21 (1969), 97-128.

19. R.N. Pederson, A proof of the Bieberbach conjecture for the sixth coefficient. Arch. Rat. Mech. Anal. 31 (1968), 331-351.

20. R.N. Pederson and M. Schiffer, A proof of the Bieberbach conjecture for the fifth coefficient. Arch. Rat. Mech. Anal. 45 (1972), 161-193.

21. Ch. Pommerenke, Relations between the coefficients of a univalent function. Invent. Math. 3 (1967), 1-15.

22. M.S. Robertson, A remark on the odd schlicht functions. Bull. Amer. Math. Soc. 42 (1936), 366-370.

Norbert Steinmetz, Mathematisches Institut I, Universität Karlsruhe, Kaiserstr. 12, D-7500 Karlsruhe, West Germany

Inequalities for Sums, Series and Integrals

Serious work

International Series of
Numerical Mathematics, Vol. 80

TAUBERIAN THEOREMS, CONVOLUTIONS AND SOME RESULTS OF D.C. RUSSELL

Matts Essén

Abstract. This paper is a comment on a paper by D.C. Russell "Tauberian-type results for convolutions of sequences" (cf. [9]). We wish to describe the relation between the results of Russell and the work of Essén on convolution inequalities (cf. [4], [5]).

In [2], E.T. Copson proved the following result:

THEOREM C. *Let* $k_i > 0$ $(i = 1,\dots,m)$, $k_1 + \dots + k_m = 1$, *and the real sequence* $\{a_n\}$ *satisfy the inequality*

$$a_{n+m} \le \sum_{i=1}^{m} k_i a_{n+m-i} \qquad (n = 0,1,2,\dots) .$$

If $\{a_n\}$ *is bounded, then it must be convergent.*

This theorem is a corollary of the discrete form of the main result in Essén [5] which holds for sequences on the integers $\underline{Z}$ (the Copson result is restricted to the nonnegative integers $\underline{Z}^+$). The following result is due to D.C. Russell [8]:

THEOREM R. *Let* $P = \{p_n\}_{0\le n\le N}$ *be real-valued and* $p(z) = \sum_{n=0}^{N} p_n z^n$ $(z \in \underline{C})$. *Let* $a = \{a_n\}_{n\ge 0}$ *be such that* $a * P$ *is real and ultimately monotone. Then in order that boundedness of* $\{a_n\}$ *shall always imply its convergence, it is necessary and sufficient that*

$$(1) \qquad p(z) \ne 0 \quad \text{in the set} \quad \Gamma = \{z \in \underline{C}: |z| = 1,\ z \ne 1\} .$$

This paper is in final form and no version of it will be submitted for publication elsewhere.

General questions of this type were studied in Essén [4]. An application is given in [5]. The purpose of this note is to point out that a more general form of Theorem R is a direct consequence of the method of proof in [5].

To explain the connection between Theorems C and R and our result, we need some notation. We consider real-valued functions on X which will be $\underline{R}$ or $\underline{Z}$. If $K \in L^1(X)$, we have

$$\hat{K}(t) = \int_{-\infty}^{\infty} K(x) e^{-ixt} dt , \quad X = \underline{R} ,$$

$$\hat{K}(e^{i\theta}) = \sum_{-\infty}^{\infty} k_n e^{in\theta} , \quad X = \underline{Z} .$$

$$P(x) = \begin{cases} \int_x^{\infty} K(y)dy , & x > 0 , \\ -\int_{-\infty}^{x} K(y)dy , & x < 0 , \end{cases} \quad X = \underline{R} ,$$

$$p_n = \begin{cases} \sum_{j=n+1}^{\infty} k_j , & n \geq 0 , \\ -\sum_{-\infty}^{n} k_j , & n \leq -1 , \end{cases} \quad X = \underline{Z} .$$

(cf. [5, p.14] and [4, p.130]). Without loss of generality, we can assume that $k_0 = 0$ (cf. [4, p.115]).

We have the following relations:

$$\hat{P}(t) = (\hat{K}(0) - \hat{K}(t))/(it) , \tag{2}$$

$$\hat{P}(e^{i\theta}) = (\hat{K}(0) - \hat{K}(e^{i\theta}))/(1 - e^{-i\theta}) . \tag{3}$$

The general theme of [4] and [5] is the study of convolution inequalities of the form $\varphi - \varphi * K \geq 0$. There, it is assumed that K is nonnegative and that either $\hat{K}(0) = 1$ when $X = \underline{R}$ or that $\hat{K}(1) = 1$ when $X = \underline{Z}$. From the proof of Theorem 1 below, it will be clear that the methods of [5] work also when K is not necessarily nonnegative.

To deduce Theorem C from Theorem R, we let K be a nonnegative sequence supported on $\underline{Z}^+$, define the sequence P as above and apply Theorem R.

REMARK. The following formal argument explains the connection between Theorem R and our convolution inequalities. Let δ be the Dirac functional on $\underline{R}$. Then we have $P' = \hat{K}(0)\delta - K$, and $\varphi * P$ is increasing if and only if $\varphi * P' = \hat{K}(0)\varphi - \varphi * K$ is nonnegative. (If $K \in L^1(\underline{Z})$, K' is defined as the sequence $\{k_n - k_{n-1}\}_{-\infty}^{\infty}$. On $\underline{Z}$, we replace $\hat{K}(0)$ by $\hat{K}(1)$ in these formulas.)

THEOREM 1. *Let* $K \in L^1(\underline{R})$ *be such that* $\int_{-\infty}^{\infty} |xK(x)|dx$ *is finite. Let* $\varphi \in L^{\infty}(\underline{R})$ *be slowly decreasing* (cf. Definition 9b, Ch. V in [10]).

a) *If* $\hat{P}(t) \neq 0$, $t \in \underline{R}$, *and* $\varphi * P$ *is increasing on* $\underline{R}^+$, *then* $\lim_{x\to\infty} \varphi(x)$ *exists*.

b) *If* $\hat{P}(t_0) = 0$ *for some* $t_0 \neq 0$, *the conclusion in* a) *is not true*.

c) *Assume that* $\hat{P}(t) \neq 0$, $t \neq 0$, *and furthermore that there exists an integer* $q \geq 2$ *such that*

$$(4) \qquad \int_{-\infty}^{\infty} x^n K(x)dx = 0\ , \quad n = 1,2,\ldots, q-1; \; A = \int_{-\infty}^{\infty} x^q K(x)dx \neq 0\ .$$

We assume also that all these integrals are absolutely convergent. If $\varphi * P$ *is increasing on* $\underline{R}^+$, *then* $\lim_{x\to\infty} \varphi(x)$ *exists*.

REMARK. It is easy to write down the analogous result on $\underline{Z}$. It should be compared to Theorem R.

Proof. As a normalization, we assume that $\hat{K}(0) = 1$. (We leave the case $\hat{K}(0) \leq 0$ to the reader.)

a) Let us first assume that φ is absolutely continuous with $\varphi' \in L^{\infty}(\underline{R})$. Then the argument in the remark preceding Theorem 1 shows that $\varphi - \varphi * K = h$ is nonnegative on $\underline{R}^+$. Hence we have

$$(5) \qquad \int_0^x h(y)dy = P * \varphi(x) - P * \varphi(0)\ ,$$

(cf. p.13 in [5]). Since $\int_{-\infty}^{\infty} |xK(x)|dx$ is finite, we know that $P \in L^1(\underline{R})$. Hence the right hand member in (5) is bounded as $x \to \infty$ and we must have $h \in L^1(\underline{R}^+)$. It follows that $\lim_{x\to\infty} \varphi * P(x)$ exists. Since $\hat{P}(t)$ is non-vanishing on $\underline{R}$ and φ is slowly decreasing, we can apply Pitt´s Tauberian theorem (cf. Theorem 10a,

Ch. V in [10]) and conclude that $\lim_{x\to\infty} \varphi(x)$ exists.

In the general case, we consider a nonnegative C^∞-function ψ with support in $[-1,1]$ and integral 1. If $\psi_n(x) = n\psi(nx)$ and $\varphi_n = \varphi * \psi_n$, $n = 1,2,\ldots$, φ_n will be absolutely continuous with $\varphi_n' \in L^\infty(\underline{R})$ and $\varphi_n * P$ will be increasing on $[1,\infty)$. Arguing as in the first part of the proof, we see that

$$\varphi_n(x) - \varphi_n * K(x) \geq 0 , \quad x \in [1,\infty) .$$

When $n \to \infty$, it is known that $\varphi_n(x)$ tends to $\varphi(x)$ at all Lebesgue points of φ. Thus a.e. on $[1,\infty)$, we have $\varphi - \varphi * K = h \geq 0$, (5) holds also in this case and the conclusion follows.

b) Choosing $\varphi(x) = \mathrm{Re}(e^{it_0x})$, we see that

$$(\varphi - \varphi * K)(x) = \mathrm{Re}\{e^{it_0x}(1 - \hat{K}(t_0)\} = 0 .$$

The last relation follows from (2) and our assumption that $t_0 \neq 0$. It is clear that $\varphi * P$ is constant and that $\varphi(x)$ does not have a limit at infinity.

c) Let $\hat{P}_q$ be defined by

$$\hat{P}_q(t) = (1 - \hat{K}(t))(it)^{-q} , \ t \neq 0; \ \hat{P}_q(0) = (-1)^{q+1} A/q!$$

It follows from (4) that $\hat{P}_q$ is continuous on $\underline{R}$. It is also clear that $\hat{P}_q$ is non-vanishing on $\underline{R}$.

Let us first assume that $\varphi \in C^\infty(\underline{R})$ with $\varphi^{(k)} \in L^\infty(\underline{R})$, $k = 0,1,\ldots$. Since $P_2' = P$, we have

$$0 \leq \varphi * P' = \varphi'' * P_2 = \varphi^{(q)} * P_q .$$

If $f = \varphi * P_q$, we know that f and its first q derivatives are in $L^\infty(\underline{R})$ and that $f^{(q)}$ is nonnegative. Consider the formula

$$f^{(q-1)}(x) = \int_0^x f^{(q)}(t)dt + f^{(q-1)}(0) .$$

Since the left hand member is bounded, the integral in the right hand member converges when $x \to \infty$ and $\lim_{x\to\infty} f^{(q-1)}(x) = b_{q-1}$ exists. We must have $b_{q-1} = 0$: it follows that

$$f^{(q-1)}(x) = -\int_x^\infty f^{(q)}(t)dt$$

is nonpositive on $\underline{R}^+$. Continuing in this way, we deduce that f' does not change sign on $\underline{R}^+$. Since $f \in L^\infty(\underline{R})$, f' is integrable and we have proved the existence of $\lim_{x\to\infty} f(x) = \lim_{x\to\infty} \varphi * P_q(x)$. Since $\hat{P}_q$ is nonvanishing on $\underline{R}$, we can apply Pitt's Tauberian theorem which gives the existence of $\lim_{x\to\infty} \varphi(x)$.

In the general case, we approximate φ by a sequence $\{\varphi_n\} = \{\varphi * \psi_n\}$ in the same way as in the proof of a): all functions in the sequence $\{\varphi_n * P_q\}$ are either increasing or decreasing and their L^∞-norms are uniformly bounded. Choosing a subsequence, we find that $\varphi * P_q$ is a monotone and bounded function on $\underline{R}^+$. Thus $\lim_{x\to\infty} \varphi * P_q(x)$ exists and gives the existence of $\lim_{x\to\infty} \varphi(x)$ in the same way as before.

REMARK. Let us in c) replace the assumption that $\varphi * P$ is increasing on $\underline{R}^+$ by the assumption that $\varphi * P$ is increasing on $\underline{R}$. Then $\varphi * P_2$ will be convex and bounded on $\underline{R}$. Thus, this function must be constant. In a standard way, it follows that φ is a polynomial of degree at most $q-1$. Hence φ must be constant.

REMARK. The function P_2 can be found in Lemma 5.3 in [4]: it is there called $-N_2$.

As a variation on this theme, we prove

THEOREM 2. *Let* $K \in L^1(\underline{R})$ *be such that* $\hat{K}(0) = 1$, $\int_{-\infty}^{\infty} |xK(x)|dx$ *is finite and* $\hat{P}(t)$ *is nonvanishing on* $\underline{R}$. *If* $\varphi \in L^\infty(\underline{R})$ *and* $\varphi * P$ *is convex on* $\underline{R}^+$, *then* $\lim_{x\to\infty} \varphi(x)$ *exists.*

THEOREM 2'. *Let* $K \in L^1(\underline{R})$ *be such that* $\hat{K}(0) = 1$ *and that the assumptions of Theorem 1c hold. If* $\varphi \in L^\infty(\underline{R})$ *and* $\varphi * P$ *is convex on* R^+, *then* $\lim_{x\to\infty} \varphi(x)$ *exists.*

We prove Theorem 2 and leave the proof of Theorem 2' to the reader.

Proof of Theorem 2. We assume first that φ is twice continuously differentiable with the first two derivatives in $L^\infty(\underline{R})$. Again using the formula $P' = \delta - K$, we see that

$$(d^2/dx^2)(\varphi * P) = \varphi' * P' = \varphi' - \varphi' * K = h \geq 0 \quad \text{on} \quad \underline{R}^+ ,$$

and that

$$\int_0^x h(y)dy = P * \varphi'(x) - P * \varphi'(0) = \varphi(x) - \varphi * K(x) - P * \varphi'(0) . \tag{6}$$

Since the right hand members are bounded and h is nonnegative on $\underline{R}^+$, we have $h \in L^1(\underline{R}^+)$ and $\lim P * \varphi'(x)$ exists. The existence of $\lim_{x\to\infty} \varphi'(x) = \alpha$ is now a consequence of Pitt's Tauberian theorem. Our assumption that $\varphi \in L^\infty(\underline{R})$ implies that $\alpha = 0$. Using (6), we find

$$\int_0^\infty h(y)dy = -P * \varphi'(0) ,$$

$$\varphi(x) - \varphi * K(x) = -\int_x^\infty h(y)dy , \quad x > 0 . \tag{7}$$

In the general case, we again approximate φ by a sequence $\{\varphi_n\} = \{\varphi * \psi_n\}$ (cf. the proof of Theorem 1a). If

$$\varphi_n(x) - \varphi_n * K(x) = -\int_x^\infty h_n(y)dy ,$$

we have

$$\left|\int_0^\infty h_n(y)dy\right| = |(P' * \varphi_n)(0)| \leq 2\|\varphi\|_\infty$$

Letting $n \to \infty$, we deduce that

$$\varphi(x) = \varphi * K(x) - g(x) , \quad x > 0 , \tag{8}$$

where g is nonnegative and decreasing to 0 as $x \to \infty$.

We claim that if (8) holds, the function φ must be slowly decreasing. To see this, we choose $\delta > 0$ and consider

$$\varphi(x+\delta) - \varphi(x) = \int_{-\infty}^\infty \varphi(y)(K(x+\delta-y) - K(x-y))dy + g(x) - g(x+\delta) .$$

It follows that

$$\liminf_{x\to\infty}\{\varphi(x+\delta) - \varphi(x)\} \geq -\|\varphi\|_\infty \int_\infty^\infty |K(\delta-y) - K(-y)|dy ,$$

and that

$$\liminf_{\delta \to 0+} \liminf_{x \to \infty} \{\varphi(x+\delta) - \varphi(x)\} \geq 0 ,$$

which is what we wanted to prove.

Since (8) holds and φ is slowly decreasing as $x \to \infty$, the existence of $\lim_{x \to \infty} \varphi(x)$ follows in the same way as before.

We would also like to discuss the main points in the proofs of Theorems 1 and 3 in Russell [9]: his argument uses classical results of Wiener and Pitt [11]. We start from the fact that if the maximal ideal space of a Banach algebra is compact and the Gelfand (or "Fourier") transform of an element in the algebra does not vanish on this space, then the element is invertible in the algebra (cf. Ch. VIII in [7]). Our Banach algebra will be either $\ell^1(\underline{Z})$ or $\ell^1_+ = \ell^1(\underline{Z}^+)$: the maximal ideal spaces are the unit circle and the closed unit disc, respectively.

Russell considers a collection Λ of classes of complex-valued sequences on $\underline{Z}$: examples are ℓ^r, $1 \leq r < \infty$, convergent sequences or sequences of bounded variation (a longer list is given in [9]). They have the property that if the norms are defined in the right way and if $\lambda \in \Lambda$ then

$$(9) \qquad \{u \in \lambda ,\ v \in \ell^1\} \Rightarrow u * v \in \lambda ,$$

$$\|u * v\|_\lambda \leq \|u\|_\lambda \, \|v\|_{\ell^1} .$$

THEOREM R1. *Let* $P \in L^1(\underline{Z})$ *be such that* $\hat{P}(e^{i\theta})$ *is non-vanishing on* $[0,2\pi)$. *If* $\lambda \in \Lambda$, *then*

$$(10) \qquad \{\varphi \in \ell^\infty ,\ \varphi * P \in \lambda\} \Rightarrow \varphi \in \lambda .$$

Proof. According to the Wiener-Lévy theorem (cf. 3.10 in Ch. VIII in [7]), there exists $v \in L^1(\underline{Z})$ such that $\hat{v}(e^{i\theta})\hat{P}(e^{i\theta}) = 1$ on $[0,2\pi)$ and thus that $v * P = \delta$. If $\varphi \in \ell^\infty(\underline{Z})$ and $T = \varphi * P$, we have also $\varphi = T * v$.

If $\varphi * P \in \lambda$, it follows from (9) that $\varphi = \varphi * P * v \in \lambda$. Thus (10) holds and the theorem is proved.

We now turn to Λ_+ which is the restriction of Λ to $\underline{Z}^+$.

There is an analogue of (9): if $\lambda_+ \in \Lambda_+$, then

(11) $\{u \in \lambda_+ ,\ v \in \ell^1_+\} \Rightarrow u * v \in \lambda_+ .$

If $P = \{p_n\}_0^\infty \in \ell^1_+$, we define $\hat{P}(z) = \sum_0^\infty p_n z^n$ in the closed unit disc.

THEOREM R3. <u>Let</u> $P \in \ell^1_+$ <u>be such that</u> $\hat{P}(z)$ <u>does not vanish on</u> $\{0 < |z| \leq 1\}$. <u>If</u> $\lambda_+ \in \Lambda_+$ <u>and</u> $\varphi = \{\varphi_n\}_0^\infty$, <u>then</u>

(12) $\varphi * P \in \lambda_+ \Rightarrow \varphi \in \lambda_+ .$

<u>Proof</u>. We write $\hat{P}(z) = z^q \hat{T}(z)$, where $\hat{T}(0) \neq 0$. It follows that $\hat{T}(z)$ is nonvanishing in the closed unit disc and that there exists $V \in \ell^1_+$ such that $\hat{T}(z)\hat{V}(z) = 1$ on $\{|z| \leq 1\}$. This means that $P * V = \delta_q$, where δ_q is the sequence $\{\delta_{k,q}\}_{k=0}^\infty$. If $\varphi * P \in \lambda_+$, then we can use (11) to deduce that

$$(\varphi * P) * V = \varphi * (P * V) = \varphi * \delta_q \in \lambda_+ ,$$

which implies that $\varphi \in \lambda_+$. The theorem is proved.

FINAL REMARKS. As an example of other classes where non-vanishing Fourier transforms have inverses in the class, we mention the non-quasianalytic weight algebras introduced by Beurling in [1]. More references are given in the remark p.283 in [3]. Spaces of this type where the elements can behave differently at ∞ and $-\infty$ are discussed in [6].

REFERENCES

1. A. Beurling, Sur les intégrales de Fourier absolument convergentes et leur application à une transformation fonctionnelle. Neuvième congrès des mathématiciens scandinaves, Helsingfors (1938), 345-366.

2. E.T. Copson, On a generalisation of monotonic sequences, Proc. Edinburgh Math. Soc. (2) 17 (1970), 159-164.

3. H.G. Diamond and M. Essén, One-sided Tauberian theorems for kernels which change sign. Proc. of the London Math. Soc. (3) 36 (1978), 273-284.

4. M. Essén, Studies on a convolution inequality. Arkiv f. matematik 5 (1963), 113-152.

5. M. Essén, Note on "A theorem on the minimum modulus of entire functions" by Kjellberg. Math. Scand. 12 (1963), 12-14.

6. M. Essén, Banach algebra methods in renewal theory. Journal d'Analyse Math. 26 (1973), 303-336.

7. Y. Katznelson, An introduction to harmonic analysis. Wiley 1968.

8. D. C. Russell, On bounded sequences satisfying a linear inequality. Proc. of the Edinburgh Math. Soc. 19 (Series II, 1974), 11-16.

9. D.C. Russell, Tauberian-type results for convolution of sequences. In: W. Walter (ed)., General Inequalities 5, Birkhäuser Verlag, Basel, 1987, pp.103-113.

10. D. Widder, The Laplace transform. Princeton Univ. Press 1946.

11. N. Wiener and H.R. Pitt, On absolutely convergent Fourier Stieltjes transforms. Duke Math. J. 4 (1938), 420-436.

Department of Mathematics
University of Uppsala
Thunbergsvägen 3
S-752 38 Uppsala, Sweden

International Series of
Numerical Mathematics, Vol. 80

ON A HARDY-LITTLEWOOD TYPE INTEGRAL INEQUALITY WITH A MONOTONIC WEIGHT FUNCTION

W.N.Everitt and A.P.Guinand

Abstract. The Hardy-Littlewood type integral inequality, which is the subject of study in this paper, is

$$\left[\int_a^\infty w(x)|f'(x)|^2dx\right]^2 \leqslant 4 \int_a^\infty w(x)|f(x)|^2dx \int_a^\infty w(x)|f''(x)|^2dx$$

where $-\infty \leqslant a < \infty$, and w is a positive, monotonic increasing function on (a,∞). The inequality is valid, with the number 4, for all complex-valued f such that f and $f'' \in L_w^2(a,\infty)$. In certain cases the number 4 is best possible and all cases of equality can be described.
The example $w(x) = x$ on $(0,\infty)$ is considered in detail and it is shown the best possible number in the inequality in this case is strictly less than 4. It is known, from simple examples, that the inequality may not hold if the monotonic increasing condition on the weight w is removed.
The proofs given in this paper are essentially in classical analysis and depend, in a curious way, on the original proof of Hardy and Littlewood from 1932, where $w(x) = 1$ on $(0,\infty)$. These proofs are compared with the later operator theoretic proof of the inequality above, first given by Kwong and Zettl in 1979, and later in 1981. Both types of proof offer an explanation of the fact that 4 is a global number for the inequality, i.e. for all intervals (a,∞) and all weights w of the kind prescribed above.

1. INTRODUCTION

In 1932 Hardy and Littlewood [4] discussed two integral inequalities which have since, particularly in recent years, been proved and generalised in a number of different ways. Suppose that f is a real-valued function on the half-line $[0,\infty)$ or the real line $(-\infty, \infty)$ such that the derivative f'' exists almost everywhere, and f and f'' are both in $L^2(0,\infty)$ or $L^2(-\infty, \infty)$ respectively; then f' is in $L^2(0,\infty)$ or $L^2(-\infty,\infty)$ and, respectively.

$$(1.1)\qquad \left[\int_0^\infty f'(x)^2dx\right]^2 \leqslant 4\int_0^\infty f(x)^2dx\int_0^\infty f''(x)^2dx$$

$$(1.2)\qquad \left[\int_{-\infty}^\infty f'(x)^2dx\right]^2 \leqslant \int_{-\infty}^\infty f(x)^2dx\int_{-\infty}^\infty f''(x)^2dx.$$

In both cases the numbers given, 4 in (1.1) and 1 in (1.2), are best possible and cases of equality are known.For details see [4, section 7] and the now classic book Inequalities by Hardy, Littlewood and Polya [5, section 7.8 and 7.9]. (Some details of the equalising functions are also given in section 2 below.)

Recently Kwong and Zettl have observed a remarkable extension of the Hardy-Littlewood inequalities (1.1) and (1.2) above; details are given in [8] and [9]. Suppose that w is any positive, monotonic increasing function on $(0,\infty)$ or $(-\infty,\infty)$; let $L_w^2(0,\infty)$ or $L_w^2(-\infty,\infty)$ denote the corresponding weighted integrable-square spaces, i.e. the collection of all Lebesgue measurable, complex-valued functions f such that, respectively,

$$(1.3)\qquad \int_0^\infty w(x)|f(x)|^2dx < \infty \quad\text{or}\quad \int_{-\infty}^\infty w(x)|f(x)|^2dx < \infty.$$

Kwong and Zettl show that for certain $f \in L_w^2(0,\infty)$ or $L_w^2(-\infty,\infty)$ such that f'' exists almost everywhere and $f'' \in L_w^2(0,\infty)$ or $L_w^2(-\infty.\infty)$ (specifically for f in the domain of a m-dissipative operator associated with differentiation on $(0,\infty)$ or $(-\infty,\infty)$) then, see again (1.1) above,

$$(1.4)\qquad \left[\int_0^\infty w(x)|f'(x)|^2dx\right]^2 \leqslant 4\int_0^\infty w(x)|f(x)|^2dx\int_0^\infty w(x)|f''(x)|^2dx$$

with a similar inequality holding on $(-\infty,\infty)$ but now see again (1.2), with in general the number 1 replaced by 4

$$(1.5)\qquad \left[\int_{-\infty}^{\infty} w(x)|f'(x)|^2dx\right]^2 \leqslant 4 \int_{-\infty}^{\infty} w(x)|f(x)|^2dx \int_{-\infty}^{\infty} w(x)|f''(x)|^2dx$$

The proof of these results by Kwong and Zettl depends upon an inequality for m-dissipative operators in Hilbert spaces given by Kato [7]. In turn Kato's inequality stems from an earlier result of Kallman and Rota [6]. Both of these operator inequalities are very general in application but, as Kwong and Zettl remark, this generality makes it difficult to determine best possible numbers in the inequalities (1.4) and (1.5) for particular cases of the weight function w. The application in [8] of the Kato inequality requires properties of dissipative and m-dissipative operators which are given in the book of Beals [1].

It is possible that the method of quadratic forms, associated with the calculus of variations, introduced by Hardy and Littlewood in [4] and subsequently extended by Everitt in [3] and Evans and Everitt in [2], would apply in certain cases in order to determine best possible constants and to characterise cases of equality. This is shown to be so in the case of (1.4) when $w(x) = x$ for all $x \in (0,\infty)$; this gives an example

$$(1.6)\qquad \left[\int_{0}^{\infty} x|f'(x)|^2dx\right]^2 \leqslant K \int_{0}^{\infty} x|f(x)|^2dx \int_{0}^{\infty} x|f''(x)|^2dx$$

for which the best possible number K is strictly less than 4, in fact it may be shown

$$(1.7)\qquad 2.350 < K < 2.351.$$

Also that there are non-null cases of equality in (1.6). These results will be reported on in section 6 below.

In the general case of (1.4) and (1.5) we give here a classical analytic proof of these inequalities which not only follows the original

Hardy-Littlewood analysis in [4], but does throw some light on the reason why the number 4 is a global upper bound for best possible constants in both inequalities, and why the number 1 in (1.2) has to be replaced by a larger number in the general case (1.5).

It is worth remarking that the general inequalities (1.4) and (1.5) cannot be studied from the properties of second-order symmetric quasi-differential expressions, see Naimark [10], as was the case in [2] and [3]. It is not possible to consider (1.4), for example, as a special case of the general inequality in [2], except when w is constant on $(0,\infty)$ in which case (1.4) reduces to (1.1).

We conclude this section with a list of notations. The results are stated in section 2; proofs and the example (1.6) are given in subsequent sections.

NOTATIONS

The symbol "$(x \in K)$" is to be read as "for all $x \in K$". R and C denote the real number and complex number number fields. Open and closed end-points of intervals of R are denoted by (and [respectively. For any interval I of R the symbol $AC_{loc}(I)$ denotes the collection of complex-valued functions on I which are absolutely continuous (Lebesgue) on all compact sub-intervals of I.

In this paper the interval I will be <u>either</u> (i) open and of the form (a, ∞) where $-\infty \leqslant a < \infty$, <u>or</u> closed and of the form $[a,\infty)$ where $-\infty < a < \infty$.

Let $f: I \to C$ and be Lebesgue measurable on I; then $L(I)$ and $L^2(I)$ denote the linear spaces of all such f which satisfy, respectively,

$$(L)\int_a^\infty |f(x)|dx < \infty \qquad (L)\int_a^\infty |f(x)|^2dx < \infty .$$

For any (weight) $w : I \to [0,\infty)$ which is Lebesgue measurable, the symbol $L_w^2(I)$ denotes the linear space of all such functions f which satisfy

$$(L)\int_a^\infty w(x)|f(x)|^2dx < \infty.$$

In these notations (L) denotes the Lebesgue integral, and is omitted from the integral when not required, i.e. all integrals are Lebesgue unless as indicated below.

Throughout this paper the weight w is monotonic increasing on I, i.e.

$$(1.8) \qquad 0 \leqslant w(x_1) \leqslant w(x_2) < \infty \qquad (x_1,x_2 \in I \text{ with } x_1 < x_2).$$

This allows the introduction of two other forms of integration over compact intervals $[\alpha,\beta] \subseteq I$:

(i) the Riemann-Stieltjes integral, restricted to the case when α and β are points of continuity of the weight w and f is continuous on $[\alpha,\beta]$,

$$(1.9) \qquad (RS)\int_{[\alpha,\beta]} f(x)dw(x); \quad \text{or} \quad (RS)\int_\alpha^\beta f(x)dw(x)$$

for definitions of this integral see Rudin [11, chapter 6] or Widder [13, Chapter V]

(ii) the Lebesgue-Stieltjes integral

$$(1.10) \qquad (LS)\int_{[\alpha,\beta]} f(x)dw(x);$$

where f is Borel measurable, and dw is to be intepreted as the regular Lebesgue-Stieltjes measure, defined on the Borel sets of R, when generated by the monotonic function w; for definitions see [11, chapter 11] and Taylor and Kingman [12, Chapters 4 and 5].

We note that when α and β are points of continuity of w, and f is continuous on $[\alpha,\beta]$ then the integrals in (1.9) and (1.10) take the same value.

Supposing (1.9) or (1.10) to exist for all compact intervals $[\alpha,\beta] \subset (a,\infty)$, and if the limit of (1.9) or (1.10) exists as, independently, $\alpha \to a+$ and $\beta \to \infty$ then this limit is denoted by, respectively

$$(1.11) \qquad (RS)\int_{\to a}^{\to\infty} f(x)dw(x); \qquad \text{or} \qquad (LS)\int_{\to a}^{\to\infty} f(x)dw(x).$$

Finally we use the Lebesgue space $L_{loc}(I)$ of those complex-valued, measurable functions f defined on the interval I such that

$$(L)\int_{[\alpha,\beta]} |f(x)|dx < \infty \qquad (\text{all } [\alpha,\beta] \subseteq I).$$

2. STATEMENT OF RESULTS

Suppose given a weight function w with the following properties

(2.1)

(i) $w; (-\infty,\infty) \to [0,\infty)$, i.e. $w(x) \geq 0$ $(x \in (-\infty,\infty)$

(ii) w is not the null function

(iii) w is monotonic increasing on $(-\infty,\infty)$, i.e.

$$0 \leq w(x_1) \leq w(x_2) \quad (x_1,x_2 \in (-\infty,\infty) \text{ with } x_1 \leq x_2).$$

We note that w is then Lebesgue measureable, and

$$(2.2) \qquad w \in L_{loc}(-\infty,\infty).$$

For such a given weight w define (uniquely) the number $a \in R \cup \{-\infty\}$, i.e. $a \geq -\infty$, by

(2.3) $$a := \inf\{x \in (-\infty,\infty) \mid w(x) > 0\}$$

and let

(2.4) $$w(a+) := \lim_{x \to a+} w(x)$$

so that $0 \leqslant w(a+) < \infty$. We shall need to distinguish the cases

(2.5) $$(\alpha)\quad w(a+) > 0 \qquad\qquad (\beta)\quad w(a+) = 0.$$

We note further that if $a > -\infty$ then $w \in L(a,x)$ $(x > a)$ and

(2.6) $$\int_a^x w(t)dt > 0 \qquad (x \in (a,\infty)).$$

Given w as above, and hence the number $a \in R \cup \{-\infty\}$, we define the linear manifold $D(w)$ of the space $L_w^2(a,\infty)$ by

(2.7) $$D(w) := \{f:(a,\infty) \to R \mid \quad \text{(i)} \quad f \text{ and } f' \in AC_{loc}(a,\infty)$$
$$\text{(ii)} \quad f \text{ and } f'' \in L_w^2(a,\infty)\}.$$

It is convenient to define the linear manifold D_ξ, for all $\xi \in R$, by

(2.8) $$D_\xi := \{f:[\xi,\infty) \to R \mid \quad \text{(i)} \quad f \text{ and } f' \in AC_{loc}(\xi,\infty)$$
$$\text{(ii)} \quad f \text{ and } f'' \in L^2(\xi,\infty)\}.$$

We recall here, for later use, the inequalities of Hardy and Littlewood [4] but as given in the book Inequalities by Hardy, Littlewood and Polya [5, section 7.8].

Theorem (Hardy and Littlewood). Let the linear manifold D_ξ of $L^2(\xi,\infty)$ be defined as in (2.8); then

(2.9) (i) $$f' \in L^2(\xi,\infty) \qquad\qquad (f \in D_\xi)$$

(ii) for any $\rho \in R$

(2.10) $$\int_\xi^\infty \{\rho^4 f(x)^2 - \rho^2 f'(x)^2 + f''(x)^2\}dx \geqslant 0 \qquad\qquad (f \in D_\xi)$$

with equality if and only if, for some $A \in R$,

(2.11) $$f(x) = A \exp[-\tfrac{1}{2}\rho(x-\xi)] \sin\{\tfrac{1}{2}\rho(x-\xi)\sqrt{3} - \pi/3\} \qquad (x \in [\xi,\infty))$$

(iii) the inequality holds

(2.12) $$\left[\int_\xi^\infty f'(x)^2 dx\right]^2 \leqslant 4 \int_\xi^\infty f(x)^2 dx \int_\xi^\infty f''(x)^2 dx \qquad (f \in D_\xi)$$

with equality if and only if for some $A \in R$ and some $\rho > 0$, f satisfies (2.11).

Proof. See [5, section 7.8].

We remark that property (i) in (2.9) is sometimes called Littlewood's principle; when a function and its second derivative behave at ∞ then its first derivative may also conform in the same way.

We may now state

Theorem 1 Let the weight w satisfy the conditions (2.1) above; let $a \in R \cup \{-\infty\}$ and $D(w) \subset L^2_w(a,\infty)$ be defined as in (2.3) and (2.7) respectively; then for all $f \in D(w)$ the following properties hold:

(2.13)

(i) for all $\xi > a$ $\quad f^{(r)} \in L^2(\xi,\infty)$ $\qquad (r=0,1,2)$

(ii) $$(\text{RS or LS}) \int_{\to a}^{\to\infty} \left\{\int_X^\infty f^{(r)}(t)^2 dt\right\} dw(x) < \infty \qquad (r=0,1,2)$$

(iii) $f' \in L^2_w(a,\infty)$ i.e. $\displaystyle\int_a^\infty w(x)|f'(x)|^2 dx < \infty$

(iv) $$\lim_{X\to\infty} w(X) \int_X^\infty f^{(r)}(t)^2 dt = 0 \qquad (r=0,1,2).$$

(v) $$\lim_{\xi\to a+} w(\xi) \int_\xi^\infty f^{(r)}(t)^2 dt \text{ exists and is finite} \qquad (r=0,1,2).$$

Case α When $w(a+) > 0$ the following additional properties hold:

(2.14) (i) $f^{(r)} \in L^2(a,\infty)$ $(r=0,1,2)$

(ii) if also $a > -\infty$ then $f(a)$ and $f'(a)$ can be defined so that f and $f' \in AC_{loc}[a,\infty)$

(iii) for all $\rho \geq 0$ and all $k \in R$

$$\int_a^\infty w(x)\left\{\rho^4 f(x)^2 - k\rho^2 f'(x)^2 + f''(x)^2\right\}dx =$$

$$w(a+)\int_a^\infty \left\{\rho^4 f(x)^2 - k\rho^2 f'(x)^2 + f''(x)^2\right\}dx +$$

$$(\text{LS or RS})\int_{\to a}^{\to\infty}\left\{\int_x^\infty\left[\rho^4 f(t)^2 - k\rho^2 f'(t)^2 + f''(t)^2\right]dt\right\}dw(x).$$

Case β When $w(a+) = 0$ the following additional properties hold

(i) $\lim_{\xi\to a+} w(\xi)\int_\xi^\infty f^{(r)}(t)^2 dt = 0$ $(r=0,1,2)$

(2.15)

(ii) for all $\rho \geq 0$ and all $k \in R$

$$\int_a^\infty w(x)\left\{\rho^4 f(x)^2 - k\rho^2 f'(x)^2 + f''(x)^2\right\}dx =$$

$$(\text{LS or RS})\int_{\to a}^{\to\infty}\left\{\int_x^\infty\left[\rho^4 f(t)^2 - k\rho^2 f'(t)^2 + f''(t)^2\right]dt\right\}dw(x).$$

Proof. We defer the proof to the next section but note here that the monotonicity of the weight w, see (2.1)(iii), is essential for these results to hold.

We remark only that the result (2.13)(iii) shows that the Littlewood principle, referred to above, extends from the Hardy-Littlewood case (2.9) to the introduction of the monotone weight w.

We pass to

Theorem 2 Let the conditions of Theorem 1 hold; then for all $\rho \geq 0$

$$\int_a^\infty w(x)\left\{\rho^4 f(x)^2 - \rho^2 f'(x)^2 + f''(x)^2\right\}dx \geq 0 \qquad (f \in D(w)). \tag{2.16}$$

For any $\rho > 0$ there is equality in (2.16) if and only if the following conditions are satisfied by the function f and the weight w:

1. If $a > -\infty$ and $w(a+) > 0$ then either f is null on (a,∞), or

   (i) w is a step function on (a,∞) with jumps occuring only at some or all of the points

$$\{a + 2n\pi/(\rho\sqrt{3}) : n = 0,1,2,\ldots\} \tag{2.17}$$

and such that

$$\int_a^\infty w(x)\exp[-\rho x]dx < \infty \tag{2.18}$$

and (ii) for some $A \in R$ and for all $x \in (a,\infty)$

$$f(x) = A\exp[-\tfrac{1}{2}\rho(x-a)]\sin\{\tfrac{1}{2}\rho(x-a) - \pi/3\}. \tag{2.19}$$

2. If $a > -\infty$ and $w(a+) = 0$, or if $a = -\infty$ then f is null on (a,∞) i.e. $f(x) = 0$ $(x \in (a,\infty))$.

Proof. See section 4 below.

Finally we state

Theorem 3. Let the weight w satisfy the conditons (2.1) above, and let a and $D(w)$ be defined as in (2.3) and (2.7) respectively; then the following integral inequality is valid

$$\left[\int_a^\infty w(x)f'(x)^2dx\right]^2 \leq 4\int_a^\infty w(x)f(x)^2dx \int_a^\infty w(x)f''(x)^2dx \qquad (f \in D(w)). \tag{2.20}$$

For any weight w satisfying (2.1) let K(w) be the best possible number, i.e. the lower bound, such that

$$(2.21)\qquad \left[\int_a^\infty w(x)f'(x)^2dx\right]^2 \leq K(w)\int_a^\infty w(x)f(x)^2dx \int_a^\infty w(x)f''(x)^2dx \qquad (f \in D(w))$$

then the following results hold

(i) for all w satisfying (2.1)

$$(2.22)\qquad 0 < K(w) \leq 4$$

(2.23) (ii) if $a > -\infty$ and $w(a+) > 0$ then $K(w) = 4$

(iii) if $a = -\infty$, or if $a > -\infty$ but $w(a+) = 0$ then it may happen that $K(w) < 4$.

If for some $\rho > 0$ the weight w satisfies the conditions in 1(i) of Theorem 2, then $K(w) = 4$ and cases of equality are given by those functions determined by 1(ii), i.e. (2.19); the converse of this statement holds i.e. if there are non-trivial cases of equality in (2.20) then w and f are so prescribed.

Proof. This is deferred to section 5 below.

Remarks 1. The inequality (2.20) is due to Kwong and Zettl; see [8] where, however, the set of functions for which the inequality is valid is the domain of the square of a m-dissipative operator in $L^2_W(a,\infty)$; this is changed in [9] to a direct definition which is equivalent to the set D(w) of this paper.

2. The result (2.23) is also due to Kwong and Zettl; see [9, Theorem 8, (2.12)].

3. Some examples to illustrate these results are:

(i) $w(x) = 0 \quad (x \in (-\infty,0)) \qquad w(x) = 1 \qquad (x \in [0,\infty))$

for which $a = 0$ and $w(a+) = 1 > 0$; this gives the Hardy-Littlewood inequality (2.12), with $\xi = 1$, and $K(w) = 4$; all cases of equality are given by (2.11) which should be compared with the results in 1 (i) of Theorem 2 above.

(ii) $w(x) = 1$ $(x \in (-\infty,\infty))$ for which $a = -\infty$ and $w(a+) = 1 > 0$; this gives the Hardy-Littlewood inequality (1.2) for which $K(w) = 1$; the only case of equality is the null function, see [5, section 7.9, Theorem 261]; these results are special and specific to this inequality and cannot be deduced from the general analysis in [8] or [9], nor from the results in Theorem 3 above.

(iii) $w(x) = 0$ $(x \in (-\infty,0])$ $w(x) = x$ $(x \in [0,\infty))$ for which $a = 0$ and $w(a+) = 0$; this is the example mentioned in the Introduction, see (1.6) above and considered in some detail in section 6 below; here $2.350 < K(w) < 2.351$ and these are non-null cases of equality.

(iv) If the monotonic increasing condition on w is replaced by a monotonic decreasing condition then the results as stated in Theorem 1 to 3 above no longer hold; for example the inequality

$$\left[\int_0^\infty e^{-x} f'(x)^2 dx\right]^2 \leqslant K \int_0^\infty e^{-x} f(x)^2 dx \int_0^\infty e^{-x} f''(x)^2 dx \tag{2.24}$$

is not valid for any number $K > 0$; to see this let $f(x) = x$ $(x \in [0,\infty))$. If the monotonicity is broken then again the inequality may fail; as an example (due to W.Walter) take

$$w(x) = 1 \left(x \in [n\pi, n\pi + \tfrac{1}{n}]\right), = 0 \quad \text{(otherwise)} \qquad (n=1,2,\ldots)$$

and

$$f(x) = \sin(x) \qquad (x \in [1,\infty);$$

then a calculation shows that

$$\int_{n\pi}^{n\pi+\frac{1}{n}} w(x)f(x)^2dx = \int_{n\pi}^{n\pi+\frac{1}{n}} w(x)f''(x)^2dx \backsim \frac{1}{3n^3} \qquad (n \to \infty)$$

and
$$\int_{n\pi}^{n\pi+\frac{1}{n}} w(x)f'(x)^2dx \backsim \frac{1}{n} \quad (n \to \infty)$$

so that f and $f'' \in L_w^2(1,\infty)$ but $f' \notin L_w^2(1,\infty)$.

(v) The inequalities in Theorems 1,2 and 3 above can all be extended to the case when the functions f are complex-valued on (a,∞); it is sufficient to replace such terms as $f(x)^2$, etc., by $|f(x)|^2$.

3. Proof of Theorem 1. We begin with a Lemma which we require at a number of places in the sections which follow.

Lemma Let the monotone function w and the number a be as given in section 2; let the function $\phi : R \to R$ satisfy $\phi \in L_{loc}(-\infty,\infty)$; let the compact interval $[\alpha,\beta] \subset (a,\infty)$ be chosen with α and β as points of continuity of w; let $\gamma \in [\alpha,\beta]$ and let $k \in R$; then the following equalities hold

$$\text{(3.1)} \qquad (L) \int_{\alpha}^{\beta} w(x)\phi(x)dx = (LS) \int_{[\alpha,\beta]} w(x)d\left[k + \int_{\gamma}^{x} \phi(t)dt\right]$$

$$\text{(3.2)} \qquad = (RS) \int_{\alpha}^{\beta} w(x)d\left[k + \int_{\gamma}^{x} \phi(t)dt\right]$$

$$\text{(3.3)} \qquad = w(\beta)\left[k + \int_{\gamma}^{\beta} \phi(t)dt\right] - w(\alpha)\left[k + \int_{\gamma}^{\alpha} \phi(t)dt\right]$$

$$- (RS) \int_{\alpha}^{\beta} \left[k + \int_{\gamma}^{x} \phi(t)dt\right] dw(x)$$

$$(3.4)\qquad = w(\beta)\left[k + \int_\gamma^\beta \phi(t)dt\right] - w(\alpha)\left[k + \int_\gamma^\alpha \phi(t)dt\right]$$

$$-(LS)\int_{[\alpha,\beta]} \left[k + \int_\gamma^x \phi(t)dt\right]dw(x).$$

<u>Proof</u> We note that $w\phi \in L_{loc}(-\infty,\infty)$.

It is sufficient to suppose that $\phi(x) \geqslant 0$ $(x \in R)$; otherwise we work separately with $\phi_+ = \max\{\phi,0\}$ and $\phi_- = -\min\{\phi,0\}$, since then $\phi = \phi_+ - \phi_-$ and we can argue separately with ϕ_+ and ϕ_-.

With $\phi \geqslant 0$ on R and given $k \in R$ define

$$(3.5)\qquad \psi(x) = k + \int_\gamma^x \phi(t)dt \qquad (x \in R)$$

so that ψ is continuous and monotonic increasing on R. Thus ψ generates a regular Lebesgue-Stieltjes measure on the Borel sets of R; see [11, section 11.4] and [12, section 4.5]. This measure is absolutely continuous with respect to Lebesgue measure, see [12, section 6.4], and (3.1) above then follows as an example of the Radon-Nikodym theory, see [12, section 6.4, example 7]. Note that the (LS) integral can be taken over the closed interval $[\alpha,\beta]$ since ϕ is continuous at the end-points α and β; see the remarks in [12, section 5.5, page 125].

The Riemann-Stieltjes integral in (3.2) is defined as in [11, section 6.2] and exists in virtue of [11, Theorem 6.9]. The identification of the (LS) and (RS) integrals in (3.1) and (3.2) follows from a straight-forward adaption of the proof given in [11, Theorem 11.22] of the identification of the (L) and (R) integrals. (The continuity of ϕ on $[\alpha,\beta]$ eases the proof; see the remark below concerning (3.4)).

The step from (3.2) to (3.3) is integration by parts; noting that w is monotonic on $[\alpha,\beta]$ and continuous at the points α and β, and that ψ, see (3.5), is monotonic and continuous on $[\alpha,\beta]$ we can appeal to the theorem in Widder [13; section 3.1, Theorem 3.2].

Finally the proof of the step from (3.3) to (3.4) again follows the argument in passing from (3.1) to (3.2) only now the Lebesgue-Stieltjes measure is generated by the monotonic function w. Here it is essential for w to be continuous at the end-points α and β for the proof in [11, Theorem 11.33] to extend, and to adapt the proof so that all partitions of $[\alpha,\beta]$ are taken at points of continuity of w. Note also that the integral in (3.4) has to be taken over the closed interval $[\alpha,\beta]$.

We pass now to the proof of Theorem 1.

From the definition of the number a in (2.3) it follows that $w(x) \geqslant w(\xi) > 0$ $(x \in [\xi,\infty))$ holds for all $\xi > a$. If now $f \in D(w)$ then

$$\int_\xi^\infty f(x)^2dx \leqslant \frac{1}{w(\xi)}\int_\xi^\infty w(x)f(x)^2dx < \infty,$$

and similarly for f'', i.e. f and $f'' \in L^2(\xi,\infty)$. It then follows from the Hardy-Littlewood inequality (2.9) that $f' \in L^2(\xi,\infty)$. These results hold for all $f \in D(w)$ and for all $\xi > a$; thus (2.13)(i) follows.

In all that follows the numbers ξ and X satisfy the following requirements

(3.6)
(i) $a < \xi < X < \infty$
(ii) the weight w is continuous at ξ and X

noting that (ii) can be satisfied for ξ close to a, and X close to ∞.

Now apply the Lemma given at the start of this section with

$$\phi = f^2 + f''^2; \quad \alpha = \gamma = \xi; \quad \beta = X; \quad k = -\int_\xi^\infty \phi(x)dx$$

and we obtain

$$(3.7)\quad \int_{\xi}^{X} w(x)\left\{f(x)^2+f''(x)^2\right\}dx = -\, w(X)\int_{X}^{\infty}\left\{f(x)^2+f''(x)^2\right\}dx + w(\xi)\int_{\xi}^{\infty}\left\{f(x)^2+f''(x)^2\right\}dx$$

$$+(\text{LS or RS})\int_{[\xi,X]}\left\{\int_{x}^{\infty}(f(t)^2+f''(t)^2)dt\right\}dw(x).$$

From the monotonicity of the positive weight w

$$w(X)\int_{X}^{\infty}\left\{f(x)^2+f''(x)^2\right\}dx \leqslant \int_{X}^{\infty} w(x)\left\{f(x)^2+f''(x)^2\right\}dx$$

and (3.7) gives

$$\int_{\xi}^{X} w(x)\left\{f(x)^2+f''(x)^2\right\}dx \geqslant -\int_{X}^{\infty} w(x)\left\{f(x)^2+f''(x)^2\right\}dx \;+$$

$$\int_{[\xi,X]}\left\{\int_{x}^{\infty}(f(t)^2+f''(t)^2)dt\right\}dw(x).$$

which, on rearrangement, gives

$$\int_{\xi}^{X}\left\{\int_{x}^{\infty}(f(t)^2+f''(t)^2)dt\right\}dw(x) \leqslant \int_{\xi}^{\infty} w(x)\left\{f(x)^2+f''(x)^2\right\}dx$$

$$\leqslant \int_{a}^{\infty} w(x)\left\{f(x)^2+f''(x)^2\right\}dx$$

Now let $\xi \to a+$ and $X \to \infty$, noting (3.6)(ii), to give

$$(3.8)\quad \int_{\to a}^{\to\infty}\left\{\int_{x}^{\infty}(f(t)^2+f''(t)^2)dt\right\}dw(x) \leqslant \int_{a}^{\infty} w(x)\left\{f(x)^2+f''(x)^2\right\}dx$$

From the inequality $2a\,b \leqslant \alpha^2 + b^2$ for $a, b \in R$, we obtain

$$(3.9)\quad \int_{\to a}^{\to\infty}\left\{\int_X^{\infty}(f(t)^2+f''(t)^2)dt\right\}dw(x) \geqslant 2\int_{\to a}^{\to\infty}\left\{\int_X^{\infty}f(t)^2dt\int_X^{\infty}f''(t)^2dt\right\}^{1/2}dw(x)$$

$$\geqslant \int_{\to a}^{\to\infty}\left\{\int_X^{\infty}f'(t)^2dt\right\}dw(x)$$

on using the Hardy-Littlewood inequality (2.12).

From (3.8) and (3.9) the required result (2.13) (ii) now follows.

Appealing again to the Lemma we now have

$$(3.10)\quad \int_{\xi}^{X}w(x)f'(x)^2dx = -\,w(X)\int_X^{\infty}f'(t)^2dt + w(\xi)\int_{\xi}^{\infty}f'(t)^2dt$$

$$+\ (\text{LS or RS})\int_{[\xi,X]}\left\{\int_X^{\infty}f'(t)^2dt\right\}dw(x)$$

$$(3.11)\quad \leqslant w(\xi)\int_{\xi}^{\infty}f'(t)^2dt + \int_{[\xi,X]}\left\{\int_X^{\infty}f'(t)^2dt\right\}dw(x).$$

Now from the monotonicity of w and the inequality (2.12)

$$(3.12)\quad \left\{w(\xi)\int_{\xi}^{\infty}f'(t)^2dt\right\}^2 \leqslant 4w(\xi)\int_{\xi}^{\infty}f(t)^2dt.w(\xi)\int_{\xi}^{\infty}f''(t)^2dt$$

$$\leqslant 4\int_a^{\infty}w(t)f(t)^2dt\int_a^{\infty}w(t)f''(t)^2dt$$

and the right-hand side is finite since $f \in D(w)$. Thus applying the results (3.10), (3.12), (3.9), (3.8) in that order we find

$$\int_{\xi}^{X} w(x)f'(x)^2dx \leqslant 2\left\{\int_a^\infty w(x)f(x)^2dx \int_a^\infty w(x)f''(x)^2dx\right\}^{1/2} + \int_a^\infty w(x)\left\{f(x)^2 + f''(x)^2\right\}dx$$

for all ξ, X satisfying (3.6). In this last result let $\xi \to a+$ and $X \to \infty$ to obtain (again using $2\,ab \leqslant a^2 + b^2$)

$$(3.13) \quad \int_a^\infty w(x)f'(x)^2dx \leqslant 2\int_a^\infty w(x)\left\{f(x)^2 + f''(x)^2\right\}dx$$

Thus $f' \in L^2_w(a,\infty)$ and the required result (2.13) (iii) now follows.

For the integers $r = 0,1,2$ we have, for all $X > a$,

$$w(X)\int_X^\infty f^{(r)}(x)^2dx \leqslant \int_X^\infty w(x)f^{(r)}(x)^2dx$$

and the right-hand side tends to the limit 0 as $X \to \infty$. This gives the required result (2.13)(iv).

From the Lemma we obtain, for $r = 0,1,2$,

$$\int_\xi^\infty w(x)f^{(r)}(x)^2dx = -\,w(X)\int_X^\infty f^{(r)}(x)^2dx + w(\xi)\int_X^\infty f^{(r)}(x)^2dx \int_{[\xi,X]}\left\{\int_X^\infty f^{(r)}(t)^2dt\right\}dw(x).$$

In this result let $X \to \infty$ to give, for all $\xi > a$,

$$(3.14) \quad \int_\xi^\infty w(x)f^{(r)}(x)^2dx = w(\xi)\int_\xi^\infty f^{(r)}(x)^2dx + \int_\xi^{\to\infty}\left\{\int_X^\infty f^{(r)}(t)^2dt\right\}dw(x).$$

From the previous results it now follows that

$$\lim_{\xi\to a+} w(\xi)\int_{\xi}^{\infty} f^{(r)}(x)^2dx$$

exists and is finite for $r = 0,1,2$. This establishes (2.12)(v).

<u>Case α</u> Suppose now, for $a \geqslant -\infty$,

$$w(a+) > 0. \tag{3.15}$$

From the condition of $f \in D(w)$ and the above results we have $f^{(r)} \in L^2_W(a,\infty)$. for $r = 0,1,2$, and so

$$w(a+)\int_a^{\infty} f^{(r)}(x)^2dx \leqslant \int_a^{\infty} w(x)f^{(r)}(x)^2dx < \infty.$$

Thus from (3.15) it follows that $f^{(r)} \in L^2(a,\infty)$ for $r = 0,1,2$. This gives the result (2.14)(i).

Suppose additionally $a > -\infty$. Then from

$$f(\xi) = f(X) - \int_{\xi}^{X} f'(x)dx \qquad f'(\xi) = f'(X) - \int_{\xi}^{X} f''(x)dx,$$

and noting, since $-\infty < a < X < \infty$, $f', f'' \in L^2(a,X)$ give $f', f'' \in L(a,X)$, it follows that $\lim_{\xi\to a+} f(\xi)$ and $\lim_{\xi\to a+} f'(\xi)$ exist and are finite. If we define $f(a)$ and $f'(a)$, respectively, by these limits then it follows that both f and $f' \in AC_{loc}[a,\infty]$; this establishes (2.14)(ii).

Returning now to the general situation with $a \geqslant -\infty$. In (3.14) let $\xi \to a+$, and using (3.15) and (2.14)(i) gives, for $r = 0,1,2$,

$$\int_a^{\infty} w(x)f^{(r)}(x)^2dx = w(a+)\int_a^{\infty} f^{(r)}(x)^2dx + \int_{\to a}^{\to\infty}\left\{\int_X^{\infty} f^{(r)}(t)^2dt\right\}dw(x). \tag{3.16}$$

Multiplying this last result by the appropriate power of ρ and, in the case $r=1$ by k, yields the result (2.14)(iii).

Case β Suppose now

$$w(a+) = 0. \tag{3.17}$$

From(2.13)(i) it follows that $f^{(r)} \in L^2(\xi,\infty)$ for all $\xi > a$ and $r = 0,1,2$. If now $f^{(r)} \in L^2(a,\infty)$ then the required result (2.15)(i) follows at once from (3.17).

Suppose then $f^{(r)} \notin L^2(a,\infty)$ i.e.

$$\lim_{\xi\to a+} \int_\xi^\infty f^{(r)}(t)^2 dt = \infty. \tag{3.18}$$

From (2.13)(v) we have

$$\lim_{\xi\to a+} w(\xi)\int_\xi^\infty f^{(r)}(t)^2 dt = L_r(\text{say}) \geqslant 0; \tag{3.19}$$

suppose $L_r > 0$. Then for ξ close to a+

$$w(\xi) \geqslant \tfrac{1}{2} L_r \left\{\int_\xi^\infty f^{(r)}(t)^2 dt\right\}^{-1} .$$

Multiply this inequality by $f^{(r)}(\xi)^2$ and integrate over the interval $[x,X]$ where X is chosen so that

$$\int_X^\infty f^{(r)}(t)^2 dt > 0.$$

Then, for all x close to a+,

$$\int_x^X w(\xi)f^{(r)}(\xi)^2 d\xi \geqslant \tfrac{1}{2} L_r \,\ell n \left\{\int_x^\infty f^{(r)}(t)^2 dt \Big/ \int_X^\infty f^{(r)}(t)^2 dt\right\}.$$

Now let $x \to a+$ and (3.18) gives a contradiction since $f^{(r)} \in L^2_w(a,\infty)$.

Thus in (3.19) it must be that $L_r = 0$, and this for $r = 0,1,2$. This establishes the result (2.15)(i).

If now in (3.14) we let $\xi \to a+$ but use (2.15)(i) we obtain, for $r = 0,1,2$,

$$(3.20)\qquad \int_a^\infty w(x)f^{(r)}(x)^2dx = \int_{\to a}^{\to\infty}\left\{\int_x^\infty f^{(r)}(t)^2dt\right\}dw(x).$$

As in Case α the required identity (2.15)(ii) now follows.

This completes the proof of Theorem 1.

4. Proof of Theorem 2. The proof of the inequality (2.16) follows from the identities (2.14)(iii) and (2.15)(ii), and the Hardy-Littlewood inequality (2.10). We can incorporate the two identities into one if we use the Lebesgue-Stieltjes integral based on the non-negative measure generated by the monotonic weight w on the Borel sets of R.

In both the cases α and β of Theorem 1 the identities (2.14)(iii) and (2.15)(ii) can be written in the single form (for all $\rho \geqslant 0$, all $k \in R$ all $f \in D(w)$)

$$(4.1)\qquad \int_a^\infty w(x)\left[\rho^4f(x)^2-k\rho^2f'(x)^2 + f''(x)^2\right]dx =$$

$$(LS)\int_{[a,\infty]}\left\{\int_x^\infty\left[\rho^4f(t)^2-k\rho^2f'(t)^2 + f''(t)^2\right]dt\right\}dw(x).$$

In (4.1) it is understood that the interval $[a,\infty)$ is to be interpreted as $(-\infty.\infty)$ in the case when $a = -\infty$. Note that when $a > -\infty$ the w-measure of the point set $\{a\}$ is $w(a+)$.

The inequality (2.16) now follows from (4.1) on taking $k=1$. The integrand on the right-hand side is then non-negative by the Hardy-

Littlewood inequality (2.10) (recall that if $f \in D(w)$ then $f \in D_\xi$ for all $\xi > a$), and the w-measure is non-negative since w is monotonic increasing on R.

We now consider the cases of equality in (2.16) as given in the statement of Theorem 2.

1. Suppose $a > -\infty$ and $w(a+) > 0$. In this case $f \in D_a$ and both terms on the right-hand side of the identity (2.14)(iii), with k=1, have to be zero. Firstly, since $w(a+) > 0$, this requires

$$\int_a^\infty \left[\rho^4 f(t)^2 - \rho^2 f'(t)^2 + f''(t)^2\right] dt = 0 \tag{4.2}$$

and this can only happen, see (2.19), if for some $A \in R$

$$f(x) = A\exp[-\rho(x-a)]\sin\{\tfrac{1}{2}\rho(x-a)\surd 3 - \pi/3\} \qquad (x \in [a,\infty)) \tag{4.3}$$

and if f given by (4.3) satisfies $f \in D(w)$; the latter is the case if (2.18) holds or, if not, that $A = 0$, i.e. f is null.

If, for the given value of $\rho > 0$, (2.18) is satisfied then secondly, equality now holds in (2.16) if and only if

$$\int_{\to a}^{\to\infty} \left\{\int_x^\infty \left[\rho^4 f(t)^2 - \rho^2 f'(t)^2 + f''(t)^2\right] dt\right\} dw(x) = 0 \tag{4.4}$$

with f given by (4.3)

Suppose that for some point $\xi \geq a$ the weight w is right-continuous and strictly increasing in the right-neighbourhood of ξ, i.e.

$$w(x) > w(\xi+) = w(\xi) \qquad (x > \xi). \tag{4.5}$$

With f given by (4.3) define $\phi: [a,\infty) \to R$ by

$$\phi(x) = \int_x^\infty \left[\rho^4 f(t)^2 - \rho^2 f'(t)^2 + f''(t)^2\right] dt \qquad (x \in [a,\infty)). \tag{4.6}$$

Recall $\phi(x) \geq 0$ $(x \in [a,\infty))$, from (2.10), and $\phi(a) = 0$. Clearly ϕ is continuous on $[a,\infty)$, so that if $\phi(\xi) > 0$ then for some $\delta > 0$ and some $\mu > 0$, $\phi(x) \geq \mu > 0$ for all $x \in [\xi,\xi+\delta]$ and

(4.7) $$\int_{\to a}^{\to\infty}\left\{\int_{x}^{\infty}\left[\rho^4 f(t)^2-\rho^2 f'(t)^2 + f''(t)^2\right]dt\right\}dw(x) \geq (w(\xi+\delta)-w(\xi))\mu > 0$$

in contradiction to (4.4), unless A=0 and f is null on $[a,\infty)$.

Otherwise $\rho(\xi) = 0$. Suppose $\phi(\tau) > 0$ for all $x \in (\xi,\xi+\delta]$ for some $\delta > 0$; then since w is increasing to the right of ξ recall (4.5) above, there exists an interval $[\alpha,\beta] \subset (\xi,\xi+\delta]$ such that $w(\beta+) > w(\alpha-)$, in which case there is again a contradiction to (4.4) as with (4.7). Alternatively we have to suppose there exists a sequence $\{x_n: n,1,2,\ldots\}$ with $\xi < x_{n+1} < x_n < \xi+\delta$ $(n=1,2\ldots)$; $\lim_{n\to\infty} x_n = \xi$, and $\phi(x_n) = 0$ $(n = 1,2,\ldots)$; however with f given by (4.3) ϕ is an analytic function on $[\xi, \xi+\delta]$ and so must be identically zero on $[\xi,\xi+\delta]$, i.e. $A = 0$ and f is null on $[a,\infty]$.

Thus for equality in (2.16) there can be no points at which w is right-continuous and strictly increasing. Similarly there are no points $\xi > 0$ at which w is left-continuous and strictly increasing.

Hence w must be a step function with a jump at the point a and, possibly, at points $\{x_n : n = ,2,\ldots\}$ with $a < x_n < x_{n+1}$ $\{n=1,2,\ldots)$. At any point x_n we have to ensure that $\phi(x_n) = 0$, otherwise (4.4) will fail due to a contribution $(w(x_n+)-w(x_n-))\phi(x_n)$. Thus the function f must satisfy, in addition to (4.3)

(4.8) $$f(x) = A_n\exp[-\tfrac{1}{2}\rho(x-x_n)]\sin\{\tfrac{1}{2}\rho(x-x_n)\surd 3-\pi/3\} \qquad (x \in [x_n,\infty)).$$

However (4.3) and (4.8) are compatible if and only if for some positive integer r

$$x_n = a + 2r\pi/(\rho\surd 3) \quad \text{and} \quad A_n\exp[-\rho(a-x_n)] = A.$$

This completes the proof of 1. of the conditions for equality in Theorem 2.

2. Suppose $a > -\infty$ and $w(a+) = 0$, or $a = -\infty$.

If $a > -\infty$ and $w(a+) = 0$, the original conditions on w, see (2.6), imply that w is continuous and strictly increasing at a. The arguments above can then be extended to show that (4.4) is satisfied if and only if f is null on $[a,\infty)$. (Define ϕ again by (4.6); if $\phi(a) = 0$ then f is given by (4.3) and (4.4) requires $A = 0$; if $\phi(a) > 0$ then a result similar to (4.7) holds to prevent this possibility.)

If $a = -\infty$ and $w(a+) > 0$ then f, f', and $f'' \in L^2(-\infty,\infty)$ and equality in (2.16) demands that (4.2) holds (with $a = \infty$). The analysis in [5, section 7.8, pages 189 and 190], in particular extend [5, (7.8.3)] to the interval $(-\infty,\infty)$. Alternatively if w is not constant on $(-\infty,\infty)$ then argue with (4.4), in the neighbourhood of $-\infty$, as above.

Finally if $a = -\infty$ and $w(a+) = 0$ then w is strictly increasing in every interval of the form $(-\infty,\xi)$ and we can adapt the argument above to show that (4.4) is only possible if f is null on $(-\infty,\infty)$.

This completes the proof of Theorem 2.

5. Proof of Theorem 3 The proof of the general inequality (2.20) follows from the inequality (2.16) as given in Theorem 2.

To see this we note beforehand, from the conditions (2.1) on the weight w, and the definition (2.3) of the number $a \geqslant -\infty$, that if $f \in D(w)$ then

$$\int_a^\infty w(x)f(x)^2\,dx = 0 \text{ if and only if } f(x) = 0 \ (x \in (a,\infty)). \tag{5.1}$$

Clearly the inequality (2.20) is satisfied if (5.1) holds.

Suppose then that $f \in D(w)$ and

(5.2) $$\int_a^\infty w(x)f(x)^2dx > 0.$$

From Theorem 2 we know that the inequality (2.16) is valid for all $f \in D(w)$ and all $\rho \geq 0$. Following the argument in [5, section 7.8, page 193] we can deduce (2.20) from (2.16) by noting that the quadratic in ρ^2 is positive semi-definite, in which case the discriminant of the quadratic is non-negative and this yields the inequality (2.20). Alternatively, in view of (5.2), given $f \in D(w)$ we choose ρ^2 in (2.16) to be

$$\rho^2 = \int_a^\infty w(x)f'(x)^2dx \left\{2\int_a^\infty w(x)f(x)^2dx\right\}^{-1} > 0$$

and (2.20) again follows.

With (2.20) established and with the definition of $K(w)$ as given in Theorem 3, the result (2.22) follows noting from (3.1) and (3.2) that $K(w)$ must be positive; for if $K(w) = 0$ then at least one of the two terms

$$\int_a^\infty w(x)f'(x)^2dx \qquad \int_a^\infty w(x)f''(x)^2dx$$

must be zero, and this leads to a contradiction on (5.2).

The result (2.23)(ii) is due to Kwong and Zettl, see [9, Theorem 8, (2.12)]. It is also possible to prove this result using the identity (2.14)(iii) and we outline the proof here. If it can be shown that, for all $f \in D(w)$,

(5.3) $$\int_a^\infty w(x)\left\{\rho^4f(x)^2-k\rho^2f'(x)^2+f''(x)^2\right\}dx \geq 0$$

<u>for some</u> $k > 0$ then the inequality (2.21) is valid with $K(w) \leq 4k^{-2}$

(the case $k=1$ yields (2.22) from (2.16)). This result follows essentially from 'completing the square' in the quadratic in ρ^2 in (5.3), (recall that (5.2) holds); for details of the resulting identity see the comparable result in Evans and Everitt [2, section 5, (5.9)]. The converse result, i.e. (5.3) follows from (2.21), holds also.

Now if, with $a > -\infty$ and $w(a+) > 0$, it is the case that (2.21) holds with $K(w) < 4$ then (5.3) is valid with $k > 1$. Returning then to the identity (2.14)(iii) we seek $f \in D(w)$ so that the first term on the right-hand side is strictly negative, which is possible since $k > 1$, and also so that the integrand of the second term is non-negative. This can be done using the technique in [9, proof of Theorem 8, page 302]. The result is that the right-hand side of the identity (2.14)(iii), and hence the left-hand side, is negative for this particular element of $D(w)$. This yields a contradiction which can only be removed by taking $k = 1$ which is then the best possible, i.e. the largest, value of k for (5.3) to hold. This in turn implies that $K(w) = 4$ in the inequality (2.21).

Note that in the argument in the previous paragraph it is essential to have $w(a+) > 0$ to make the right-hand side of (2.14)(iii) strictly negative, and to have $a > -\infty$ since otherwise the construction taken from [9] is not possible. It should also be remarked that the w-measure gives positive measure of the point set $\{a\}$, i.e. $w(a+) - w(a-) = w(a+) > 0$, which is an alternative observation on the reason for the weight w forcing the number $K(w)$ up to the bound 4 in these circumstances.

Finally when w satisfies the conditions 1(i) of Theorem 2, i.e. (2.17) and (2.18), then the condition (2.23)(ii) is satisfied and $K(w) = 4$. If also condition 1(ii) of Theorem 2 is satisfied then from the identity in [2, section 5, (5.19)], appropriately adapted, we see that f as given by (2.19) is a non-trivial case of equality in (2.20). Conversely from [2, section 5, (5.19)] if there is a non-trivial case of equality in (2.20) then there has to be a non-trivial case of equality in (2.16) and, from Theorem 2, the conditions on w and f are then circumscribed.

This completes the proof of Theorem 3.

6. <u>The example $w(x) = x$</u> In this section we give some details of the case when the weight w is given by

(6.1) $$w(x) = 0 \quad (x \in (-\infty, 0]) \qquad w(x) = x \quad (x \in [0,\infty))$$

for which, in the notations of section 2

$$a = 0 \quad \underline{\text{and}} \quad w(a+) = 0.$$

For this example we shall write, in terms of the earlier notations

$$K(w) = K_0 \quad D(w) = D_0 \quad L^2_w(a,\infty) = L^2_0(0,\infty)$$

so that

(6.2) $$L^2_0(0,\infty) = \left\{f : (0,\infty) \to C \;\Big|\; \int_0^\infty x|f(x)|^2 dx < \infty\right\}$$

(6.3) $$D_0 = \Big\{f : (0,\infty) \to R \;\Big|\; \text{(i)} \;\; f \text{ and } f' \in AC_{loc}(0,\infty)$$
$$\text{(ii)} \;\; f \text{ and } f'' \in L^2_0(0,\infty)\Big\}.$$

The best possible inequality (2.21)then takes the form

(6.4) $$\left[\int_0^\infty xf'(x)^2 dx\right]^2 \leqslant K_0 \int_0^\infty xf(x)^2 dx \int_0^\infty xf''(x)^2 dx \qquad (f \in D_0)$$

where $0 < K_0 \leqslant 4$. We shall show in fact that $2.350 < K_0 < 2.351$ and that there is a family of cases of equality in (6.4).

We begin with

Lemma For all $f \in D_0$

(6.5) (i) $f \in AC_{loc}[0,\infty)$ (ii) f and $f' \in L^2(0,\infty)$.

Proof. For any $\epsilon > 0$, integrating by parts,

$$(6.6) \qquad \int_\epsilon^1 xf(x)f'(x)dx = \Big[\tfrac{1}{2}x\, f(x)^2\Big]_\epsilon^1 - \tfrac{1}{2}\int_\epsilon^1 f(x)^2 dx$$

Also integrating f' and multiplying by $x^{1/2}$

$$x^{1/2}f(x) = x^{1/2}f(1) - x^{1/2}\int_x^1 f'(t)dt$$

i.e.

$$x^{1/2}|f(x)| \leqslant x^{1/2}\,|f(1)| + x^{1/2}\int_x^1 |f'(t)|dt$$

$$\leqslant x^{1/2}\,|f(1)| + \left\{x \ln(^1/x)\int_x^1 tf'(t)^2 dt\right\}^{1/2}.$$

Thus from (6.6) we see that $f \in L^2(0,1)$. A similar argument shows that $f' \in L^2(0,1)$. Since, clearly, f and $f' \in L^2(1,\infty)$ we obtain (6.5)(ii).

With $f' \in L^2(0,1)$ we have $f' \in L(0,1)$ and so $f \in AC[0,1]$ and (6.5)(i) now follows.

To study the inequality (6.4) we follow the method of Hardy and Littlewood in [4] and given also in [5, section 7.8, proof (1) of Theorem 260], but adapted as in [3] and [2, section 5].

We consider the functional

$$(6.7) \qquad J_{\rho,k}(f,g) = \int_0^\infty x\{\rho^4 f(x)g(x) - k\rho^2 f'(x)g'(x) + f''(x)g''(x)\}dx$$

defined for all $f, g \in D_0$, all $\rho \geqslant 0$ and all $k \geqslant 0$. We write $J_{\rho,k}(f)$ for $J_{\rho,k}(f,f)$. If for some $k > 0$ it can be shown that

(6.8) $$J_{\rho,k}(f) \geqslant 0 \qquad (f \in D_0,\ \rho > 0)$$

then the quadratic form arguments used in the proof of Theorem 3 show that inequality (6.4) holds with K_0 replaced by $K = 4k^{-2}$. We look then for the largest (the upper bound) value of k, say k_0, for which (6.8) can be established, and then the best possible inequality (6.4) holds with

(6.9) $$K_0 = 4k_0^{-2} .$$

The cases of equality in (6.4) then remain to be determined.

If, as in [5, section 7.8], we apply the ideas of the calculus of variations to the study of (6.8) then we find the Euler-Lagrange equation of the functional $J_{\rho,k}$ is

(6.10) $$(xY''(x))'' + k\rho^2(xY'(x))' + \rho^4 xY(x) = 0 \qquad (x \in (0,\infty))$$

which is to be studied for $k > 0$ and $\rho > 0$. This linear fourth-order differential equation has singular points at ∞, and also at 0 where the coefficient of $Y^{(4)}$ takes the value zero. We have to look for solutions of this equation which are regular at the point 0, i.e. $Y^{(r)} \in AC_{loc}[0,\infty)$ for $r = 0,1,\ldots,4$, and such that $Y^{(r)}(x) = O(e^{-\epsilon x})$ $(x\to\infty)$ for some $\epsilon > 0$ and $r = 0,1,\ldots,4$. Note that such a solution would also satisfy $Y \in D_0$.

Suppose such a solution to exist then repeated integration by parts gives, for any $f \in D_0$,

(6.10a) $$\begin{aligned} J_{\rho,k}(f,Y) &= \int_0^\infty f(x)\{\rho^4 xY(x) - k\rho^2(xY'(x))' + (xY''(x))''\}dx \\ &\qquad + f(0)Y''(0) \\ &= f(0)Y''(0) \end{aligned}$$

(recall from the Lemma given above that $f \in AC_{loc}[0,\infty)$ so that $f(0)$ is well-defined).

Now let $g \in D_0$ with $g(0) = 0$; then

$$\int_0^\infty xg'(x)^2 dx = [xg(x)g'(x)]_0^\infty - \int_0^\infty \{xg''(x) + g'(x)\}g(x)dx$$

$$= - \int_0^\infty xg''(x)g(x)dx - [\tfrac{1}{2}g(x)^2]_0^\infty$$

$$= - \int_0^\infty xg''(x)g(x)dx.$$

Thus for such $g \in D_0$

$$(6.11) \quad J_{\rho,k}(g) = \int_0^\infty x\{\rho^4 g(x)^2 + k\rho^2 g(x)g''(x) + g''(x)^2\}dx$$

$$= \int_0^\infty x(g''(x) + \tfrac{1}{2}k\rho^2 g(x))^2 dx + \rho^4\left(1 - \frac{k^2}{4}\right)\int_0^\infty xg(x)^2 dx$$

and so for $k \in (0,2]$ and $\rho > 0$

$$(6.12) \qquad J_{\rho,k}(g) \geq 0 \qquad (g \in D_0 \text{ with } g(0) = 0)$$

with equality if and only if $g(x) = 0$ $(x \in [0,\infty))$.

Suppose then we can find $k_0 \in (0,2]$ such that for all $\rho > 0$ there is a solution Y_{ρ,k_0} of (6.10) satisfying

(i) $\quad Y_{\rho,k_0} \in AC_{loc}[0,\infty) \qquad (r=0,1,\ldots 4)$

(6.13) (ii) $\quad$ for some $\epsilon > 0$ $\;Y_{\rho,k_0}(x) = 0(e^{-\epsilon x}) \quad (x\to\infty;\ r=0,1,\ldots,4)$

(iii) $\quad Y_{\rho,k_0}(0) \neq 0 \qquad$ (iv) $\quad Y''_{\rho,k_0}(0) = 0$

then

$$(6.14) \qquad J_{\rho,k_0}(Y_{\rho,k_0}) = 0 \qquad (\rho \in (0,\infty)).$$

Given any $f \in D_0$ we can find $\alpha \in R$ so that

(6.15) $$f(x) = \alpha Y_{\rho,k_0}(x) + g(x) \qquad (x \in [0,\infty))$$

where $g \in D_0$ and $g(0) = 0$; in fact $\alpha = f(0)/Y_{\rho,k_0}(0)$.

With the representation (6.15) for f we obtain

$$J_{\rho,k_0}(f) = \alpha^2 J_{\rho,k_0}(Y_{\rho,k_0}) + 2\alpha J_{\rho,k_0}(Y_{\rho,k_0},g) + J_{\rho,k_0}(g).$$

Now using (6.13) and then (6.10a) we find $J_{\rho,k_0}(Y_{\rho,k_0},g) = 0$, and then using (6.14) gives

$$J_{\rho,k_0}(f) = J_{\rho,k_0}(g).$$

From (6.12) this yields

(6.8) $$J_{\rho,k_0}(f) \geqslant 0 \qquad (f \in D_0;\ \rho \in (0,\infty))$$

with equality if and only if g is null on $[0,\infty)$, i.e. if an only if,

(6.16) $$f(x) = \alpha Y_{\rho,k_0}(x) \qquad (x \in [0,\infty))$$

for some $\rho > 0$.

We shall show that there is a unique $k_0 \in (0,2)$ for which there is a solution Y_{ρ,k_0}, for all $\rho > 0$, such that both (6.13) and (6.14) hold. This number k_0 gives the upper bound of $k \in (0,2]$ for which the inequality (6.8) is valid; in turn this gives $K_0 = 4k_0^{-2}$ for the best possible inequality (6.4). All the cases of equality in (6.4) are then determined by (6.16) but now for all $\rho > 0$.

It remains then to establish the existence of the special solution Y_{ρ,k_0} with the properties (6.13) and (6.14).

Solutions of the differential equation (6.10) can be obtained by the method of contour integration; these solutions take the form

(6.17) $$Y_{\rho,k}(x) = \int_\Gamma \frac{e^{-xz}}{\sqrt{(z^4 + k\rho^2 z^2 + \rho^4)}}\, dz$$

where Γ is a simple closed contour in the complex z-plane chosen so that the $\surd(.)$ factor remins single-valued on making a circuit of Γ. The quartic $z^4 + k\rho^2 z^2 + \rho^4$ has simple zeros at the four points

$$\tfrac{1}{2}\rho\{\pm \surd(2-k) \pm i\surd(2+k)\}$$

for $k \in (0,2)$. If we take for Γ a simple closed contour, with positive orientation, enclosing the two points $\tfrac{1}{2}\rho\{\surd(2-k) \pm i\surd(2+k)\}$ such that re $[z] > 0$ $(z \in \Gamma)$ then we obtain a solution which satisfies (6.13)(i), (ii) and (iii), with ϵ in (ii) given by ϵ $\tfrac{1}{2}\rho(2-k) - \delta$ for any $\delta \in (0, \tfrac{1}{2}\rho\surd(2-k))$; this solution is valid for all $\rho > 0$ and all $k \in (0,2)$. Further consideration of the other contour integral solutions shows that such a solution is unique up to linear independence.

By deforming the loop contour Γ onto the imaginary axis of the complex z-plane we can represent this solution in the form

$$(6.18) \qquad Y_{\rho,k}(x) = \int_0^\infty \frac{\cos(xt)}{\surd(t^4 - k\rho^2 t^2 + \rho^4)}\,dt \qquad (x \in [0,\infty);\ \rho \in (0,\infty))$$

$$= \frac{1}{\rho}\int_0^\infty \frac{\cos(\rho xt)}{\surd(t^4 - kt^2 + 1)}\,dt.$$

We can rewrite (6.18) in the form

$$(6.19) \quad Y_{\rho,k}(x) = \frac{1}{\rho}\int_0^\infty \frac{\cos(\rho xt)}{t^2 + 1}\,dt + \frac{(2+k)}{\rho}\int_0^\infty \frac{t^2\cos(\rho xt)}{(t^2+1)\surd(P).(t^2+1+\surd(P))}\,dt$$

$$= \frac{\pi}{2\rho}\,e^{-\rho x} + \frac{(2+k)}{\rho}\int_0^\infty \cdots\cdots\cdots\, dt$$

where $P = P(t,k) = t^4 - kt^2 + 1$ $(t \in [0,\infty);\ k \in (0,1))$.

where Γ is a simple closed contour in the complex z-plane chosen so that the $\surd(.)$ factor remains single-valued on making a circuit of Γ. The quartic $z^4 + k\rho^2 z^2 + \rho^4$ has simple zeros at the four points

$$\tfrac{1}{2}\rho\{\pm\surd(2-k) \pm i\surd(2+k)\}$$

for $k \in (0,2)$. If we take for Γ a simple closed contour, with positive orientation, enclosing the two points $\tfrac{1}{2}\rho\{\surd(2-k) \pm i\surd(2+k)\}$ such that $re\,[z] > 0 \; (z \in \Gamma)$ then we obtain a solution which satisfies (6.13)(i), (ii) and (iii), with ϵ in (ii) given by $\epsilon\ \tfrac{1}{2}\rho(2-k) - \delta$ for any $\delta \in (0, \tfrac{1}{2}\rho\surd(2-k))$; this solution is valid for all $\rho > 0$ and all $k \in (0,2)$. Further consideration of the other contour integral solutions shows that such a solution is unique up to linear independence.

By deforming the loop contour Γ onto the imaginary axis of the complex z-plane we can represent this solution in the form

$$(6.18) \qquad Y_{\rho,k}(x) = \int_0^\infty \frac{\cos(xt)}{\surd(t^4 - k\rho^2t^2 + \rho^4)}\,dt \qquad (x \in [0,\infty);\ \rho \in (0,\infty))$$

$$= \frac{1}{\rho}\int_0^\infty \frac{\cos(\rho xt)}{\surd(t^4 - kt^2 + 1)}\,dt.$$

We can rewrite (6.18) in the form

$$(6.19) \qquad Y_{\rho,k}(x) = \frac{1}{\rho}\int_0^\infty \frac{\cos(\rho xt)}{t^2+1}\,dt + \frac{(2+k)}{\rho}\int_0^\infty \frac{t^2\cos(\rho xt)}{(t^2+1)\surd(P).(t^2+1+\surd(P))}\,dt$$

$$= \frac{\pi}{2\rho}e^{-\rho x} + \frac{(2+k)}{\rho}\int_0^\infty \cdots\cdots\cdots\ dt$$

where $P = P(t,k) = t^4 - kt^2 + 1 \; (t \in [0,\infty);\ k \in (0,1))$.

Then the inequality (6.4) holds with $K_0 = 4k_0^{-2}$ and all cases of equality are determined by

$$f(x) = \alpha Y_{\rho,k_0}(x) \qquad (x \in [0,\infty))$$

for all $\alpha \in R$ and all $\rho > 0$.

Finally numerical calculations, based on the integral representation (6.20) of $I(k)$, show that

$$1{\cdot}30445 < k_0 < 1{\cdot}30446$$

and this gives in turn, from $K_0 = 4k_0^{-2}$,

$$2{\cdot}35070 < K_0 < 2{\cdot}35075.$$

In particular note that $K_0 < 4$ as recorded in (1.7). This confirms the statement made in (2.23)(iii) of Theorem 3.

Acknowledgements Both authors thank M.K. Kwong and A. Zettl for helpful discussions concerning their results in the two papers quoted in the references to this paper. W.N. Everitt thanks W. Walter for advice and for providing the counterexample at the end of section 2.

REFERENCES

1. R.Beals, Topics in operator theory. Chicago University Press, 1971.
2. W.D.Evans and W.N.Everitt, A return to the Hardy-Littlewood integral inequality. Proc. Royal Soc. Lond. A 380 (1982), 447-486.
3. W.N.Everitt, On an extension to an integro-differential inequality of Hardy, Littlewood and Polya. Proc. Royal Soc. Edinb. A 69 (1971/72), (295-333.
4. G.H.Hardy and J.E.Littlewood, Some integral inequalities connected with the calculus of variations. Q.J.Math.(2) 3 (1932), 241-252.

5. G.H.Hardy, J.E.Littlewood and G.Polya, Inequalities. Cambridge University Press, 1934.

6. R.R.Kallman and G-C.Rota, On the inequality $||f'||^2 \leqslant 4\ ||f||\ ||f''||$. Inequalities II (Ed. Oved Shisha) 187-192; Academic Press, New York and London, 1970.

7. T.Kato, On an inequality of Hardy, Littlewood and Polya. Adv. Math. 1 (1971), 217-218.

8. M.K.Kwong and A.Zettl, An extension of the Hardy-Littlewood inequality. Proc. Amer.Math. Soc. 77 (1979), 117-118.

9. M.K.Kwong and A.Zettl, Norm inequalities of product form in weighted L^p spaces. Proc. Royal Soc. Edinb. A89, 293-307, 1981.

10. M.A.Naimark, Linear differential operators II. Ungar, New York, 1968.

11. W.Rudin, Principles of mathematical analysis. McGraw-Hill Kogakusha, Tokyo, 1976.

12. S.J.Taylor, and J.F.C.Kingman, Introduction to measure and probability. Cambridge University Press, 1966.

13. D.V.Widder, Advanced calculus. Prentice-Hall, New York, 1947.

W.N.Everitt,
Department of Mathematics,
University of Birmingham,
P.O.BOX 363,
BIRMINGHAM,
England. U.K.

A.P. Guinand,
Department of Mathematics,
Trent University,
Peterborough,
ONTARIO,
Canada.

International Series of
Numerical Mathematics, Vol. 80

CONTRIBUTIONS TO INEQUALITIES II

Alexander Kovacec

Abstract. We give a class of functions of N nonnegative variables for which the problem to maximize them on the compact set of all N-tuples $\underline{x} = (x_1, x_2, \dots, x_N)$ with $x_i \geq 0$ $(1 \leq i \leq N)$, $\Sigma\, x_i = a$ leads naturally to a dynamic programming approach. For the case $N \nearrow \infty$, we prove, roughly speaking, that in case of homogeneity the "maximizing sequences" $(a_1, a_2, \dots)$ of the functions in question tend to be close to geometric progressions.

1. INTRODUCTION

For a nonnegative real a define the compact set $K_N(a)$ by $K_N(a) := \{\underline{x} \in \mathbb{R}^N\colon\ \underline{x} \geq \underline{0},\ \sum x_i = a\}$, and let $\Delta \subseteq \mathbb{R}^2$ be defined by $\Delta := \{(x,y) \in \mathbb{R}^2\colon\ 0 \leq x \leq y\}$. Fix a continuous function $g\colon \Delta \to \mathbb{R}$ and define $U_N\colon \mathbb{R}_+^N \to \mathbb{R}$ by

$$(1) \qquad U_N(\underline{x}) := \sum_{i=1}^{N} g\left(x_i,\ \sum_{j=i}^{N} x_j\right) .$$

The main body of this note centers around the observation that the problem of maximizing U_N on $K_N(a)$ can be reduced to that of maximizing U_{N-1} on $K_{N-1}(b)$ for some $b < a$ (Lemma 1 below). In many cases the reduction process can actually be carried through; it leads to new inequalities with best possible constants (Theorem 3) of which only infinite versions seem to be known (Example 1). In other cases where one encounters difficulties in maximizing $U_N(\underline{x})$ on $K_N(a)$, it may still be possible to find $\sup\{U_\infty(\underline{x})\colon \underline{x} \in K_\infty(a)\}$ by a completely different direct approach which does not rely on the determination of $\sup\{U_N(\underline{x})\colon \underline{x} \in K_N(a)\}$ and a subsequent limit

---

This paper is in final form and no version of it will be submitted for publication elsewhere.

procedure (Theorem 4, Example 2). For this, homogeneity of U is required.

This paper is a continuation of [1], but needs none of the results there. It is much more closely related (in spirit) to (parts of) papers of Wang (see [4] and references) and the dynamic programming approach in general. (The reader may get a striking impression of certain manifestations of that principle by a glance on the paper of Redheffer [3] (where it is - however - not mentioned explicitly).)

2. RESULTS AND PROOFS

The following lemma contains our main observation.

LEMMA 1. _With the notations, definitions and assumptions on_ a, $K_N(a)$, Δ _and_ g _as given above, if we define_

$$F_N(a) := \sup\{U_N(\underline{x}) : \underline{x} \in K_N(a)\}\ , \tag{2}$$

then there holds the recurrence relation

$$F_N(a) = \sup_{0 \le x \le a} \{g(x,a) + F_{N-1}(a-x)\}\ . \tag{3}$$

Proof. One computes with $\underline{x}' := (x_2, \ldots, x_N)$ that

$$F_N(a) = \sup\{g(x_1,a) + \sum_{i=2}^{N} g\Big(x_i, \sum_{j=i}^{N} x_j\Big) : 0 \le x_1 \le a,\ \underline{x}' \in K_{N-1}(a-x_1)\}$$

$$= \sup_{0 \le x \le a} \sup\{g(x,a) + \sum_{i=2}^{N} g\Big(x_i, \sum_{j=i}^{N} x_j\Big) : \underline{x}' \in K_{N-1}(a-x)\}$$

$$= \sup_{0 \le x \le a} \{g(x,a) + \sup\{\sum_{i=2}^{N} g\Big(x_i, \sum_{j=i}^{N} x_j\Big) : \underline{x}' \in K_{N-1}(a-x)\}\}$$

$$= \sup_{0 \le x \le a} \{g(x,a) + \sup\{U_{N-1}(\underline{x}) : \underline{x} \in K_{N-1}(a-x)\}\}\ ,$$

which yields the result.

The following lemma will not be used. We state it for completeness.

LEMMA 2. *If* $(a_1,a_2,\dots,a_N)$ *is a point that maximizes the function* $\underline{x}\mapsto U_N(\underline{x})$, $\underline{x}\in K_N(a)$, *then, for* $1\le i\le N-1$, a_i *maximizes* $x\mapsto g\left(x,a-\sum_{l=1}^{i-1}a_l\right)+F_{N-1}\left(\left(a-\sum_{l=1}^{i-1}a_l\right)-x\right)$ $\left(0\le x\le a-\sum_{l=1}^{i-1}a_l\right)$, *while* $a_N = a-\sum_{l=1}^{N-1}a_l$. *Conversely, a point so determined maximizes* $\underline{x}\mapsto U_N(\underline{x})$, $(\underline{x}\in K_N(a))$.

Proof. The proof follows directly from a glance at the proof of Lemma 1: To find the maximizing point, we first vary the first coordinate x_1 (called x there) in order to maximize $x\mapsto g(x,a)+F_{N-1}(a-x)$. If a_1 is the point that maximizes this function, then $F_N(a)=g(a_1,a)+F_{N-1}(a-a_1)$. To find $F_{N-1}(a-a_1)$ we maximize the function $x\mapsto g(x,a-a_1)+F_{N-2}(a-a_1-x)$. This yields a point a_2. The proof now follows from a clear inductive procedure and by noting that $a=\sum_{i=1}^{N}a_i$ yields the condition on a_N.

THEOREM 3. (a) *Let* $r>0$ *and define a function* $f\colon [1/1+r,\infty)\to[1/1+r,\infty)$ *by* $f(c)=1-r/c^{1/r}(1+r)^{1+1/r}$. *Then for every* $N\in\mathbb{N}$ *and nonnegative reals* $x_1,\dots,x_N$ *there holds the inequality*

$$(4)\qquad (f\circ\dots\circ f)(1)\left(\sum_{i=1}^{N}x_i\right)^{1+r}\le x_1\left(\sum_{i=1}^{N}x_i\right)^r+x_2\left(\sum_{i=2}^{N}x_i\right)^r+\dots+x_N\left(\sum_{i=N}^{N}x_i\right)^r .$$

(b) *Let* $p>1$ *and define a function* $f\colon [0,p/p-1]\to[0,p/p-1]$ *by* $f(c)=1+(p-1)^{p-1}\left(\frac{c}{p}\right)^p$. *Then for every* $N\in\mathbb{N}$ *and nonnegative reals* $x_1,\dots,x_N$ *there holds the inequality*

$$(5)\qquad \frac{x_1}{\sqrt[p]{x_1+x_2+\dots+x_N}}+\frac{x_2}{\sqrt[p]{x_2+\dots+x_N}}+\dots+\frac{x_N}{\sqrt[p]{x_N}}\le(f\circ\dots\circ f)(1)\left(\sum_{i=1}^{N}x_i\right)^{1-1/p} .$$

In both parts ((a) and (b)), $f\circ\dots\circ f$ *denotes the* (N-1)-*fold iterate of* f, *and* $(f\circ\dots\circ f)(1)$ *is the largest (in (a)) respectively smallest (in (b)) possible constant.*

Proof. To gain more insight let us consider the qualitative behaviour of $x\mapsto xa^r+c(a-x)^{1+r}$ $(c>0,\ 0\le x\le a)$. It turns out that strict extrema in the interior of $[0,a]$ are attained only in the cases $r>0$, $c>\frac{1}{1+r}$ and $-1<r<0$, $c<\frac{1}{1+r}$. In fact, qualitatively we have the following pictures for the mentioned functions:

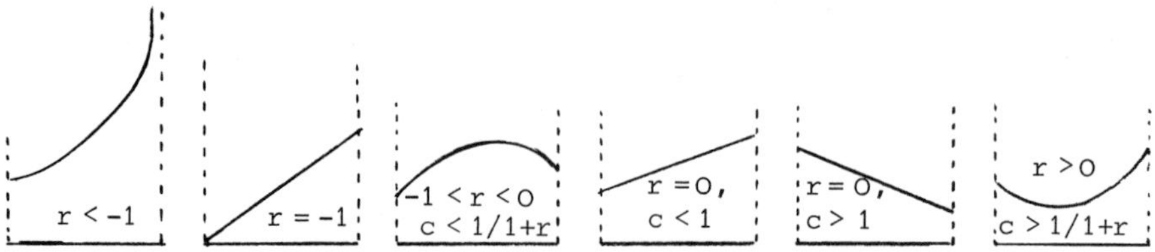

The ideas resulting from these considerations can easily be made rigourous to obtain by standard techniques with $f(c) := 1 - \left(\frac{r}{c}\right)^{1/r}(1+r)^{-1-1/r}$ that

$$(6)\qquad \inf_{0\le x\le a} \{xa^r + c(a-x)^{1+r}\} = f(c)a^{1+r} \qquad \left(r>0,\ c>\frac{1}{1+r}\right),$$

$$(7)\qquad \sup_{0\le x\le a} \{xa^r + c(a-x)^{1+r}\} = f(c)a^{1+r} \qquad \left(-1<r<0,\ c<\frac{1}{1+r}\right).$$

Also one easily convinces oneself that, if c satisfies one of the conditions required, then f(c) satisfies the same condition. For the proof of (a) put $g(u,v) := uv^r$ $(r>0)$ and note that by definition of F_N, we have $F_1(a) = g(a,a) = a^{1+r}$. Therefore, by the inf-version of Lemma 1, since $1 > \frac{1}{1+r}$, we have by (3), (6) that $F_2(a) = \inf_{0\le x\le a}\{xa^r + 1\cdot(a-x)^{1+r}\} = f(1)a^{1+r}$. From this and since $f(1) > \frac{1}{1+r}$ we have again by (3) and (6) that $F_3(a) = \inf\{xa^r + f(1)(a-x)^{1+r}\} = f(f(1))a^{1+r}$. This suffices to see how inductively the iterated f's come about and so, by definition of F_N,

$$F_N(a) = (f\circ\ldots\circ f)(1)a^{1+r} = (f\circ\ldots\circ f)(1)\left(\sum_{i=1}^{N} x_i\right)^{1+r}$$

$$\le \sum_{i=1}^{N} g\left(x_i, \sum_{j=i}^{N} x_j\right) = \sum_{i=1}^{N} x_i\left(\sum_{j=i}^{N} x_j\right)^{1+r}.$$

This proves (a). (b) follows by obvious modifications applied to the function

$$(8)\qquad g(u,v) = \begin{cases} uv^r, & v>0, \\ 0, & v=0, \end{cases}$$

which is continuous on Δ (although $-1<r<0$), and replacing r by $-1/p$ with $p>1$ afterwards. The statement on the constant is a consequence of the method.

REMARK. Easy but tedious additional thoughts show that very slight modifications of the recursive functional approach of Lemma 1 can be used also to find upper and lower bounds for $\sum_{i=1}^{N} x_i(\sum_{j=i}^{N} x_j)^r$ in terms of $\sum x_i$ for other r than those investigated. For example, for $r=-1$ we get the sharp inequality

$$\frac{x_1}{x_1+x_2+\dots+x_N}+\frac{x_2}{x_2+\dots+x_N}+\dots+\frac{x_{N-1}}{x_{N-1}+x_N}+\frac{x_N}{x_N}\le N .$$

For the following theorem we need some notation. $K(a)=K_\infty(a)$ will denote the set of _all_ sequences $\underline{x}=(x_1,x_2,\dots)$ with $x_i\ge 0$, $\sum_{i=1}^{\infty} x_i=a$. We think of $K_N(a)$ as imbedded into $K_\infty(a)$, i.e., $K_N(a)=\{\underline{x}:\ \underline{x}=(x_1,\dots,x_N,0,0,\dots),\ x_i\ge 0,\ \sum_{i=1}^{N} x_i=a\}$. We define $G(a)=\{a(1-q)(1,q,q^2,\dots):\ 0\le q<1\}$, the set of all _geometric_ sequences of sum $a>0$. Finally, let $U(x) := \sum_{i=1}^{\infty} g\left(x_i, \sum_{j\ge i} x_j\right)$.

THEOREM 4. _If_ $g:\Delta\to\mathbb{R}_+$ _is a continuous function, homogeneous of positive degree with_ $g(0,u)=0$ (_whenever_ $(0,u)\in\Delta$), _then_

$$\sup_{\underline{x}\in K(a)} \sum_{i=1}^{\infty} g\left(x_i,\sum_{j\ge i}x_j\right) = \sup_{\underline{x}\in G(a)} \sum_{i=1}^{\infty} g\left(x_i,\sum_{j\ge i}x_j\right) . \tag{9}$$

Proof. The proof rests on the following

LEMMA. _Given_ $a\in\mathbb{R}_+$, $N\in\mathbb{N}$, _for any_ $\underline{x}\in K_N(a)$ _there exists_ $\underline{q}\in G(a)$ _such that_ $U(\underline{x})\le U(\underline{q})$.

To see this, note for $\underline{x}\in K_N(a)$ that $U(\underline{x}) = \sum_{i=1}^{N} g\left(x_i,\sum_{j=i}^{N}x_j\right)$. Since the function $\underline{x}\mapsto U(\underline{x})$, $\underline{x}\in K_N(a)$, is continuous, there is an $\underline{m}=(m_1,\underline{m}')=(m_1,\dots,m_N,0,\dots)\in K_N(a)$ such that

$$U(\underline{x}) \le \sup_{\underline{x}\in K_N(a)} U(\underline{x}) = U(\underline{m}) = g\left(m_1,\sum_{i\ge 1} m_i\right)+U(\underline{m}') . \tag{10}$$

Obviously, $\underline{m}'\in K_{N-1}(a-m_1)\subseteq K_N(a-m_1)$, hence, defining $q:=(a-m_1)/a$,

$$U(\underline{m}') \le \sup_{\underline{x}\in K_N(a-m_1)} U(\underline{x}) = U\left(\frac{a-m_1}{a}\,\underline{m}\right) = q^r\cdot U(\underline{m}) , \tag{11}$$

where the first equality follows from the definition of $\underline{m}$ and the homogeneity of U, which also yields the second equality. Note also the inequality for $U(\underline{m})$ that results from (10) and (11) to obtain for every $N \in \mathbb{N}$ the inequality

$$U(\underline{x}) \le g\Big(m_1, \sum_{i\ge 1} m_i\Big)+q^r U(m) \le g\Big(m_1, \sum_{i\ge 1} m_i\Big)+q^r \Big(g\Big(m_1, \sum_{i\ge 1} m_i\Big)+ +q^r U(m)\Big) \;\ldots$$

$$\le \ldots \le g\Big(m_1, \sum_{i\ge 1} m_i\Big) \sum_{j=0}^{N} (q^j)^r \le \sum_{j=0}^{\infty} g(q^j m_1, q^j a) .$$

For the last inequality we used homogeneity of g of degree $r > 0$, $g(u,v) \ge 0$, $a = \sum_{i\ge 1} m_i$, and a limit procedure. The lemma follows.

The rest of the proof is now easy: Because of $K(a) \supseteq G(a)$ it suffices to show

$$\sup_{\underline{x}\in K(a)} U(x) \le \sup_{\underline{x}\in G(a)} U(x) . \tag{12}$$

Case 1. For every $\underline{x} \in K(a)$ we have $U(\underline{x}) < \infty$. Choose $\underline{x} \in K(a)$, put $s := U(\underline{x})$. For every $\varepsilon > 0$, we find an N such that $U(x_1, x_2, \ldots, x_N, 0, 0, \ldots) \ge s-\varepsilon$. Let $\mu := \sum_{i\ge N+1} x_i$. By the lemma we find $\underline{q} \in G(a-\mu)$ with $U(q) \ge U(x_1, \ldots, x_N, 0, \ldots) \ge s-\varepsilon$. Then $\underline{q}' := \frac{a}{a-\mu}\underline{q} \in G(a)$ and $U(\underline{q}') \ge s-\varepsilon$. Since ε was arbitrary, this proves (12).

Case 2. There exists an $\underline{x} \in K(a)$ such that $U(\underline{x}) = \infty$. Then, for such an $\underline{x}$, for every $M \in \mathbb{R}_+$ we find an $N \in \mathbb{N}$ such that $M \le U(x_1, \ldots, x_N, 0, 0, \ldots) < \infty$. From now on arguments as in the last case can be used to yield that for every $\varepsilon > 0$, $M \in \mathbb{R}_+$, we find $\underline{q} \in G(a)$ such that $U(\underline{q}) \ge M-\varepsilon$. As $M \in \mathbb{R}_+$ is arbitrary, this yields $\sup_{\underline{x}\in G(a)} U(\underline{x}) = \infty$ in this case, which again proves (12).

REMARK. In case we add the assumption that the left-hand side supremum of (9) is actually attained, the proof would be simpler. Example 1 below shows, however, that this need not be the case, while Example 2 shows that this can be the case.

EXAMPLE 1. Based on Theorem 3 and Theorem 4, respectively, one can give two rather different proofs of the inequality

$$(13) \qquad \sum_{i=1}^{\infty} \frac{x}{\sqrt[p]{\sum_{j\geq i} x_i}} \leq \frac{p}{p-1} \left(\sum_{i=1}^{\infty} x_i\right)^{1-1/p} \qquad (p>1,\ x_i \geq 0)$$

and show that the constant at the right-hand side is best possible. In fact one obtains this inequality easily by applying a limit procedure ($N\to\infty$) to inequality (5) of Theorem 3, or by defining the function $g\colon \Delta \to \mathbb{R}_+$, by $g(u,v) = u/\sqrt[p]{v}$ if $0 \leq u < v$ and $= 0$ if $0 = u \leq v$, and determining $\sup_{\underline{x}\in G(a)} \sum g(x_i, \sum_{j\geq i} x_j)$. Since $\underline{x} = (x_1, x_2, \ldots) \in G(a)$ iff $x_i = a(1-q)q^{i-1}$ for some $0 \leq q < 1$ and since then $\sum_{j\geq i} x_j = a(1-q)q^{i-1} \sum_{j\geq 0} q^j = aq^{i-1}$, one obtains

$$\sup_{x\in G(a)} \sum g(x_i, \sum_{j\geq i} x_j) = \sup_{0\leq q<1} \sum_{i=1}^{\infty} \frac{a(1-q)q^{i-1}}{\sqrt[p]{aq^{i-1}}}$$

$$= a^{1-1/p} \sup_{0\leq q<1} (1-q)/1-q^{1-1/p}$$

$$= \frac{p}{p-1}\, a^{1-1/p} .$$

Now Theorem 4 yields the second proof of (13). (This provides an answer to an (unstarred) Monthly-Problem [2].)

EXAMPLE 2. The following example shows that Theorem 4 is not only of theoretical interest. Define

$$g(u,v) = \begin{cases} \dfrac{u}{v^{1/3}} + \dfrac{u^{1/3}}{v^{2/3}} & (0 < u \leq v) \\ 0 & (0 = u \leq v) . \end{cases}$$

Then g is continuous on Δ and homogeneous of degree $\frac{2}{3} > 0$. The problem of determining the functions $F_N(a)$ yields computational troubles. Nevertheless, $\lim_{N\to\infty} F_N(a) = \sup_{\underline{x}\in K(a)} \sum_{i=1}^{\infty} g(x_i, \sum_{j\geq i} x_j)$ can be established directly by means of Theorem 4. Namely, this value is equal to

$$\sup_{\underline{x}\in G(a)} \sum_{i=1}^{\infty} g\left(x_i, \sum_{j\geq i} x_j\right) = a^{2/3} \sup_{0\leq q<1} \frac{(1-q)(1+(1-q)^{1/3})}{1-q^{2/3}},$$

as a small calculation reveals. A numerical calculation of the right-hand side yields the bound $2.43\,a^{2/3}$, in other words, we have the inequality

$$\sum_{i=1}^{\infty} \frac{x_i}{\left(\sum_{j\geq i} x_j\right)^{1/3}} + \sum_{i=1}^{\infty} \frac{x_i^{4/3}}{\left(\sum_{j\geq i} x_j\right)^{2/3}} \leq 2.43 \left(\sum_{i=1}^{\infty} x_i\right)^{2/3}.$$

(Separate estimates of the two infinite sum expressions would only yield the constant ≈ 2.63 on the right-hand side.)

Finally we want to remark that inequalities of the form (13) go back to Dini. Apart from [2] the reader is referred to the articles of M.J. Pelling [AMM 84, 138] and P. Biler [AMM 92, 347].

REFERENCES

1. A. Kovacec, Two contributions to inequalities. In: W. Walter (ed.), General Inequalities 4 (pp.37-46), Birkhäuser Verlag, Basel, 1984.

2. D. Oberlin and P. Novinger, Problem E 2996. Amer. Math. Monthly 90 (1983), 334; solution in 93 (1986), 303.

3. R.M. Redheffer, Recurrent Inequalities. Proc. London Math. Soc. 17 (1967), 683-699.

4. C.L. Wang, Inequalities and Mathematical Programming. In: E.F. Beckenbach and W. Walter (eds.), General Inequalities 3 (pp.149-169), Birkhäuser Verlag, Basel, 1983.

Alexander Kovacec, Institut für Mathematik, Universität Wien, Strudlhofgasse 4, A-1090 Wien, Austria

International Series of
Numerical Mathematics, Vol. 80

ON SOME DISCRETE QUADRATIC INEQUALITIES

László Losonczi

Abstract. Inequalities of the form

$$\alpha \sum_{j=0}^{n} |x_j|^2 \le \sum |x_j \pm x_{j+k}|^2 \le \beta \sum_{j=0}^{n} |x_j|^2$$

are studied, where $x_0,\dots,x_n$ are real or complex variables, α, β are constants, $1 \le k \le n$, the summation (with respect to j) in the middle term can be understood in four different ways (see introduction) and either the plus or the minus sign is taken. The best constants α, β are found in all cases. This is based on the determination of eigenvalues of suitable Hermitean matrices.

1. INTRODUCTION

Let n be a fixed natural number, $1 \le k \le n$, $x_0,\dots,x_n$ real or complex numbers. To simplify our notations we introduce the following summation symbols:

(i) $$\sum{}^1 = \sum_{j=0}^{n-k} ,$$

(ii) $$\sum{}^2 = \sum_{j=0}^{n} \text{ with } x_{n+1} = \dots = x_{n+k} = 0 ,$$

(iii) $$\sum{}^3 = \sum_{j=-k}^{n-k} \text{ with } x_{-k} = \dots = x_{-1} = 0 ,$$

(iv) $$\sum{}^4 = \sum_{j=-k}^{n} \text{ with } x_{-k} = \dots = x_{-1} = 0 = x_{n+1} = \dots = x_{n+k} .$$

This paper is in final form and no version of it will be submitted for publication elsewhere.

Our aim is to study inequalities of the form

$$(1)\qquad \alpha_i^{\pm}\sum_{j=0}^{n}|x_j|^2 \le \sum{}^i|x_j\pm x_{j+k}|^2 \le \beta_i^{\pm}\sum_{j=0}^{n}|x_j|^2 ,$$

where $\alpha_i^{\pm}$, $\beta_i^{\pm}$ $(i=1,2,3,4)$ are constants and either the + or the - sign is taken. It is easy to see that the cases $i=2$ and $i=3$ are the same apart from the notation of the variables x_j. Hence there are 6 different cases in (1) corresponding to $i=1,2$ or $3,4$ and the + or - sign.

K. Fan, O. Taussky and J. Todd [3] considered (1) in the case $k=1$ with the - sign and determined the best constants. Their inequalities are the discrete analogues of Wirtinger's inequality; see, e.g., G.H. Hardy, J.E. Littlewood and G. Pólya [5], p.184.

In connection with extremal properties of non-negative trigonometric polynomials G. Szegö [8], E. Egerváry and O. Szász [2] proved that

$$(2)\qquad -\gamma\sum_{j=0}^{n}|x_j|^2 \le \sum_{j=0}^{n-k}(x_j\bar{x}_{j+k}+\bar{x}_j x_{j+k}) \le \gamma\sum_{j=0}^{n}|x_j|^2 \quad (x_j\in\mathbb{C})$$

holds with the constant

$$\gamma = 2\cos\pi/([\tfrac{n}{k}]+2) ,$$

which is optimal. Here $[x]$ denotes the greatest integer not exceeding x. (2) is clearly related to (1). In the case $k=1$, (2) was proved by L. Fejér [4].

Other inequalities related to (1) were studied by G.V. Milovanovic and I.Z. Milovanovic [6] who determined the best constants A, B in the inequalities

$$(3)\quad A\sum_{j=0}^{n}p_jx_j^2 \le \sum_{j=-1}^{n-1}r_j(x_j-x_{j+1})^2 \le B\sum_{j=0}^{n}p_jx_j^2 \quad (x_{-1}=0,\ x_j\in\mathbb{R}) ,$$

where $p_0,\dots,p_n$, $r_{-1},\dots,r_{n-1}$ are given sequences of positive numbers. They also investigated a variation of (3) where the summation in the middle term runs from $j=-1$ to $j=n$ with $x_{-1}=x_{n+1}=0$.

In Section 2 we find the value of a tri-diagonal determinant depending on three parameters. For some special values of the parameters the factorization of this determinant can also be given. In Section 3 we use the results of Section 2 to find the best constants in the inequality (1).

2. EVALUATION AND FACTORIZATION OF SOME TRI-DIAGONAL DETERMINANTS

Our first aim in this section is to evaluate the determinant of order n+1,

$$(4) \qquad D = D(u,v,w) = \begin{vmatrix} u & & & 1 & & & & & \\ & \ddots & & & \ddots & & & & \\ & & u & & & \ddots & & & \\ 1 & & & v & & & \ddots & & \\ & \ddots & & & \ddots & & & \ddots & \\ & & \ddots & & & v & & & 1 \\ & & & \ddots & & & w & & \\ & & & & \ddots & & & \ddots & \\ & & & & & 1 & & & w \end{vmatrix},$$

where both u and w appear in the main diagonal k times and the number 1 appears (twice) n+1-k times. Elements not indicated are zero. Thus the elements d_{ij} of D are given by

$$d_{ii} = \begin{cases} u & \text{if } i = 1,\dots,k\ , \\ v & \text{if } i = k+1,\dots,n-k+1\ , \\ w & \text{if } i = n-k+2,\dots,n+1 \end{cases}$$

and

$$d_{ij} = \begin{cases} 1 & \text{if } |i-j| = k\ , \\ 0 & \text{if } 0 < |i-j| \neq k\ . \end{cases}$$

We assume that $k \geq 1$, $n+1 \geq 2k$. Since n and k are fixed, our notation does not indicate the dependence of D on n and k.

Let $\alpha_1 = u \neq 0$. Multiplying the rows $1,\dots,k$ of D by $-1/\alpha_1$ and adding them to the rows $k+1,\dots,2k$, respectively, the first k elements "1" in the lower triangle of D will disappear, and in the rows $k+1,\dots,2k$ the diagonal element v will be replaced by

$$\alpha_2 = v - \frac{1}{\alpha_1}\ .$$

Assuming that $\alpha_2 \neq 0$ we multiply the rows $k+1,\dots,2k$ of D by $-1/\alpha_2$

and add them to the rows 2k+1,...,3k, respectively. Again, the next k elements 1 will vanish in the lower triangle, and the diagonal element v in the rows 2k+1,...,3k will change to

$$\alpha_3 = v - \frac{1}{\alpha_2} .$$

Let $n+1 = kq+r$ with integer r, $0 \le r < k$.

If $r = 0$, repeating the above process q-1 times the lower triangle of D will contain only zeros, and in the diagonal we shall have

$$\text{(5)} \qquad \alpha_1 = u, \alpha_2 = v - \frac{1}{\alpha_1}, \dots, \alpha_{q-1} = v - \frac{1}{\alpha_{q-2}}, \tilde{\alpha}_q = w - \frac{1}{\alpha_{q-1}},$$

each value occurring k times. Hence

$$\text{(6)} \qquad D = \left(\prod_{i=1}^{q-1} \alpha_i^k\right) \tilde{\alpha}_q^k .$$

If $0 < r < k$, then after q-2 steps we shall have in the diagonal $\alpha_1, \alpha_2, \dots, \alpha_{q-1}$, each appearing k times, $\alpha_q = v - \frac{1}{\alpha_{q-1}}$ appearing r times and $\tilde{\alpha}_q = w - \frac{1}{\alpha_{q-1}}$ appearing k-r times, while the last r elements in the diagonal will be w. We still have r elements 1 in the lower triangle. Multiplying the rows n+2-k-r,...,n+2-k-1 of D by $-1/\alpha_q$ and adding them to the rows n+2-r,...,n+1, respectively, all elements of the lower triangle will become zero, and the last r elements of the diagonal will be

$$\text{(7)} \qquad \tilde{\alpha}_{q+1} = w - \frac{1}{\alpha_q} .$$

Hence

$$\text{(8)} \qquad D = D(u,v,w) = \left(\prod_{i=1}^{q-1} \alpha_i^k\right) \alpha_q^r \tilde{\alpha}_q^{k-r} \tilde{\alpha}_{q+1}^r .$$

Since for $r = 0$, (8) changes over into (6), (8) is valid in general.

REMARK. During the above calculation we had to assume that $\alpha_1 \neq 0, \dots, \alpha_q \neq 0$. But D is clearly a polynomial in u, v, w; therefore after performing the multiplications on the right hand side of (8) we must have a polynomial in u, v, w. This polynomial is identically equal to D(u,v,w), since the excluded values of u, v, w (by $\alpha_i \neq 0$) are of measure zero in $\mathbb{R}^3$ or C^3.

Now we are ready to prove

THEOREM 1. *The following identity holds*

(9) $$D(u,v,w) = (wT_{q-1}-T_{q-2})^{k-r}(wT_q-T_{q-1})^r \sin^{-k}\theta ,$$

where $D(u,v,w)$ *is the determinant of order* $n+1$ *defined by* (4),

(10) $$v = 2\cos\theta ,$$

(11) $$T_i = T_i(\theta,u) = u\sin i\theta - \sin(i-1)\theta \quad (i = 0,1,\dots,q)$$

and $n = kq+r$, $0 \le r < k$, $k \ge 1$, $n+1 \ge 2k$.

Proof. We claim that the quantities α_i defined by (5) can be obtained as

(12) $$\alpha_i = \frac{T_i}{T_{i-1}} \qquad (i = 1,\dots,q) .$$

(12) is proved by induction. For $i = 1$, (12) is clearly valid as

$$\alpha_1 = u = \frac{u\sin\theta}{\sin\theta} = \frac{T_1}{T_0} .$$

Supposing that (12) holds for i, we have by (5) and (11)

$$\alpha_{i+1} = v - \frac{1}{\alpha_i} = 2\cos\theta - \frac{T_{i-1}}{T_i}$$

$$= \frac{1}{T_i}\Big(u[\sin(i+1)\theta + \sin(i-1)\theta] - [\sin i\theta + \sin(i-2)\theta]$$

$$- u\sin(i-1)\theta + \sin(i-2)\theta\Big) .$$

Writing $2\cos\theta\sin i\theta$ and $2\cos\theta\sin(i-1)\theta$ for the brackets, respectively, we get $\alpha_{i+1} = T_{i+1}/T_i$, i.e., (12) holds for $i+1$ too.

By (12) we have

(13) $$\prod_{i=1}^{q-1}\alpha_i = \frac{T_{q-1}}{T_0} = \frac{T_{q-1}}{\sin\theta} .$$

Calculating α_q, $\tilde\alpha_q$, $\tilde\alpha_{q+1}$ by using (11), (12), we easily obtain (9) from (8).

We need another form of D which can be used more conveniently for our purpose.

THEOREM 2. *For the determinant* D_1 *given by*

(14) $$D_1(v,a,b) := D(v+a,v,v+b)$$

we have

(15) $$D_1(v,a,b) = R_{q-1}(\theta,a,b)^{k-r} R_q(\theta,a,b)^r \sin^{-k}\theta ,$$

<u>where</u>

(16) $$R_i(\theta,a,b) = \sin(i+2)\theta + (a+b)\sin(i+1)\theta + ab\sin i\theta \quad (i = q-1,q)$$

<u>and</u> $v = 2\cos\theta$, $n+1 = kq+r$, $0 \le r < k$, $k \ge 1$, $n+1 \ge 2k$.

<u>Proof.</u> Repeated application of the identity $2\cos\theta\sin j\theta = \sin(j+1)\theta + \sin(j-1)\theta$ and some simple calculations show that

(17) $$\begin{aligned} wT_i - T_{i-1} &= (v+b)T_i - T_{i-1} \\ &= (2\cos\theta+b)[(2\cos\theta+a)\sin i\theta - \sin(i-1)\theta] \\ &\qquad - [(2\cos\theta+a)\sin(i-1)\theta - \sin(i-2)\theta] \\ &= R_i(\theta,a,b) . \end{aligned}$$

Now we obtain (15) from (9) and (17).

We remark that in the case $k = 1$ (15) was proved by D.E. Rutherford [7].

We can decompose $D_1(v,a,b)$ into linear factors of v in the following 6 cases:

(18) $$\begin{aligned} &a = b = 1;\ a = b = -1;\ a = b = 0;\ a = 1,\ b = -1\ (\text{or } a = -1,\ b = 1); \\ &a = 0,\ b = 1\ (\text{or } a = 1,\ b = 0);\ a = 0,\ b = -1\ (\text{or } a = -1,\ b = 0). \end{aligned}$$

This decomposition is given by

THEOREM 3. <u>Let</u> $D_1(v,a,b)$ <u>be the determinant of order</u> $n+1$ <u>given by</u> (14) <u>and assume that</u> $n+1 = kq+r$, $0 \le r < k$, $k \ge 1$, $n+1 \ge 2k$. <u>Then the following identities hold</u>:

(19) $$D_1(v,1,1) = (v+2)^k \prod_{j=1}^{q-1} \left(v - 2\cos\frac{j\pi}{q}\right)^{k-r} \prod_{j=1}^{q} \left(v - 2\cos\frac{j\pi}{q+1}\right)^r ,$$

(20) $$D_1(v,-1,-1) = (v-2)^k \prod_{j=1}^{q-1} \left(v - 2\cos\frac{j\pi}{q}\right)^{k-r} \prod_{j=1}^{q} \left(v - 2\cos\frac{j\pi}{q+1}\right)^r ,$$

(21) $$D_1(v,0,0) = \prod_{j=1}^{q} \left(v - 2\cos\frac{j\pi}{q+1}\right)^{k-r} \prod_{j=1}^{q+1} \left(v - 2\cos\frac{j\pi}{q+2}\right)^r ,$$

(22) $$D_1(v,1,-1) = \prod_{j=0}^{q-1}\left(v - 2\cos\frac{(2j+1)\pi}{2q}\right)^{k-r} \prod_{j=0}^{q}\left(v - 2\cos\frac{(2j+1)\pi}{2(q+1)}\right)^{r},$$

(23) $$D_1(v,0,1) = \prod_{j=1}^{q}\left(v - 2\cos\frac{2j\pi}{2q+1}\right)^{k-r} \prod_{j=1}^{q+1}\left(v - 2\cos\frac{2j\pi}{2q+3}\right)^{r},$$

(24) $$D_1(v,0,-1) = \prod_{j=0}^{q-1}\left(v - 2\cos\frac{(2j+1)\pi}{2q+1}\right)^{k-r} \prod_{j=0}^{q}\left(v - 2\cos\frac{(2j+1)\pi}{2q+3}\right)^{r}.$$

Proof. The proofs are based on the factorization of R_q (and R_{q-1}) given by

(25) $$R_q(\theta,1,1) = 2\sin(q+1)\theta(1+\cos\theta),$$

(26) $$R_q(\theta,-1,-1) = 2\sin(q+1)\theta(1-\cos\theta),$$

(27) $$R_q(\theta,0,0) = \sin(q+2)\theta,$$

(28) $$R_q(\theta,1,-1) = 2\cos(q+1)\theta,$$

(29) $$R_q(\theta,0,1) = 2\sin\frac{(2q+3)\theta}{2}\cos\frac{\theta}{2},$$

(30) $$R_q(\theta,0,-1) = 2\cos\frac{(2q+3)\theta}{2}\sin\frac{\theta}{2}.$$

Since the proofs of (19) - (24) are similar, we prove one of them, say (23).

Substituting (29) into (15) we get

$$D_1(v,0,1) = \left(\frac{\sin\frac{(2q+1)\theta}{2}}{\sin\frac{\theta}{2}}\right)^{k-r}\left(\frac{\sin\frac{(2q+3)\theta}{2}}{\sin\frac{\theta}{2}}\right)^{r}.$$

The zeros of the right hand side are $\theta = 2j\pi/(2q+1)$ ($j = 1,\dots,q$), each with multiplicity $k-r$, and $\theta = 2j\pi/(2q+3)$ ($j = 1,\dots,q+1$), each with multiplicity r. Thus the zeros of $D_1(v,0,1)$ are $2\cos\frac{2j\pi}{2q+1}$ ($j = 1,\dots,q$) and $2\cos\frac{2j\pi}{2q+3}$ ($j = 1,\dots,q+1$), each with multiplicity $k-r$ and r, respectively. Since the main coefficient of the polynomial $D_1(v,0,1)$ is one, we immediately get (23).

REMARK. Using (25) - (30) we can easily obtain trigonometric formulae for $D_1(v,a,b)$ in the cases (18). Using these forms we can express $D_1(v,a,b)$ by means of the Chebychev polynomials C_j

and U_j of the first and second kind. These are defined by

$$C_j(\cos\theta) = \cos j\theta\,, \qquad U_j(\cos\theta) = \frac{\sin(j+1)\theta}{\sin\theta}\,.$$

For example

$$D_1(v,0,1) = U_{2q}^{k-r}\left(\frac{\sqrt{v+2}}{2}\right)U_{2q+2}^{r}\left(\frac{\sqrt{v+2}}{2}\right)\,,$$

since

$$\frac{\sin\frac{(2q+1)\theta}{2}}{\sin\frac{\theta}{2}} = U_{2q}\left(\cos\frac{\theta}{2}\right) = U_{2q}\left(\sqrt{\frac{1+\cos\theta}{2}}\right) = U_{2q}\left(\frac{\sqrt{v+2}}{2}\right)\,.$$

3. DISCRETE QUADRATIC INEQUALITIES

Let $A = (a_{ij})$ be an Hermitean matrix of order n+1 with eigenvalues $\lambda_o \geq \lambda_1 \geq \dots \geq \lambda_n$ and let $z_o, z_1, \dots, z_n$ be the corresponding linearly independent eigenvectors. Then

$$(31)\qquad \lambda_n\langle x,x\rangle \leq \langle Ax,x\rangle \leq \lambda_o\langle x,x\rangle$$

holds for every vector x in the real or complex unitary (n+1)-space. Here $\langle\cdot,\cdot\rangle$ denotes the usual inner product

$$\langle x,y\rangle = \sum_{j=0}^{n} x_j\bar{y}_j$$

for $x = (x_o,\dots,x_n)^T$, $y = (y_o,\dots,y_n)^T$. The equality $\lambda_n\langle x,x\rangle = \langle Ax,x\rangle$ holds if and only if $x = 0$ or x is an eigenvector corresponding to λ_n (if $\lambda_n < \lambda_{n-1}$, then x is a scalar multiple of z_n). Similarly the equality $\langle Ax,x\rangle = \lambda_o\langle x,x\rangle$ holds if and only if $x = 0$ or x is an eigenvector corresponding to λ_o; see e.g. [1].

Let us return now to the inequality (1). With the above notation (1) can be written as

$$(32)\qquad \alpha_i^{\pm}\langle x,x\rangle \leq \langle A_i^{\pm}x,x\rangle \leq \beta_i^{\pm}\langle x,x\rangle$$

($i = 1,2,3,4$ and either the + or - sign is taken), where $A_i^{\pm}$ are Hermitean matrices given later. It is clear that the best constants $\alpha_i^{\pm}$ and $\beta_i^{\pm}$ in (1) or (32) are the minimal and maximal eigenvalues of $A_i^{\pm}$, respectively.

Our results concerning the inequality (1) can be summarized as follows.

THEOREM 4. *Let* n, k *be fixed natural numbers*, $1 \le k \le n$. *The inequalities*

$$\alpha_i^{\pm} \sum_{j=0}^{n} |x_j|^2 \le \sum{}^{i} |x_j \pm x_{j+k}|^2 \le \beta_i^{\pm} \sum_{j=0}^{n} |x_j|^2 \tag{1}$$

($i = 1,2,3,4$, *and with* + *or* −) *hold for every real or complex* (n+1)-*vector* $x = (x_0,\dots,x_n)^T$. *The best constants are*

$$\alpha_1^+ = \alpha_1^- = 0 , \qquad \beta_1^+ = \beta_1^- = 4\cos^2\frac{\pi}{2([\frac{n}{k}]+1)} ; \tag{33}$$

$$\begin{cases} \alpha_2^+ = \alpha_2^- = \alpha_3^+ = \alpha_3^- = 4\sin^2\dfrac{\pi}{2(2[\frac{n}{2}]+3)} , \\ \\ \beta_2^+ = \beta_2^- = \beta_3^+ = \beta_3^- = 4\cos^2\dfrac{\pi}{2[\frac{n}{k}]+3} ; \end{cases} \tag{34}$$

$$\alpha_4^+ = \alpha_4^- = 4\sin^2\frac{\pi}{2([\frac{n}{k}]+2)} , \qquad \beta_4^+ = \beta_4^- = 4\cos^2\frac{\pi}{2([\frac{n}{k}]+2)} . \tag{35}$$

Proof. 1. *$n+1 \ge 2k$.* Consider the square matrix of order n+1

$$C^{\pm}(v,a,b) = \left(\begin{array}{ccc|ccc|ccc} v+a & & & \pm 1 & & & & & \\ & \ddots & & & \ddots & & & & \\ & & v+a & & & \ddots & & & \\ \hline \pm 1 & & & v & & & \ddots & & \\ & \ddots & & & \ddots & & & \ddots & \\ & & \ddots & & & v & & & \pm 1 \\ \hline & & & \ddots & & & v+b & & \\ & & & & \ddots & & & \ddots & \\ & & & & & \pm 1 & & & v+b \end{array}\right) , \tag{36}$$

where in the main diagonal we have k times v+a, then n+1−2k times v, then k times v+b. Note that

$$\det C^+(v,a,b) = D_1(v,a,b) , \quad \det C^-(v,a,b) = (-1)^{n+1} D(-v,-a,-b).$$

The matrices $A_i^{\pm}$ which generate the Hermitean form

$$\langle A_i^{\pm}x,x\rangle = \sum{}^{i} |x_j \pm x_{j+k}|^2$$

are given by

$$(37)\qquad \begin{cases} A_1^{\pm} = C^{\pm}(2,-1,-1) , & A_3^{\pm} = C^{\pm}(2,0,-1) , \\ A_2^{\pm} = C^{\pm}(2,-1,0) , & A_4^{\pm} = C^{\pm}(2,0,0) . \end{cases}$$

Hence

$$(38)\qquad \det(A_1^{+}-\lambda E) = D_1(2-\lambda,-1,-1) , \quad \det(A_1^{-}-\lambda E) = (-1)^{n+1}D_1(\lambda-2,1,1),$$

$$(39)\qquad \det(A_2^{+}-\lambda E) = D_1(2-\lambda,-1,0) = D_1(2-\lambda,0,-1) = \det(A_3^{+}-\lambda E) ,$$

$$(40)\ \det(A_2^{-}-\lambda E) = (-1)^{n+1}D_1(\lambda-2,1,0) = (-1)^{n+1}D_1(\lambda-2,0,1) = \det(A_3^{-}-\lambda E),$$

$$(41)\qquad \det(A_4^{+}-\lambda E) = D_1(2-\lambda,0,0) , \quad \det(A_4^{-}-\lambda E) = (-1)^{n+1}D_1(\lambda-2,0,0),$$

where E is the unit matrix of order n+1.

The maximal and minimal eigenvalues of $A_i^{\pm}$ can now be determined with the help of Theorem 3. Since the proofs of (33), (34), (35) are similar, we present only the

Proof of (33). The number α_1^{+} is the smallesteigenvalue of A_1^{+}, hence by (38)

$$\alpha_1^{+} = 2-\lambda_1^{+} ,$$

where λ_1^{+} is the greatest zero of $D_1(v,-1,-1)$. From (20) follows $\lambda_1^{+} = 2$ and $\alpha_1^{+} = 0$.

Again by (38) the smallest eigenvalue α_1^{-} of A_1^{-} has the form

$$\alpha_1^{-} = \lambda_1^{-}+2 ,$$

where λ_1^{-} is the smallest zero of $D_1(v,1,1)$. From (19) we have $\lambda_1^{-} = -2$ and $\alpha_1^{-} = 0$.

The number β_1^{+} is the largest eigenvalue of A_1^{+}, thus by (38)

$$\beta_1^{+} = 2-\mu_1^{+} ,$$

where μ_1^{+} is the smallest zero of $D_1(v,-1,-1)$. From (20) follows

$$\mu_1^{+} = \begin{cases} 2\cos\frac{(q-1)\pi}{q} = -2\cos\frac{\pi}{q} & \text{if } r=0 , \\ 2\cos\frac{q\pi}{q+1} = -2\cos\frac{\pi}{q+1} & \text{if } r>0 . \end{cases} ,$$

Using the equation $n+1 = kq+r$ we can easily check that

$$[\tfrac{n}{k}] = \begin{cases} q-1 & \text{if } r = 0 , \\ q & \text{if } r > 0 , \end{cases}$$

therefore

$$\beta_1^+ = 2-\mu_1^+ = 2+2\cos\frac{\pi}{[\frac{n}{k}]+1} = 4\cos^2\frac{\pi}{2([\frac{n}{k}]+1)} .$$

For the largest eigenvalue β_1^- of A_1^- we get by (38)

$$\beta_1^- = \mu_1^- + 2 ,$$

where μ_1^- is the largest zero of $D_1(v,1,1)$. From (19) we have

$$\mu_1^- = 2\cos\frac{\pi}{[\frac{n}{k}]+1}$$

and

$$\beta_1^- = \beta_1^+ .$$

2. If <u>n+1 < 2k</u>, (37) is no longer valid. In this case

$$A_1^{\pm} = \left(\begin{array}{ccc|ccc|ccc} 1 & & & & & & \pm 1 & & \\ & \ddots & & & & & & \ddots & \\ & & 1 & & & & & & \pm 1 \\ \hline & & & 0 & & & & & \\ & & & & \ddots & & & & \\ & & & & & 0 & & & \\ \hline \pm 1 & & & & & & 1 & & \\ & \ddots & & & & & & \ddots & \\ & & \pm 1 & & & & & & 1 \end{array}\right) . \tag{42}$$

Here we have $n+1-k$ elements 1 in the main diagonal, then $2k-n-1$ zeros and again $n+1-k$ times 1. In the lower and upper triangle we have $n+1-k$ elements ± 1.

$A_2^{\pm}$ has the same form as $A_1^{\pm}$, but in its main diagonal 1 appears k times, then 2 appears $n+1-k$ times. In the diagonal of $A_3^{\pm}$, 2 appears $n+1-k$ times, then 1 k times. The main diagonal of $A_4^{\pm}$ contains $n+1$ elements 2.

To find the eigenvalues of $A_i^{\pm}$ $(i = 1,2,3,4)$ we need the value of the determinant

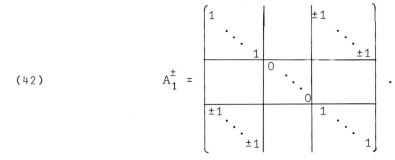

$$D_2(v,a,b) = \left|\begin{array}{c|c|c} \begin{matrix} v+a & & \\ & \ddots & \\ & & v+a \end{matrix} & & \begin{matrix} 1 & & \\ & \ddots & \\ & & 1 \end{matrix} \\ \hline & \begin{matrix} v & & \\ & \ddots & \\ & & v \end{matrix} & \\ \hline \begin{matrix} 1 & & \\ & \ddots & \\ & & 1 \end{matrix} & & \begin{matrix} v+b & & \\ & \ddots & \\ & & v+b \end{matrix} \end{array}\right| \tag{43}$$

D_2 is a determinant of order n+1, and its main diagonal consists of v+a (n+1-k times), v (2k-n-1 times) and v+b (n+1-k times). In the lower and upper triangle all elements are zero except for the n+1-k elements 1. Multiplying the first n+1-k lines of D_2 by -1/v+a and adding them to the last n+1-k lines the lower triangle of D_2 will have only zeros, and D_2 will be the product of the new diagonal elements:

$$\begin{aligned} D_2(v,a,b) &= (v+a)^{n+1-k} v^{2k-n-1} \left(v + b - \frac{1}{v+a}\right)^{n+1-k} \\ &= [(v+a)(v+b)-1]^{n+1-k} v^{2k-n-1} . \end{aligned} \tag{44}$$

By (42) and (44) we get

$$\det (A_1^+ - \lambda E) = (-1)^{2k-n-1} \lambda^k (\lambda-2)^{n+1-k} ,$$

hence $\alpha_1^+ = 0$, $\beta_1^+ = 2$. Similarly,

$$\det (A_1^- - \lambda E) = (-1)^{n+1} \lambda^k (\lambda-2)^{n+1-k} ,$$

therefore $\alpha_1^- = 0$, $\beta_1^- = 2$. Since now $[\frac{n}{k}] = 1$, (33) is valid again. The formulas (34), (35) can be proved in a similar way using (44).

REFERENCES

1. R. Bellman, Introduction to matrix analysis. McGraw-Hill, New York, 1960.

2. E. Egerváry and O. Szász, Einige Extremalprobleme im Bereiche der trigonometrischen Polynome. Math. Z. 27 (1928), 641-692.

3. K. Fan, O. Taussky and J. Todd, Discrete analogs of inequalities of Wirtinger. Monatsh. Math. 59 (1955), 73-79.

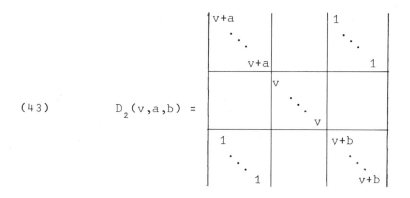

4. L. Fejér, Über trigonometrische Polynome. J. Reine Angew. Math. 146 (1915), 53-82.

5. G.H. Hardy, J.E. Littlewood and G. Pólya, Inequalities. Cambridge Univ. Press, Cambridge, 2nd Edition, 1952.

6. G.V. Milovanovic and I.Z. Milovanovic, On discrete inequalities of Wirtinger's type. J. Math. Anal. Appl. 88 (1982), 378-387.

7. D.E. Rutherford, Some continuant determinants arising in physics and chemistry, I. Proc. Royal Soc. Edinburgh, Sect. A 62 (1947), 229-236.

8. G. Szegö, Koeffizientenabschätzungen bei ebenen und räumlichen harmonischen Entwicklungen. Math. Ann. 96 (1926-27), 601-632.

László Losonczi, Department of Mathematics, Kossuth Lajos University, H-4010 Debrecen, Pf.12, Hungary

International Series of
Numerical Mathematics, Vol. 80

SOME INEQUALITIES FOR GEOMETRIC MEANS

E.R. Love

Abstract. This paper is mainly concerned with a discrete inequality bearing some resemblance to recent inequalities of Heinig and of Cochran and Lee. These are typified by

$$\sum_{m=1}^{\infty} m^{k-1}\left(\prod_{n=1}^{m} x_n^{n^{p-1}}\right)^{p/m^p} \le e^{k/p} \sum_{m=1}^{\infty} m^{k-1} x_m$$

under appropriate conditions. The products on the left are replaced, in this paper, by geometric means with more general weights, and the factors m^{k-1} on both sides by factors r^{-m} for suitably small r. Some inequalities having an analogous character are first discussed, since they led the way into this study.

1. INTRODUCTION

J.A. Cochran and C.-S. Lee [1: Theorem 1] found the inequality

$$(1) \qquad \int_0^{\infty} x^{k-1} \exp\left(\frac{p}{x^p}\int_0^x t^{p-1}\log f(t)\,dt\right)dx \le e^{k/p}\int_0^{\infty} x^{k-1} f(x)\,dx,$$

where $p > 0$, k is real, f(t) is positive and $t^{p-1}\log f(t)$ locally integrable, in $[0,\infty)$. This generalizes an inequality of Knopp [2: Theorem 335], which is the case $p = 1 = k$ of (1), and also some inqualities of Heinig [3: Theorems 1 and 2].

At GI 5 I remarked that the exponential in the left side of (1) is a geometric mean, and proved the following generalization of (1).

This paper is in final form, and no version of it will be submitted for publication elsewhere.

Define a geometric mean of f on (0,x):

$$Gf(x) = \exp\left\{\int_0^x w(t/x)\log f(t)\,dt \Big/ \int_0^x w(t/x)\,dt\right\} .$$

If w(x) and f(x) are positive and measurable, σ(x) is submultiplicative,

$$\rho(x) = x^{-1}\sigma(x^{-1}) , \qquad \int_0^1 w(t)\rho(t)^{1/n}dt < \infty$$

for all positive integers $n \ge N$, _and_ $\tau(x)$ _is decreasing, then_

$$(2) \qquad \int_0^\infty Gf(x)\sigma(x)\tau(x)dx \le G\rho(1)\int_0^\infty f(x)\sigma(x)\tau(x)dx \quad .$$

This generalizes (1), which is the case of (2) in which

$$w(t) = t^{p-1} , \quad \sigma(x) = x^{k-1} , \quad \tau(x) = 1 .$$

A proof of (2) will appear in [4], together with discussion of best possible constants.

Cochran and Lee [1: Theorem 2] also gave a discrete analogue of (1), and showed that the constant in it, like that in (1), is best possible. This inequality is

$$(3) \qquad \sum_{m=1}^\infty m^{k-1}\left(\prod_{n=1}^m x_n^{n^{p-1}}\right)^{p/m^p} \le e^{k/p}\sum_{m=1}^\infty m^{k-1}x_m \quad ,$$

where $p \ge 1$, $k \ge 1$ and $0 \le x_n \le 1$. This also improved on a result of Heinig [3: Theorem 8].

A discrete analogue of (2) will appear in [4]; but it is less satisfactory than (2) in that conditions which seem unnatural and heavy are needed to make it coincide with even a special case of (3).

The inequality proved in this paper was found in the course of attempts to improve this situation. It is neither a discrete analogue of (1) or (2), nor a generalization of (3); but it seems worth recording as a companion to (3) having its own independent interest.

Notation. Let $w_{mn} > 0$ and $x_n > 0$ for all integers m and n such that $0 \le n \le m$. Let $\alpha(t)$ be strictly increasing and right-continuous in [0,1], $\alpha(0) = 0$, and $\lambda(t)$ be positive and measurable on (0,1).

Define the following geometric and arithmetic means, if they exist.

$$G_w x_m = \exp\left\{\sum_{n=0}^{m} w_{mn}\log x_n \Big/ \sum_{n=0}^{m} w_{mn}\right\}$$

$$= \left\{\prod_{n=0}^{m} x_n^{w_{mn}}\right\}^{1/W_m}, \qquad \text{where } W_m = \sum_{n=0}^{m} w_{mn},$$

$$G_\alpha\lambda = \exp\left\{\int_0^1 \log\lambda(t)d\alpha(t) \Big/ \int_0^1 d\alpha(t)\right\},$$

$$A_\alpha\lambda = \int_0^1 \lambda(t)d\alpha(t) \Big/ \int_0^1 d\alpha(t).$$

2. THEOREM

Let $0 < r < 1$, $C = \alpha(1)$, $\mu(t) = t^{C-1}$, $\nu(t) = t^{-C}$,

(4) $$w_{mo} = \alpha(r^m) \quad \text{and} \quad w_{mn} = \alpha(r^{m-n}) - \alpha(r^{m-n+1})$$

for $0 < n \le m$, *together with* $w_{oo} = \alpha(1)$. *If also*

(5) $$r \le \int_0^1 t^C d\alpha(t) \Big/ \int_0^1 t^{C-1} d\alpha(t),$$

the denominator integral being supposed convergent, then

(6) $$\sum_{m=1}^{\infty} r^{-m} G_w x_m \le A_\alpha\mu G_\alpha\nu \sum_{m=0}^{\infty} r^{-m} x_m.$$

If instead $r \le \xi$, *where* ξ *is the unique root in* $(0,1)$ *of*

(7) $$\xi = \int_\xi^1 t^{C-1}(1-t)d\alpha(t) \Big/ \int_0^1 t^{C-1}(1-t)d\alpha(t),$$

then (6) *can be strengthened by replacing the lower terminal of summation on the left by* $m = 0$.

Observe that the factor $A_\alpha\mu$ can be omitted from both (6) and its strengthened form whenever $C \ge 1$, because then $A_\alpha\mu \le 1$. Compare the special case $C = 1$ which follows.

COROLLARY. *Let* $0 < r < 1$, $\alpha(1) = 1$, $\nu(t) = 1/t$, *and let* w_{mn} *be as in* (4). *If*

$$r \le \int_0^1 t\, d\alpha(t),$$

then

$$\sum_{m=1}^{\infty} r^{-m} G_w x_m \le G_\alpha \nu \sum_{m=0}^{\infty} r^{-m} x_m \quad . \tag{8}$$

If instead $r \le \eta$, *where* η *is the unique root in* $(0,1)$ *of*

$$\eta = \int_\eta^1 (1-t)\,d\alpha(t) \Big/ \int_0^1 (1-t)\,d\alpha(t) \ ,$$

then (8) *can be strengthened in the same way as before, replacing* $m=1$ *by* $m=0$.

3. PROOF OF THEOREM

Let f be the real function: $f(t) = x_0$ for $0 < t \le 1$, $f(t) = x_n$ for $r^{-(n-1)} < t \le r^{-n}$ and positive integral n; and let $g(t) = \log f(t)$. Both f and g are left-continuous. Since

$$\sum_{n=0}^{m} w_{mn} = \alpha(1) = C \ ,$$

$$\begin{aligned}
\sum_{m=1}^{\infty} r^{-m} G_w x_m &= \sum_{m=1}^{\infty} r^{-m} \exp\left(\frac{1}{C} \sum_{n=0}^{m} w_{mn} \log x_n\right) \\
&= \sum_{m=1}^{\infty} r^{-m} \exp\left(\int_0^{r^m} \log x_0 \,\frac{d\alpha(t)}{C} + \sum_{n=1}^{m} \int_{r^{m-n+1}}^{r^{m-n}} \log x_n \,\frac{d\alpha(t)}{C}\right) \\
&= \sum_{m=1}^{\infty} r^{-m} \exp\left(\int_0^{r^m} g(r^{-m}t)\frac{d\alpha(t)}{C} + \sum_{n=1}^{m} \int_{r^{m-n+1}}^{r^{m-n}} g(r^{-m}t)\frac{d\alpha(t)}{C}\right) \\
&= \sum_{m=1}^{\infty} r^{-m} \exp\left(\int_0^1 g(r^{-m}t)\frac{d\alpha(t)}{C}\right) \\
&= \sum_{m=1}^{\infty} r^{-m} G_\alpha \nu \exp\left(\int_0^1 \{\log f(r^{-m}t) - \log \nu(t)\} \,\frac{d\alpha(t)}{C}\right) \\
&= G_\alpha \nu \sum_{m=1}^{\infty} r^{-m} \exp\left(\int_0^1 \log \frac{f(r^{-m}t)}{\nu(t)} \,\frac{d\alpha(t)}{C}\right) \\
&\le G_\alpha \nu \sum_{m=1}^{\infty} r^{-m} \int_0^1 \frac{f(r^{-m}t)}{\nu(t)} \,\frac{d\alpha(t)}{C} \ ,
\end{aligned}$$

the last step by the inequality of arithmetic and geometric means. Now writing

(9) $$\beta(s) = \int_0^s \frac{1}{\nu(t)} \frac{d\alpha(t)}{C} = \frac{1}{C}\int_0^s t^C d\alpha(t) ,$$

$$\sum_{m=1}^{\infty} r^{-m} G_w x_m \le G_\alpha \nu \sum_{m=1}^{\infty} r^{-m} \int_0^1 f(r^{-m}t) d\beta(t)$$

(10) $$= G_\alpha \nu \left(\sum_{m=1}^{\infty} r^{-m} \int_0^{r^m} x_o \, d\beta(t) + \sum_{m=1}^{\infty} r^{-m} \sum_{n=1}^{m} \int_{r^{m-n+1}}^{r^{m-n}} x_n d\beta(t) \right) .$$

The convergence of the denominator integral in (5) ensures that

(11) $$\beta(s)/s \to 0 \quad \text{as} \quad s \to 0 ,$$

because

$$0 < C \frac{\beta(s)}{s} = \int_0^s \frac{t^C}{s} d\alpha(t) \le \int_0^s t^{C-1} d\alpha(t) .$$

The term in (10) involving x_o can be written, using (11) and (9),

$$G_\alpha \nu \frac{x_o}{1-r} \sum_{m=1}^{\infty} (r^{-m} - r^{-m+1}) \beta(r^m)$$

$$= G_\alpha \nu \frac{x_o}{1-r} \left(\sum_{m=1}^{\infty} r^{-m} \{\beta(r^m) - \beta(r^{m+1})\} - \beta(r) \right)$$

$$= G_\alpha \nu \frac{x_o}{1-r} \left(\sum_{m=1}^{\infty} r^{-m} \int_{r^{m+1}}^{r^m} d\beta(t) - \beta(r) \right)$$

$$\le G_\alpha \nu \frac{x_o}{1-r} \left(\sum_{m=1}^{\infty} \int_{r^{m+1}}^{r^m} t^{-1} d\beta(t) - \beta(r) \right)$$

(12) $$= G_\alpha \nu \frac{x_o}{1-r} \left(\int_0^r t^{-1} d\beta(t) - \int_0^r d\beta(t) \right) .$$

The terms in (10) not involving x_o can be written

$$G_\alpha \nu \sum_{m=1}^{\infty} r^{-m} \sum_{n=1}^{m} x_n \int_{r^{m-n+1}}^{r^{m-n}} d\beta(t)$$

$$= G_\alpha \nu \sum_{n=1}^{\infty} x_n \sum_{m=n}^{\infty} r^{-m} \int_{r^{m-n+1}}^{r^{m-n}} d\beta(t)$$

$$= G_\alpha \nu \sum_{n=1}^{\infty} r^{-n} x_n \sum_{l=0}^{\infty} r^{-l} \int_{r^{l+1}}^{r^l} d\beta(t) , \quad \text{by } l = m-n ,$$

$$\leq G_\alpha \nu \sum_{n=1}^{\infty} r^{-n} x_n \sum_{l=0}^{\infty} \int_{r^{l+1}}^{r^l} t^{-1}\, d\beta(t)$$

(13) $$= G_\alpha \nu \sum_{n=1}^{\infty} r^{-n} x_n \int_0^1 t^{-1}\, d\beta(t) .$$

Together (10), (12) and (13) give

$$\sum_{m=1}^{\infty} r^{-m} G_w x_m \leq G_\alpha \nu \sum_{n=0}^{\infty} r^{-n} x_n \int_0^1 t^{-1}\, d\beta(t) -$$

(14) $$- G_\alpha \nu \frac{x_0}{1-r} \left\{ (1-r) \int_0^1 t^{-1}\, d\beta(t) - \int_0^r (t^{-1}-1)\, d\beta(t) \right\} .$$

The bracket B of integrals in this last line is

(15) $$B = \int_0^r d\beta(t) + \int_r^1 t^{-1}\, d\beta(t) - r \int_0^1 t^{-1}\, d\beta(t)$$

$$\geq \int_0^r d\beta(t) + \int_r^1 d\beta(t) - r \int_0^1 t^{-1}\, d\beta(t)$$

(16) $$= \frac{1}{C} \left\{ \int_0^1 t^C\, d\alpha(t) - r \int_0^1 t^{C-1}\, d\alpha(t) \right\} \geq 0$$

using (9) and (5). So by (14), (16) and (9),

$$\sum_{m=1}^{\infty} r^{-m} G_w x_m \leq G_\alpha \nu \sum_{n=0}^{\infty} r^{-n} x_n \frac{1}{C} \int_0^1 t^{C-1}\, d\alpha(t) ;$$

this is (6), the first conclusion of the theorem.

For the other conclusion we have, by the inequality of arithmetic and geometric means,

(17) $$\frac{1}{G_\alpha \nu} = G_\alpha(\nu^{-1}) \leq A_\alpha(\nu^{-1}) = \frac{1}{C} \int_0^1 t^C\, d\alpha(t) = \beta(1) .$$

Also $G_w x_0 = x_0$; so by (14),

(18) $$\sum_{m=0}^{\infty} r^{-m} G_w x_m \leq A_\alpha \mu G_\alpha \nu \sum_{n=0}^{\infty} r^{-n} x_n - G_\alpha \nu \frac{x_0}{1-r} \left\{ B - \frac{1-r}{G_\alpha \nu} \right\} ,$$

B having the meaning assigned in (15).

Now by (15) and (17),

$$B - \frac{1-r}{G_\alpha \nu} \geq \int_0^r d\beta(t) + \int_r^1 t^{-1}\, d\beta(t) - r \int_0^1 t^{-1}\, d\beta(t) - (1-r) \int_0^1 d\beta(t)$$

$$= \int_r^1 t^{-1}\, d\beta(t) - r \int_0^1 (t^{-1}-1)\, d\beta(t) - \int_r^1 d\beta(t)$$

$$= \int_r^1 (t^{-1}-1)d\beta(t) - r\int_0^1 (t^{-1}-1)d\beta(t) \;. \tag{19}$$

The last line decreases strictly as r increases, is positive for sufficiently small r and negative for $r = 1$. Being continuous it has a unique zero, ξ, in $(0,1)$, and is positive for $r < \xi$. Further ξ satisfies

$$\xi = \int_\xi^1 (t^{-1}-1)d\beta(t) \Big/ \int_0^1 (t^{-1}-1)d\beta(t) \;.$$

This equation is equivalent to (7), and thus (18) and (19) give the required conclusion, namely that

$$\sum_{m=0}^{\infty} r^{-m} G_w x_m \le A_\alpha \mu G_\alpha \nu \sum_{n=0}^{\infty} r^{-n} x_n \tag{20}$$

for $0 < r \le \xi$ where ξ is given by (7).

Remark. The weight factors r^{-m} in (20) differ considerably from their counterparts m^{k-1} in (3), although both tend to infinity with m. However the generic resemblance of these two inequalities supports the view, already partly substantiated in [4], that there may be a synthesis of them.

4. ACKNOWLEDGMENT

The early part (preceding (9)) of the above proof owes much to the method used by Heinig [3: upper half of p.709] and by Cochran and Lee [1: upper half of p.5].

REFERENCES

1. J.A. Cochran and C.-S. Lee, Inequalities related to Hardy's and Heinig's. Math. Proc. Cambridge Phil. Soc. 96 (1984),1-7.

2. G.H. Hardy, J.E. Littlewood and G. Pólya, Inequalities. Cambridge, 1934.

3. H.P. Heinig, Some extensions of Hardy's inequality. SIAM J. Math. Anal. 6 (1975), 698-713.

4. E.R. Love, Inequalities related to those of Hardy and of Cochran and Lee. To appear in Math. Proc. Cambridge Phil.Soc.

E.R. Love, Department of Mathematics, University of Melbourne, Parkville, Victoria 3052, Australia.

International Series of
Numerical Mathematics, Vol. 80

ESTIMATION OF AN INTEGRAL

Ray Redheffer

Abstract. Different techniques are used to get upper and lower bounds of the integral

$$J(a;\lambda,\theta) = \int_a^b \frac{ds}{s^\lambda - b} \quad \text{with} \quad b = \theta a^\lambda ,$$

where $0 < \lambda < 1$, $0 < \theta < 1$, $0 < a < b < 1$. These bounds are used to study the asymptotic behavior of J as $a \to 0$ (with λ and θ fixed). The problem arises in the study of the delay differential equation $\rho'(t) = \rho(t)^\lambda + \rho(t-\mu(t))$.

The object of our investigation is the integral

$$\text{(1)} \qquad J(a) = J(a;\lambda,\theta) = \int_a^b \frac{ds}{s^\lambda - b} \qquad (0 < a < b = \theta a^\lambda) ,$$

where λ and θ are real parameters with $0 < \lambda < 1$, $0 < \theta < 1$. The condition $b < a^\lambda$ is needed to make the integral converge and, if b is allowed to be arbitrarily close to a^λ, the value of J can be arbitrarily large. This is the reason for introducing θ. Although the condition $0 < \lambda < 1$ is not intrinsic to the problem, it is satisfied in the application which is outlined next.

The integral J arises in connection with the delay differential equation

$$\rho'(t) = \rho^\lambda - \rho(t-\mu(t)) , \quad 0 \le \mu(t) \le t ,$$

which in turn plays a role in a comprehensive investigation of nonlinear delay systems of parabolic type. For this application we have $0 < \lambda < 1$. The problem of getting upper bounds which tend to 0 with a is easy, but to show that it is easy is, after all, a legitimate goal. We present a variety of elementary techniques, each of which gives a different insight. The idea is to have fun, not just to estimate an integral.

Let

$$E(a) = E(a;\lambda,\theta) = \frac{\theta^{1-\lambda}}{1-\lambda} a^{\lambda(1-\lambda)} . \tag{2}$$

We seek upper bounds for J(a) of the form CE(a), E(a)+F(a), or at worst CE(a)+F(a) where C is independent of a and $F = o(E)$ as $a \to 0$. Clearly $J(a) < \theta/(1-\theta)$ for all relevant a, but this is not a useful estimate. The side conditions imply $0 < a < \theta^{1/(1-\lambda)}$ and whenever an inequality such as $J(a) < E(a)$ is stated without qualification, it is understood that the inequality holds for all a in this range.

First we show that $J(a) \sim E(a)$ as $a \to 0$. Let $s = at$ to get

$$J = a^{1-\lambda} \int_1^c \frac{dt}{t^\lambda - \theta} = a^{1-\lambda} I \qquad (c = \theta a^{\lambda - 1}) ,$$

where I is defined by the equation. The condition $c > 1$ holds automatically. Let us choose m satisfying $1 < m < c$ and note that the integrand in I admits the upper bounds

$$\frac{1}{1-\theta} \text{ on } (1,m) \qquad \text{and} \qquad \frac{m^\lambda}{m^\lambda - \theta} \frac{1}{t^\lambda} \text{ on } (m,c) .$$

This gives

$$I < \frac{m-1}{1-\theta} + \frac{m^\lambda}{m^\lambda - \theta} \frac{\theta^{1-\lambda} a^{-(1-\lambda)^2}}{1-\lambda} .$$

For any $K > 1$ we can choose m so large that $m^\lambda/(m^\lambda - \theta) < K$. We can then take a so small that $c > m$ holds for this m. The result is an upper bound for I which implies

$$\limsup_{a \to 0} \frac{J(a)}{E(a)} \le K .$$

Since K is as close as we please to 1, it can be replaced by 1. On the other hand

$$J(a) > \int_a^b \frac{ds}{s^\lambda} = E(a) - \frac{a^{1-\lambda}}{1-\lambda} . \tag{3}$$

This gives an estimate for lim inf J(a)/E(a) which together with the above implies

$$\lim_{a \to 0} \frac{J(a)}{E(a)} = 1 .$$

The choice $m = 1$ in the calculations above gives an estimate for I which, when multiplied by $a^{1-\lambda}$, yields

$$J(a) < \frac{E(a)}{1-\theta} . \tag{4}$$

Although the coefficient of E(a) exceeds 1, the estimate has the advantage of being both simple and applicable for all values of the parameters.

The change of variable $t = s^\lambda$ gives

$$J = \frac{1}{\lambda} \int_{a^\lambda}^{b^\lambda} \frac{t^\mu - b^\mu + b^\mu}{t-b} \, dt , \qquad \lambda\mu = 1-\lambda , \tag{5}$$

where the extra terms b^μ in the numerator permit the use of the mean-value theorem. Namely,

$$\frac{t^\mu - b^\mu}{t-b} = \mu x^{\mu-1} , \qquad b < x < t . \tag{6}$$

Throughout this paper we define

$$L(a) = \frac{1}{\lambda} \int_{a^\lambda}^{b^\lambda} \frac{dt}{t-b} = \frac{1}{\lambda} \log \frac{b^\lambda - b}{a^\lambda - b}$$

and we note that

$$0 < L(a) < \frac{1}{\lambda} \log \frac{1}{1-\theta} + (1-\lambda) \log \frac{1}{a} .$$

If $\mu \ge 1$, corresponding to $\lambda \le 1/2$, then $\mu x^{\mu-1} \le \mu t^{\mu-1}$ and hence

$$J(a) < \frac{1-\lambda}{\lambda} E(a) + a^{1-\lambda} L(a) , \quad \lambda \le \frac{1}{2} . \tag{7}$$

As $\theta \to 1$ this is much stronger than (4) because $1-\theta$ occurs under the log. In fact, we could let $1-\theta = a^N/N$ where N is an arbitrarily large constant and still get a reasonable estimate as $a \to 0$.

If $\lambda > 1/2$ the maximum of $\mu x^{\mu-1}$ in (6) occurs when $x = b$. The resulting estimate involves a to the wrong power, $(1-\lambda)^2$ instead of $\lambda(1-\lambda)$, and will not be developed here.

The calculation (3) shows why E(a) is closely related to J(a) and will now be exploited more fully. Let us note that

$$\frac{1}{s^\lambda - b} = \frac{1}{s^\lambda} + \frac{b}{s^\lambda(s^\lambda - b)} . \tag{8}$$

We integrate from a to b using (3) for the first term and setting $s = at$ in the second. The result is

$$J(a) + \frac{a^{1-\lambda}}{1-\lambda} = E(a) + \theta a^{1-\lambda} H(a) \ , \tag{9}$$

where

$$H(a) = \int_1^c \frac{dt}{t^\lambda (t^\lambda - \theta)} \ , \qquad c = \theta a^{\lambda - 1} \ .$$

Since $t^\lambda - \theta > t^\lambda(1-\theta)$ for $t > 1$ we get

$$\begin{cases} J(a) < E(a) + \frac{1}{1-\theta} \frac{1}{2\lambda - 1} a^{1-\lambda} & (\lambda > \frac{1}{2}) \\ J(a) < E(a) + \frac{1}{1-\theta} \frac{1}{1-2\lambda} a^{2\lambda(1-\lambda)} & (\lambda < \frac{1}{2}) \ . \end{cases} \tag{10}$$

The integral for $\lambda = 1/2$ is elementary and will be considered later. However, the value involves $a^{1-\lambda} \log a$, and hence singular behavior of the above estimates as $\lambda \to 1/2$ is inevitable.

The foregoing results enable us to put the problem of estimation into sharp focus. The estimate (7) has a favorable θ-dependence, but it is valid only for a limited range of λ and the coefficient of E exceeds 1. The estimates (10) together cover the whole range of λ, except $\lambda = 1/2$, and the coefficient of E is 1, but the θ-dependence is unfavorable. Thus the central problem is seen as follows: Is there an estimate $J < E + o(E)$ which covers the whole range $0 < \lambda < 1$ and in which the θ-dependence as $\theta \to 1$ is described by $\log(1-\theta)$? The answer is yes and a solution is given by

$$\begin{cases} J(a) < E(a) + a^{1-\lambda} L(a) + \frac{1}{\lambda} a^{2\lambda(1-\lambda)} & (\lambda \le \frac{1}{3}) \\ J(a) < E(a) + a^{1-\lambda} L(a) + \frac{1-2\lambda}{\lambda^2} a^{(1-\lambda)^2} & (\frac{1}{3} \le \lambda \le \frac{1}{2}) \\ J(a) < E(a) + a^{1-\lambda} L(a) & (\frac{1}{2} \le \lambda) \ . \end{cases} \tag{11}$$

For proof, integrate (8) and set $s^\lambda = t$ in the second expression to get

$$J(a) + \frac{a^{1-\lambda}}{1-\lambda} = E(a) + \frac{b}{\lambda} \int_a^b \frac{t^{\mu-1} - b^{\mu-1}}{t-b} \, dt + b^\mu L(a) \ .$$

Then estimate the integral by (6) with $\mu - 1$ replacing μ.

Let us inquire next whether $J(a) < E(a)$ holds for any significant class of values (θ, λ). By (9) this holds if, and only if,

(12) $$\theta H(a) \le \frac{1}{1-\lambda}\ .$$

Since H(a) increases as a decreases, the condition $J(a) < E(a)$ for all relevant a is equivalent to the same condition as $a \to 0$. When $\lambda \le 1/2$ the integral defining H(a) diverges as $c \to \infty$ and we conclude that $J(a) > E(a)$ for small a. The same conclusion follows from (11). But if $\lambda > 1/2$ the integral converges and we get

(13) $$\theta \int_1^\infty \frac{dt}{t^\lambda(t^\lambda-\theta)} \le \frac{1}{1-\lambda}$$

as a necessary and sufficient condition for $J(a) < E(a)$.

Equation (13) cannot hold for θ arbitrarily close to 1 because the integral tends to ∞ as $\theta \to 1$. The same conclusion follows from (11), in view of the behavior of L(a). On the other hand $t^\lambda - \theta > t^\lambda(1-\theta)$ as noted previously, and applying this to (13) we get the following:

(14) $\frac{1}{\lambda} < 2$ is necessary and $\theta + \frac{1}{\lambda} \le 2$ is sufficient

for $J(a) < E(a)$ when a ranges over all values permitted by the constraint.

It is of some interest to compare (14) with the exact criterion given by (13). Expanding the second factor of the integrand in powers of θ/t^λ we find that (13) holds if, and only if, $\lambda > 1/2$ and

(15) $$\frac{1}{1-\lambda} \ge \frac{\theta}{2\lambda-1} + \frac{\theta^2}{3\lambda-1} + \frac{\theta^3}{4\lambda-1} + \cdots\ .$$

The criterion (14) is obtained when the denominators $j\lambda-1$ on the right are all replaced by $2\lambda-1$.

The technique leading to (15) can be applied to the original integral J. Let us write

$$\frac{1}{s^\lambda - b} = \frac{1}{s^\lambda}\,\frac{1}{1 - b/s^\lambda}$$

and expand the second factor on the right in geometric series. The remainder R_m after m-1 terms satisfies

(16) $$b^{m-1}s^{-\lambda m} < R_m < \frac{1}{1-\theta}\, b^{m-1}s^{-\lambda m}$$

and the general term of the series is $b^{j-1}s^{-\lambda j}$, $j = 1,2,\dots,m-1$.

For $j\lambda \neq 1$

$$(17) \qquad \int_a^b b^{j-1} s^{-\lambda j} ds = \frac{\theta^{j(1-\lambda)} a^{\lambda(1-\lambda)j} - \theta^{j-1} a^{1-\lambda}}{1-\lambda j}$$

and, using this with $j = 1,2,\dots,m$, we get an asymptotic expansion of which $E(a)$ is the leading term. Specifically, let $F_o = 0$ and

$$F_m(t) = \sum_{j=1}^{m} \frac{t^j}{1-\lambda j}, \qquad \lambda j \neq 1, \qquad m \geq 1 .$$

Then the summation of (17) is expressible with ease by $F_{m-1}(t)$ and the error is estimated by (16). Instead of writing it out, let us consider an interesting special case. If $\lambda(1-\lambda)m < 1-\lambda$, terms with $1-\lambda$ as exponent are negligible compared to the other terms as $a \to 0$. Dropping these, we get an expression which still has the correct asymptotic behavior and which still provides a true upper bound. Namely,

$$(18) \qquad J(a) < F_{m-1}(\theta^{1-\lambda} a^{\lambda(1-\lambda)}) + \frac{a^{\lambda(1-\lambda)m}}{(1-\theta)(1-\lambda m)} \quad (\lambda m < 1) .$$

Note that the first term is a polynomial with $E(a)$ as principal part.

Estimating the remainder by the technique used for (11) instead of by (16) we get an expansion in which θ as $\theta \to 1$ occurs only in the form $\log(1-\theta)$. Also, at the cost of additional complication, the condition $\lambda m < 1$ can be dropped. Since no new ideas are involved, details are not given here.

The case $\lambda j = 1$ which leads to trouble in the asymptotic expansion is perhaps the simplest case of all as far as J itself is concerned. If $1/\lambda$ is an integer then $\mu = (1/\lambda)-1$ in (5) is also an integer and by a short calculation

$$(19) \qquad J(a) = F_\mu(\theta^{1-\lambda} a^{\lambda(1-\lambda)}) - a^{1-\lambda}\theta^{-1} F_\mu(\theta) + \theta^\mu a^{1-\lambda} L(a) .$$

The central term involving $a^{1-\lambda}$ has a lower order of magnitude than the other terms and can be dropped to get a simpler but still asymptotically correct upper bound. When $\lambda = 1/2$ the result (19) agrees with that obtained by direct calculation, as it should. Dropping such terms as can be dropped we get

$$J(a) < E(a) + \theta a^{1/2}\left(2 \log\frac{1}{1-\theta} + \log\theta + \frac{1}{2}\log\frac{1}{a}\right) \qquad \left(\lambda = \frac{1}{2}\right),$$

where $E(a) = 2\theta^{1/2}a^{1/4}$. This supplements (10) and is in good agreement with (11).

Ray Redheffer, Department of Mathematics, University of California at LA, Los Angeles, CA 90024, U.S.A.

International Series of
Numerical Mathematics, Vol. 80

TAUBERIAN-TYPE RESULTS FOR CONVOLUTION OF SEQUENCES

Dennis C. Russell

Abstract. Copson (1970) proved a theorem expressible as: If p is a finite strictly decreasing positive sequence, if a is a real semi-infinite bounded sequence, and if $a*p$ is monotone, then a is convergent (the monotonicity of $a*p$ is equivalent to a linear inequality on the a_n). Later Borwein and Russell independently published necessary and sufficient conditions for the conclusion, related to the distribution of the zeros of the generating power series (or polynomial) $p(z)$. Here we use bi-infinite sequences and a sequence space λ with the property $(u\in\lambda,\ v\in\ell^1) \Rightarrow u*v \in \lambda$, and we ask for a theorem of the form $a*p \in \lambda \Rightarrow a \in \lambda$, provided that the tauberian condition $a \in \ell^\infty$ holds. While (for λ = bv) this already includes all previous results, the theorem can be further improved for semi-infinite sequences.

0. THE BACKGROUND

In [2], E.T.Copson considered, for real weights r_j ($j = 1,\ldots,N$; N fixed) and a real infinite sequence (a_n), the linear inequality

$$(0.1)\qquad a_{n+N} \geq \sum_{j=1}^{n} r_j\, a_{n+N-j} \qquad (n = 0,1,\ldots)$$

and he showed that if $r_j > 0$ ($j = 1,\ldots,N$), $r_1+\ldots+r_N = 1$, then boundedness of (a_n) necessarily implies its convergence. The direction of the inequality in (0.1) is clearly immaterial, since we may replace a_n by $-a_n$ and reach the same conclusion. For $N = 1$ we obtain the elementary result that a bounded monotone sequence is convergent. Copson's condition $r_j>0$ is not necessary: boundedness of (a_n) still implies convergence in the example

$$(0.2)\qquad a_{n+3} \geq 4a_{n+2} - 5a_{n+1} + 2a_n .$$

This paper is in final form and no version of it will be submitted for publication elsewhere.

On the other hand, $r_j \geq 0$ is not sufficient: in the example

$$a_{n+4} \geq \tfrac{1}{2}(a_{n+2} + a_n) \,, \tag{0.3}$$

boundedness of (a_n) implies only the separate convergence of the odd and even subsequences, and not necessarily of the whole sequence. By making the notational change

$$p_0 = 1,\ p_j = 1 - r_1 - \dots - r_j \ (j = 1,\dots,N-1),\ p_j = 0 \ (j \geq N), \tag{0.4}$$

we may express Copson's theorem in the following form:

THEOREM C. Let

$$1 = p_0 > p_1 > \dots > p_N = p_{N+1} = \dots = 0. \tag{0.5}$$

If the real sequence $a := (a_n)_{n \geq 0}$ is bounded and such that

$$t_n := (a * p)_n := \sum_{j=0}^{n} a_{n-j}\, p_j \tag{0.6}$$

is monotone, then (a_n) is convergent.

Shortly afterwards, D.Borwein [1] and I [7] independently reached similar conclusions when seeking necessary and sufficient conditions on (p_j) (or on (r_j)) in order that the conclusions of Copson's theorem should remain valid. My result [7, Theorem 3] can be written as follows:

THEOREM R. Let $p := (p_n)_{0 \leq n \leq N}$ be real-valued, and denote $p(z) := \sum_{n=0}^{N} p_n z^n \ (z \in \mathbb{C})$. Let $a := (a_n)_{n \geq 0}$ be such that $a * p$ is real and ultimately monotone. Then in order that boundedness of (a_n) shall always imply its convergence, it is necessary and sufficient that

$$p(z) \neq 0 \ \text{ in the set } \ \Gamma := \{\, z : |z| = 1,\ z \neq 1 \,\}. \tag{0.7}$$

For example, the polynomial corresponding to (0.2) is $p(z) = (1-z)(1-2z)$, which has no zero in Γ; while the polynomial corresponding to (0.3) is $p(z) = (1+z)(1+\frac{1}{2}z^2)$, which does have a zero in Γ.

Borwein's generalization of Theorem C [1, Theorem 1] read:

THEOREM B. Let $p := (p_n)_{n\geq 0}$ be complex-valued with $\sum_{n=0}^{\infty}|p_n| < \infty$, and let

$$p(z) := \sum_{n=0}^{\infty} p_n z^n \neq 0 \quad \text{on} \quad |z| = 1. \tag{0.8}$$

If the sequence $a := (a_n)_{n\geq 0}$ is bounded and such that $a * p$ is real and ultimately monotone, then (a_n) is convergent.

A companion theorem [1, Theorem 2] showed that when $p(z)$ is analytic on $|z| \leq 1$ and $p(1) \neq 0$, then (0.8) is also necessary for the conclusion of Theorem B. Hypothesis (0.5) of Theorem C implies (0.8), as pointed out in slightly more generality in Lemma 2 below.

1. THE OBJECTIVE AND THE MAIN RESULTS

In order to generalize Theorem B we shall consider complex-valued bi-infinite sequences $u := (u_n)_{n\in\mathbb{Z}}$ belonging to a general sequence space λ ; or if $(u_n)_{n\geq 0}$ is semi-infinite, we shall designate the sequence space by λ_+ . Some of the standard Banach sequence spaces in common use are defined as follows:

(1.1)

$u = (u_n)_{n\in\mathbb{Z}}$	$u = (u_n)_{n\geq 0}$	property and norm of (u_n)
ℓ^r $(1 \leq r < \infty)$	ℓ^r_+	$\lVert u\rVert := \sum\|u_n\|^r < \infty$
ℓ^∞	ℓ^∞_+	$\lVert u\rVert := \sup \|u_n\| < \infty$
c	c_+	$\exists \lim u_n$; ℓ^∞-norm
c^0	c^0_+	$\exists \lim u_n = 0$; ℓ^∞-norm
bv	bv_+	$\lVert u\rVert := \sum\|u_n - u_{n-1}\| < \infty$
cs	cs_+	$\sum u_n$ convergent, with $\lVert u\rVert := \sup_{n_1,n_2}\left\|\sum_{n_1}^{n_2} u_n\right\|$.

These spaces satisfy the strict inclusions $(1 < r < \infty)$:

$$\ell^1 \subset \ell^r \subset c^0 \subset c \subset \ell^\infty;\ \ell^1 \subset bv \subset c \subset \ell^\infty;\ \ell^1 \subset cs \subset c^0 \subset \ell^\infty.$$

The set of all these spaces (excluding ℓ^∞) will be denoted by

$$\Lambda := \{\ \ell^1,\ \ell^r\ (1< r<\infty),\ cs,\ c^0,\ bv,\ c\ \} \tag{1.2}$$

with Λ_+ for the analogous set of spaces of semi-infinite sequences. The convolution $u * v$ of two sequences u and v is defined by

$$(u * v)_n := \sum_{j \in \mathbb{Z}} u_{n-j} v_j \quad (n \in \mathbb{Z}).$$

It is easy to verify:

LEMMA 1. *If* $\lambda \in \Lambda \cup \ell^\infty$ (*as defined in* (1.2)), *then*

(1.3) $$[u \in \lambda,\ v \in \ell^1] \Rightarrow u * v \in \lambda$$

and, in each case, $\|u * v\|_\lambda \le \|u\|_\lambda \|v\|_{\ell^1}$.

We shall use (1.3) as a characterization of admissible sequence spaces λ .

Since, in particular,

$$[a \in \ell_+^\infty ,\ p \in \ell_+^1] \Rightarrow a * p \in \ell_+^\infty$$

the conclusion of Theorem B (where $a * p$ is real) takes the form

$$[a \in \ell_+^\infty ,\ a * p \text{ monotone \& bounded}] \Rightarrow a \in c_+ .$$

Now a weaker hypothesis and stronger conclusion here (and also allowing all sequences to be complex-valued) would be

$$[a \in \ell_+^\infty ,\ a * p \in bv_+] \Rightarrow a \in bv_+$$

and this suggests that our objective should be to match the hypothesis to the conclusion by the use of a more general sequence space λ, in the form

(1.4) $$[a \in \ell^\infty ,\ a * p \in \lambda] \Rightarrow a \in \lambda .$$

In making the extension to bi-infinite sequences we effectively rule out the use of methods based on complex power series (such as the original proof of Theorem B) since we are then forced to work entirely on the boundary of the unit disc. To revert to the semi-infinite case we merely take all terms of negative index to be zero. In fact, it turns out to be more illuminating to work with Stieltjes convolutions of functions, which not only leads to functional analogues, but by specialization to step-functions yields the results for sequences.

Having in mind (1.3) which, with the pre-condition $p \in \ell^1$, would give $a \in \lambda \Rightarrow a * p \in \lambda$, statement (1.4) would be a corrected converse of this which is immediately recognizable as a tauberian-type theorem with tauberian condition $a \in \ell^\infty$; and in fact we shall be able to prove our first result in precisely that form:

THEOREM 1. Let $p := (p_n)_{n\in\mathbb{Z}} \in \ell^1$, $p(z) := \sum_{n\in\mathbb{Z}} p_n z^n$, $a := (a_n)_{n\in\mathbb{Z}}$. Let λ be a space of bi-infinite complex-valued sequences with the property

$$(1.5) \qquad [u \in \lambda,\ v \in \ell^1] \Rightarrow u * v \in \lambda .$$

If $p(z) \neq 0$ on $|z| = 1$ then

$$(1.6) \qquad [a \in \ell^\infty,\ a * p \in \lambda] \Rightarrow a \in \lambda .$$

In particular, by Lemma 1, $\lambda \in \Lambda$ (as defined in (1.2)) gives a number of examples of well-known sequence spaces λ for which Theorem 1 holds ($\lambda = \ell^\infty$ is also possible, but Theorem 1 is trivial in that case).

As a converse to Theorem 1, we can construct counter-examples as follows:

THEOREM 2. Let $q := (q_n)_{n\in\mathbb{Z}} \in \ell^1$, $p_n := \zeta q_n - q_{n-1}$, where $|\zeta| = 1$. Let $\lambda \in \Lambda$. Then

$$(1.7) \qquad \exists\, a \in \ell^\infty \setminus \lambda \ \text{ such that } \ a * p \in \lambda .$$

REMARK 1. The hypotheses on p and q in Theorem 2 imply that $p \in \ell^1$ and that the corresponding power series are related by $p(z) = (\zeta - z)q(z)$, the series being convergent at least on $|z| = 1$. Thus the hypothesis is that $p(z)$ has at least one zero on $|z| = 1$. See Remark 4 below concerning the restriction of Theorem 2 to the semi-infinite case, which we designate as Theorem 2_+.

In the case of Theorem 1_+ (the restriction of Theorem 1 to semi-infinite sequences) it turns out that if we strengthen the hypothesis on $p(z)$ then we can even remove the hypothesis $a \in \ell^\infty_+$ (thus the tauberian theorem then becomes a mercerian theorem).

THEOREM 3. *Let* $p := (p_n)_{n\ge 0} \in \ell^1_+$, $p(z) := \sum_{n=0}^{\infty} p_n z^n$, $a := (a_n)_{n\ge 0}$. *Let* $\lambda_+ \subseteq \ell^\infty_+$ *be a space of semi-infinite complex-valued sequences with the property*

$$[u \in \lambda_+ ,\ v \in \ell^1_+] \Rightarrow u * v \in \lambda_+ . \tag{1.8}$$

If $p(z) \neq 0$ *on* $0 < |z| \le 1$ *then*

$$a * p \in \lambda_+ \ \Rightarrow\ a \in \lambda_+ . \tag{1.9}$$

We can apply Theorem 3 to obtain an improved form of Theorem C, if we use

LEMMA 2. *Let* $p_0 > p_1 \ge p_2 \ge \ldots \ge 0$, $(p_n) \in \ell^1_+$. *Then* $p(z) \neq 0$ *on* $|z| \le 1$.

Proof of Lemma 2. By hypothesis, $r_n := p_n - p_{n-1} \le 0$ $(n \ge 1)$, $r_0 := p_0 > 0$. Thus $(1-z)p(z) = \sum_0^\infty r_n z^n = r_0 - \sum_1^\infty |r_n| z^n$, and $z=1$ gives $r_0 = \sum_1^\infty |r_n|$. Hence for $z = \rho e^{i\theta}$ $(0 \le \rho \le 1, 0 < \theta < 2\pi)$ we have

$$\mathrm{Re}[(1-z)p(z)] = \sum_1^\infty |r_n|(1-\rho^n \cos n\theta) \ge |r_1|(1-\rho\cos\theta) > 0 .$$

Together with $p(1) \ge p_0 > 1$, this gives us $p(z) \neq 0$ on $|z| \le 1$. □

REMARK 2. When $p(z)$ is a polynomial, Lemma 2 is the Eneström-Kakeya Theorem (e.g. see [5, III.22]).

Now using Lemma 2 in Theorem 3, and taking $\lambda_+ = bv_+$, we obtain:

COROLLARY 3. *Let* $p_0 > p_1 \ge p_2 \ge \ldots \ge 0$, $(p_n) \in \ell^1_+$. *Then*

$$a * p \text{ monotone \& bounded} \Rightarrow a * p \in bv_+ \Rightarrow a \in bv_+ \Rightarrow a \in c_+ .$$

Even in the case where (p_n) is a finite sequence this is an improved version of Theorem C, since the boundedness of (a_n) is not explicitly assumed but is absorbed into the hypothesis on $a * p$.

2. HILFSSÄTZE

We use the standard Lebesgue space $L^1(\mathbb{R})$ and the space $BV(\mathbb{R})$ of (complex-valued) functions of bounded variation on the whole real line. It is well-known (e.g. see [6, p.53]) that any $k(\cdot) \in BV(\mathbb{R})$ is decomposable into

$$k = k_1 + k_2 + k_3$$

where $k_1 \in AC(\mathbb{R})$ (the absolutely continuous component), k_2 is a step-function, and k_3 is the "singular component", namely a non-constant continuous function of bounded variation with zero derivative almost everywhere. When k_3 is absent we denote, following Pitt [3, p.511; 4, p.114]:

(2.1) $k(\cdot) \in V^* :\Leftrightarrow k(\cdot) \in BV(\mathbb{R})$ and k has no singular component.

If $k(\cdot) \in BV(\mathbb{R})$ we also denote its Laplace-Stieltjes transform by

$$K(\omega) := \int_{\mathbb{R}} e^{-\omega x}\, dk(x) \qquad (\omega = \sigma + i\tau \in \mathbb{C}) \tag{2.2}$$

whenever the integral converges absolutely. In particular, the Fourier-Stieltjes transform of $k(\cdot)$ is

$$K(i\tau) := \int_{\mathbb{R}} e^{-i\tau x}\, dk(x) \qquad (k \in BV(\mathbb{R}),\ \tau \in \mathbb{R}). \tag{2.3}$$

Following Wiener's classical paper on tauberian theorems [8] where he needed the non-vanishing of the Fourier transform of an L^1-function, in an extension of this work Wiener and Pitt jointly [9] considered the question "When is the reciprocal of a Fourier-Stieltjes transform also a Fourier-Stieltjes transform ?" Obviously they needed $K(i\tau) \neq 0$ $(\forall\, \tau \in \mathbb{R})$, but actually they needed more, namely that $|K(i\tau)|$ should be bounded away from zero (their theorem also allows k to have a singular component, provided this is "not too large"). If we combine their result [9, Theorem 1] with a theorem of Pitt on the inversion of a convolution [3, Theorem 3;or 4, V.Theorem 9 (p.115)] we obtain the following:

LEMMA 3. <u>If</u> $k(\cdot) \in V^*$ <u>and</u> $\inf_{\tau \in \mathbb{R}} |K(i\tau)| > 0$ <u>then</u> $\exists\, h(\cdot) \in BV(\mathbb{R})$ <u>such that</u>

$$[K(i\tau)]^{-1} = \int_{\mathbb{R}} e^{-i\tau y}\, dh(y) \qquad (\forall\, \tau \in \mathbb{R}) \tag{2.4}$$

<u>If, further,</u> $a(\cdot)$ <u>is bounded and Borel measurable on</u> $\mathbb{R}$, <u>and if</u>

$$t(x) := \int_{\mathbb{R}} a(x-y)\, dk(y) \qquad [\ t = a * dk\ ] \tag{2.5}$$

<u>then</u>

$$a(x) = \int_{\mathbb{R}} t(x-y)\, dh(y)\ . \qquad [\ a = t * dh\ ] \tag{2.6}$$

REMARK 3. The hypothesis $\inf_{\tau \in \mathbb{R}} |K(i\tau)| > 0$ implies that $k(\cdot)$ cannot merely be $AC(\mathbb{R})$; for if $k(\cdot) \in AC(\mathbb{R})$ then $\lim_{|\tau| \to \infty} |K(i\tau)| = 0$ by the Riemann-Lebesgue Lemma. Hence $k(\cdot)$ must have at least one jump.

When the integral transforms are one-sided (namely over $\mathbb{R}_+$ instead of $\mathbb{R}$), Pitt shows [3, Theorem 6; or 4, V. Theorem 13 (p.120)] that a strengthening of the condition on $K(\omega)$ allows us to shift the boundedness of $a(\cdot)$ from hypothesis to conclusion:

LEMMA 4. Let $k(\cdot) \in V^*$ and $k(\cdot)$ be constant on $(-\infty, 0)$, and let

$$\inf_{\sigma \geq 0} |K(\sigma + i\tau)| > 0 . \tag{2.7}$$

Let $a(\cdot)$ be Borel measurable, $a(x) = 0$ for $x < 0$, and define

$$t(x) := \int_0^x a(x-y)\, dk(y) . \tag{2.8}$$

Then $t(\cdot)$ bounded implies $a(\cdot)$ bounded.

3. PROOFS OF THE THEOREMS

Proof of Theorem 1. Given $p := (p_j)_{j \in \mathbb{Z}} \in \ell^1$, define $k(x) := \sum_{j \leq x} p_j$ $(\forall x \in \mathbb{R})$. Then $k(\cdot) \in V^*$ (as defined in (2.1)) and

$$p(e^{-i\tau}) = \sum_{j \in \mathbb{Z}} p_j\, e^{-i\tau j} = \int_{\mathbb{R}} e^{-i\tau x}\, dk(x) =: K(i\tau) .$$

Hence, since $K(i\tau)$ is continuous and periodic,

$$p(z) \neq 0 \text{ on } |z| = 1 \Rightarrow K(i\tau) \neq 0 \text{ on } [0, 2\pi] \Rightarrow \inf_{\tau \in \mathbb{R}} |K(i\tau)| > 0 .$$

Let $a \in \ell^\infty$ and define

$$a(x) := a_n \quad \text{for} \quad n - \tfrac{1}{2} \leq x < n + \tfrac{1}{2} \quad (\forall n \in \mathbb{Z}) ;$$

then $a(\cdot)$ is bounded and Borel measurable, and

$$t(x) := \int_{\mathbb{R}} a(x-y)\, dk(y) = \sum_{j \in \mathbb{Z}} a_{n-j} p_j =: t_n$$

for $n - \frac{1}{2} \leq x < n + \frac{1}{2}$ $(\forall n \in \mathbb{Z})$. The hypotheses of Lemma 3 therefore hold, and so

$$\exists\, h(\cdot) \in BV(\mathbb{R}) \text{ such that } a(x) = \int_{\mathbb{R}} t(x-y)\, dh(y) .$$

Define $v_j := \int_{j-\frac{1}{2}}^{j+\frac{1}{2}} dh(y)$, so that $v \in \ell^1$ and

$$a_n = a(n) = \int_{\mathbb{R}} t(n-y)\, dh(y) = \sum_{j \in \mathbb{Z}} t_{n-j} \int_{j-\frac{1}{2}}^{j+\frac{1}{2}} dh(y) = (t * v)_n .$$

But then, by (1.5), $t := a * p \in \lambda$ and $v \in \ell^1$ together imply that $a = t * v \in \lambda$, and we obtain (1.6) as required. □

Proof of Theorem 2. For any sequence $u = (u_n)$, denote $\Delta_\zeta u := (\zeta u_n - u_{n-1})$, $\Delta := \Delta_1$. Thus, given $q \in \ell^1$, we have $p := \Delta_\zeta q \in \ell^1$ (see also Remark 1). Note also that whenever $a \in \ell^\infty$ and $q \in \ell^1$ we have, by partial summation,

$$t := a * p = a * \Delta_\zeta q = q * \Delta_\zeta a .$$

(i) If $|\zeta| = 1$ but $\zeta \neq 1$ choose $a_n = \zeta^{-n-1}$ $(n \geq 0)$, $a_n = 0$ $(n < 0)$. Then $(\Delta_\zeta a)_n = 0$ $(n \neq 0)$, $(\Delta_\zeta a)_o = 1$, and consequently

$$t = a * p = q * \Delta_\zeta a = q .$$

The sequence (a_n) chosen thus satisfies

(3.1) $$a \in \ell^\infty \setminus c , \quad a * p \in \ell^1 .$$

(ii) Suppose $\zeta = 1$.

$(ii)_1$ Choose $a_n = \sin\log n$ $(n \geq 1)$, $a_n = 0$ $(n \leq 0)$; then $a \in \ell^\infty \setminus c$. But now $\Delta a_n = O(1/n)$; so $\Delta a \in c^o$ and hence (since $q \in \ell^1$) $t = q * \Delta a \in c^o$. Also $\Delta^2 a_n = O(n^{-2})$, so $\Delta^2 a \in \ell^1$ and hence (since $q \in \ell^1$) $\Delta t = q * \Delta^2 a \in \ell^1$, namely $t \in bv$. Our chosen sequence therefore now satisfies

(3.2) $$a \in \ell^\infty \setminus c , \quad a * p \in c^o \cap bv .$$

$(ii)_2$ Choose $a_n = n^{-1/r}$ $(n \geq 1)$ for a fixed r in $1 \leq r < \infty$, $a_n = 0$ $(n \leq 0)$. Now $a \in \ell^\infty \setminus \ell^r$, while $\Delta a \in \ell^1$ and so we obtain $t = q * \Delta a \in \ell^1$. Thus for this (a_n) we have

(3.3) $$1 \leq r < \infty : \; a \in \ell^\infty \setminus \ell^r , \; a * p \in \ell^1 .$$

Thus if $\lambda \in \Lambda$ (that is, if λ is any of the spaces listed in (1.2)) we have shown that (1.7) holds, namely that if $|\zeta| = 1$ then

(3.4) $$\exists\, a \in \ell^\infty \setminus \lambda \text{ such that } a * p \in \lambda .$$

For: (3.1) ⇒ (3.4) for any $\lambda \in \Lambda$.
(3.2) ⇒ (3.4) for $\lambda \in \{c^o, c, bv\}$.

(3.3) ⇒ (3.4) for $\lambda = \ell^r$ $(1 \le r < \infty)$, and since $a_n \ge 0$ in this example, the case $r = 1$ also deals with $\lambda = cs$. □

REMARK 4. Since the sequences chosen in the proof of Theorem 2 were all semi-infinite, and provide counter-examples whether (p_n), (q_n) are bi-infinite or semi-infinite, we obtain a corresponding Theorem 2_+ for the case where all the sequences are semi-infinite. However, Theorem 2_+ deals only with counter-examples for which $p(z)$ has a zero on $|z| = 1$ and hence provides a converse only to Theorem 1_+. To complete a converse of Theorem 3 (in which the hypothesis is that $p(z) \neq 0$ on $0 < |z| \le 1$) we would need to add the case where $p(z)$ may have a zero in $0 < |z| < 1$. It is enough to prove:

<u>Let</u> $q := (q_n)_{n\ge 0} \in \ell^1_+$ <u>and</u> $p_n := \zeta q_n - q_{n-1}$ $(n \ge 0,\ q_{-1} = 1)$, $0 < |\zeta| < 1$.
<u>Then</u> $\exists\, a \notin \ell^\infty$ <u>with</u> $a * p \in \ell^1$.

<u>Proof</u>. Choose $a_n = \zeta^{-n-1}$ $(n \ge 0)$, for which $a * p = q$. □

<u>Proof of Theorem 3</u>. The hypothesis on $p(z)$ as stated allows $p(0) = 0$. However, if p_m is the first non-vanishing p_n and $p(z) = z^m q(z)$, then $p_{n+m} = q_n$ $(n \ge 0)$ and also $(a * p)_{n+m} = (a * q)_n$. The hypothesis on $p(z)$ would then give $q \in \ell^1_+$ and $q(z) \neq 0$ on $0 \le |z| \le 1$, and we could replace p by q throughout our proof. Thus we may assume without loss of generality that $p(z) \neq 0$ on $0 \le |z| \le 1$.

As in the proof of Theorem 1, given $p := (p_j)_{j\ge 0} \in \ell^1_+$, define $k(x) := \sum_{j\le x} p_j$ $(\forall\, x \in \mathbb{R})$, so that $k(\cdot) \in V^*$ and $k(x) = 0$ on $(-\infty, 0)$.
Thus
$$p(e^{-\omega}) = \sum_{j=0}^{\infty} p_j e^{-\omega j} = \int_{\mathbb{R}} e^{-\omega x}\, dk(x) =: K(\omega)\,.$$
Now $p(z)$ is continuous on the compact unit disc and hence
$$p(z) \neq 0 \text{ on } |z| \le 1 \;\Rightarrow\; \inf_{|z|\le 1} |p(z)| > 0 \;\Rightarrow\; \inf_{\operatorname{Re}\omega \ge 0} |K(\omega)| > 0\,.$$
Given $a = (a_n)_{n\ge 0}$, define $a(x) := a_n$ $(n \le x < n+1;\ n=0,1,\ldots)$, $a(x) := 0$ $(x < 0)$; then

$$t(x) := \int_0^x a(x-y)\, dk(y) = \sum_{j=0}^{n} a_{n-j}\, p_j =: t_n$$

for $n \le x < n+1$, $n = 0,1,\ldots$. Hence, by Lemma 4, we now have

$$t := a * p \in \lambda_+ \subseteq \ell_+^\infty \quad \Rightarrow \quad a \in \ell_+^\infty .$$

But now the full hypotheses of Lemma 3 hold (in the one-sided case) and hence (as in the last part of the proof of Theorem 1) $\exists\, v \in \ell_+^1$ with $a = t * v$; then, by (1.8), $t \in \lambda_+$ and $v \in \ell_+^1$ imply $a = t * v \in \lambda_+$. □

REFERENCES

1. D. Borwein, Convergence criteria for bounded sequences. Proc.Edin.Math.Soc.(2) 18 (1972), 99-103.

2. E.T. Copson, On a generalisation of monotonic sequences. Proc.Edin.Math.Soc.(2) 17 (1970), 159-164.

3. H.R. Pitt, Mercerian theorems. Proc.Camb.Phil.Soc. 34 (1938), 510-520.

4. H.R. Pitt, Tauberian Theorems. Oxford University Press, 1958.

5. G. Pólya and G. Szegö, Aufgaben und Lehrsätze aus der Analysis. Springer-Verlag, Berlin: 4.Auflage, 1970.

6. F. Riesz and B. Sz.-Nagy, Functional Analysis. Ungar, New York, 1971.

7. D.C. Russell, On bounded sequences satisfying a linear inequality. Proc.Edin.Math.Soc.(2) 19 (1974), 11-16.

8. N. Wiener, Tauberian theorems. Ann.of Math. 33 (1932), 1-100.

9. N. Wiener and H.R. Pitt, On absolutely convergent Fourier-Stieltjes transforms. Duke Math.J. 4 (1938), 420-436.

Added June 1986:The reader should also see (these Proc.)

10. Matts Essén, Tauberian theorems, convolutions and some results of D.C. Russell. General Inequalities 5 (Birkhäuser, ed. W.Walter).

Dennis C. Russell, Department of Mathematics, York University, Toronto-Downsview, Ontario M3J 1P3, Canada.

Inequalities in Analysis and Applications

The bungalows

International Series of
Numerical Mathematics, Vol. 80

WEIGHTED INEQUALITIES FOR MAXIMAL FUNCTIONS IN SPACES OF HOMOGENEOUS TYPE WITH APPLICATIONS TO NON-ISOTROPIC FRACTIONAL INTEGRALS

Kenneth F. Andersen

Abstract. Given a space (Ω,d,μ) of homogeneous type, necessary and sufficient conditions are obtained which ensure that given a non-negative weight function $U(x)$ there is a non-negative weight function $V(x)$ which is finite a.e. and the fractional maximal function operator is bounded from $L^p(Vd\mu)$ to $L^q(Ud\mu)$. The dual problem and the analogous problems for non-isotropic fractional integrals are also solved.

1. INTRODUCTION

Let (Ω,d,μ) be a space of homogeneous type. Thus, Ω is a point set, d is a non-negative, symmetric quasi-distance function on $\Omega\times\Omega$ satisfying $d(x,y) = 0$ if and only if $x = y$ and $d(x,y) \leq K(d(x,z) + d(z,y))$ for some constant K independent of $x,y,z \in \Omega$; μ is a positive Borel measure on a σ-algebra containing the balls $B = B(x,r) = \{y \in \Omega: d(y,x) < r\}$ and which satisfies the doubling property $\mu(B(x,2r)) \leq D\mu(B(x,r))$ for some constant D independent of $x \in \Omega$ and $r > 0$. Let $0 \leq \alpha < 1$ and define the fractional maximal function operator M_α by

$$(M_\alpha f)(x) = \sup_{r>0} [\mu(B(x,r))]^{\alpha-1}\int_{B(x,r)} |f(y)|d\mu(y) .$$

The purpose of this paper is to study inequalities of the form

$$(1) \qquad \left(\int_\Omega |M_\alpha f|^q \, Ud\mu\right)^{1/q} \leq C\left(\int_\Omega |f|^p \, Vd\mu\right)^{1/p}$$

This paper is in final form and no version of it will be submitted for publication elsewhere.

where C is an absolute constant independent of f. The case $\alpha = 0$ has been most widely studied; thus, for example, it is known [1] (see also [3], [7]) that (1) holds for $\alpha = 0$, $1 < p = q < \infty$ with $U = V$ if and only if U satisfies the A_p condition, namely

$$\left(\int_B U d\mu\right)^{1/p}\left(\int_B U^{-1/(p-1)} d\mu\right)^{1/p'} \leq C\mu(B)$$

for some constant C and all balls B. Here and throughout, $1/p + 1/p' = 1$. More generally, (1) holds with $0 < \alpha < 1$, $1 < p < 1/\alpha$, $1/q = 1/p - \alpha$, $V = U^{p/q}$ if and only if U satisfies the condition $A_{1+(q/p')}$. When the restriction $V = U^{p/q}$ is dropped, a useful characterization of the weights U, V for which (1) holds remains an open problem. However, the main results of this paper, Theorems 1 and 2, provide characterizations of those U (respectively V) for which there is V (respectively U) such that (1) holds. An application to non-isotropic fractional integrals in R^n is given as Theorem 3.

THEOREM 1. *Let* $1 < p < \infty$, $0 \leq \alpha < 1/p$, $1/q = 1/p - \alpha$ *and suppose* $U(x) \geq 0$ *a.e. on* Ω. *There is* $V(x)$ *with* $0 \leq V(x) < \infty$ *a.e. such that*

$$(2) \qquad \left(\int_\Omega |M_\alpha f|^q \, U d\mu\right)^{1/q} \leq \left(\int_\Omega |f|^p \, V d\mu\right)^{1/p}$$

if and only if for some $x_0 \in \Omega$

$$(3) \qquad \int_\Omega \frac{U(x)d\mu(x)}{[1+\mu(B(x_0,d(x,x_0)))]^{(1-\alpha)q}} < \infty.$$

THEOREM 2. *Let* $1 < p < \infty$, $0 \leq \alpha < 1/p$, $1/q = 1/p - \alpha$ *and suppose* $0 \leq V(x) < \infty$ *a.e. on* Ω. *There is* $U(x)$ *with* $U(x) > 0$ *a.e. such that (2) holds if and only if for some* $x_0 \in \Omega$

$$(4) \qquad \limsup_{r \to \infty} \, [\mu(B(x_0,r))]^{(\alpha-1)p'} \int_{B(x_0,r)} V(x)^{-1/(p-1)} d\mu(x) < \infty.$$

REMARK 1. The particular choice of x_0 in conditions (3) and (4) is of no importance since these conditions are satisfied by some x_0 if and only if they are satisfied by all x_0. This follows easily from the following fact: If $c > 1$, $r/c \leq s \leq cr$ and $B(x,r) \cap B(y,s) \neq \emptyset$, then there are constants A, B depending only on c,K and D such that

$$(5) \qquad A\mu(B(x,r)) \leq \mu(B(y,s)) \leq B\mu(B(x,r)).$$

To see this one need only observe that $B(x,r)$ is contained in $B(y,ks)$ where $k = K^2(2c+1)$ so the first inequality of (5) follows from the doubling condition; reversing the roles of $B(x,r)$ and $B(y,s)$ yields the second inequality.

An important example to which our results apply is the space R^n equipped with a quasi-distance function defined as follows. For each i, $1 \leq i \leq n$, let ϕ_i be a continuous increasing function mapping $[0,\infty)$ onto itself satisfying for some constant C,

$$\phi_i(2t) \leq C\phi_i(t) \quad \text{and} \quad \phi_i^{-1}(2t) \leq C\phi_i^{-1}(t).$$

If $x = (x_1,\ldots,x_n) \in R^n$ and

$$\rho_\beta(x) = \begin{cases} \left(\sum_{i=1}^{n} [\phi_i^{-1}(|x_i|)]^\beta\right)^{1/\beta}, & 1 \leq \beta < \infty \\ \max\{\phi_i^{-1}(|x_i|)\}, & \beta = \infty \end{cases}$$

then $d(x,y) = \rho_\beta(x-y)$ defines a quasi-distance on R^n for which Lebesgue measure satisfies the doubling condition. Since for our purposes these quasi-distances are all equivalent for $1 \leq \beta \leq \infty$, we take $\beta = 2$ for convenience and write ρ in place of ρ_2. Choosing $\phi_i(t) = t^{\gamma_i}$, $\gamma_i > 0$, $\gamma = \Sigma\, \gamma_i$, we may define the non-isotropic fractional integral of order α, $0 < \alpha < 1$, by

$$(I_\alpha f)(x) = \int_{R^n} [\rho(x-y)]^{\gamma(\alpha-1)} f(y)dy .$$

For these we have the following analogue of Theorems 1 and 2.

THEOREM 3. <u>Let</u> $0 < \alpha < 1$, $1 < p < 1/\alpha$, $1/q = 1/p - \alpha$.

(a) <u>If</u> $U(x) \geq 0$ <u>a.e. there is</u> $V(x)$ <u>with</u> $0 \leq V(x) < \infty$ <u>a.e. such that</u>

$$(6) \qquad \left(\int_{R^n} |(I_\alpha f)(x)|^q\, U(x)dx\right)^{1/q} \leq \left(\int_{R^n} |f(x)|^p\, V(x)dx\right)^{1/p}$$

<u>if and only if</u>

$$(7) \qquad \int_{R^n} \frac{U(x)dx}{[1+\rho(x)]^{\gamma(1-\alpha)q}} < \infty.$$

(b) <u>If</u> $0 \leq V(x) < \infty$ <u>a.e. there is</u> $U(x)$ <u>with</u> $U(x) > 0$ <u>a.e. such that (6) holds if and only if</u>

$$(8) \qquad \int_{R^n} \frac{V(x)^{-1/(p-1)}dx}{[1+\rho(x)]^{\gamma(1-\alpha)p'}} < \infty.$$

For the special case $\gamma_i = 1$, $1 \leq i \leq n$, and $\alpha = 0$ Theorem 1 was obtained by W.-S. Young [10] and independently by A. Gatto and C. Gutiérrez [6] while Theorem 2 in that context is due to L. Carleson and P. Jones [4], see also J. Rubio de Francia [8]. The case $0 < \alpha < 1$ was treated by the author in [2].

The proof of Theorem 1 follows closely that of Gatto and Gutiérrez for the Euclidean case and is based on the following Lemma which is of interest in its own right. Fix x_0 and set

$$(L_\alpha f)(x) = \sup_{0<r<\frac{1+d(x,x_0)}{K+1}} [\mu(B(x,r))]^{\alpha-1} \int_{B(x,r)} |f(y)|d\mu(y)$$

and

$$f^*(x) = \sup_{\substack{x \in B(y,r) \\ 0<r<5K^2[1+d(x,x_0)]}} [\mu(B(y,r))]^{-1} \int_{B(y,r)} |f(z)|d\mu(z).$$

As the proof will show, the Lemma remains true if L_α is replaced by M_α and in the definition of f^* the restriction $r < 5K^2[1+d(x,x_0)]$ is deleted; in particular with $g = 1$ and $p = 1$ it follows that M_α is of weak type $(1,1/(1-\alpha))$.

LEMMA. *Let* $1 \leq p < \infty$, $0 \leq \alpha < 1/p$, $1/q = 1/p - \alpha$ *and suppose* $g > 0$ *a.e.* *If* $E_\lambda = \{x:(L_\alpha f)(x) > \lambda\}$, *then there are constants* C *independent of* f *and* g *such that*

(9) $$\left(\int_{E_\lambda} g\,d\mu\right)^{1/q} \leq C\lambda^{-1}\left(\int_\Omega |f|^p\,[g^*]^{1-\alpha p}d\mu\right)^{1/p}$$

and if $p > 1$,

(10) $$\left(\int_\Omega |L_\alpha f|^q\,g\,d\mu\right)^{1/q} \leq C\left(\int_\Omega |f|^p\,[g^*]^{1-\alpha p}d\mu\right)^{1/p}.$$

2. PROOF OF LEMMA

Fix a ball $B \in \Omega$ and set $E_\lambda(B) = B \cap E_\lambda$. For each $x \in E_\lambda(B)$ there is $r = r(x) < [1+d(x,x_0)]/[K+1]$ so that

(11) $$\int_{B(x,r)} |f|\,d\mu > \lambda[\mu(B(x,r))]^{1-\alpha}.$$

Theorem 1.2 of [5] shows that there is a pairwise disjoint sequence $B_j = B(x_j,r_j)$ of these balls for which the balls $B_j^\# = B(x_j,5Kr_j))$ cover $E_\lambda(B)$. Thus, from (11)

(12) $$\begin{aligned}\int_{B_j^\#} g\,d\mu &\leq \left(\int_{B_j^\#} g\,d\mu\right)\left(\lambda^{-1}[\mu(B_j)]^{\alpha-1}\int_{B_j} |f|\,d\mu\right)^q \\ &\leq \lambda^{-q}\left([\mu(B_j)]^{-1}\int_{B_j^\#} g\,d\mu\right)\left(\int_{B_j} |f|^p d\mu\right)^{q/p}\end{aligned}$$

by Hölder's inequality and the relation $1/q = 1/p - \alpha$. Now for $y \in B_j$ it follows that $5Kr_j \leq 5K^2[1+d(y,x_0)]$ since

$$\begin{aligned}(K+1)r_j < 1+d(x_j,x_0) &\leq 1+K[d(x_j,y)+d(y,x_0)] \\ &\leq 1+K[r_j+d(y,x_o)].\end{aligned}$$

The doubling property of μ then shows that the first bracketed term on the right of (12) does not exceed a constant multiple of $g^*(y)$. Thus, there is a constant C depending only on K and D such that

$$\int_{B_j^\#} g\,d\mu \leq C\lambda^{-q}\left(\int_{B_j} |f|^p\,[g^*]^{p/q}d\mu\right)^{q/p}$$

and summing over j then yields

$$\int_{E_\lambda(B)} g\,d\mu \le C\lambda^{-q} \sum_j \Big(\int_{B_j} |f|^p [g^*]^{p/q} d\mu\Big)^{q/p}$$

$$\le C\lambda^{-q} \Big(\sum_j \int_{B_j} |f|^p [g^*]^{p/q} d\mu\Big)^{q/p}$$

$$\le C\lambda^{-q} \Big(\int_\Omega |f|^p [g^*]^{p/q} d\mu\Big)^{q/p}$$

where we used the fact that $q/p > 1$ to obtain the second inequality. Since B is arbitrary this implies (9).

Observe that for fixed α, (9) shows that the sublinear operator $(Tf)(x) = \big(L_\alpha[f(g^*)^\alpha]\big)(x)$ satisfies

$$\Big(\int_{\{x:(Tf)(x) > \lambda\}} g\,d\mu\Big)^{1/q} \le C\lambda^{-1}\Big(\int_\Omega |f|^p g^* d\mu\Big)^{1/p}$$

so (10) follows from the Marcinkiewicz Interpolation Theorem.

3. PROOF OF THEOREM 1

To prove the necessity of (3), fix x_0 and let $E_n = \{x \in B(x_0,n): V(x) \le n\}$. Since $V(x) < \infty$ a.e. it follows that $\mu(E_n) > 0$ for some $n = n_0$. Let χ denote the characteristic function of E_{n_0}. For $x \in B(x_0,2n_0)$, $B(x_0,n_0) \subset B(x,3Kn_0)$ so

$$(M_\alpha\chi)(x) \ge [\mu(B(x,3Kn_0))]^{\alpha-1}\int_{B(x,3Kn_0)} \chi\,d\mu \tag{13}$$

$$\ge C[\mu(B(x_0,n_0))]^{\alpha-1}\mu(E_{n_0})$$

where we have used (5) to obtain the last inequality. On the otherhand, if $x \notin B(x_0,2n_0)$, then $B(x_0,n_0) \subset B(x,2Kd(x,x_0))$ so we find, again by the definition of M_α and (5), that

$$(M_\alpha\chi)(x) \ge C[\mu(B(x_0,d(x,x_0)))]^{\alpha-1}\mu(E_{n_o}). \tag{14}$$

Upon taking $f = \chi$ in (2) and using the estimates (13) and (14) we obtain (3).

For the sufficiency of (3), observe first that the function $h(x) = [1+d(x,x_0)]^{-\beta}$ belongs to $L(\Omega,d\mu)$ whenever $2^\beta > D$ since

$$\int_\Omega h\,d\mu = \int_{d(x,x_0)<1} h\,d\mu + \sum_1^\infty \int_{2^{j-1}\leq d(x,x_0)<2^j} h\,d\mu$$

$$\leq \mu(B(x_0,1)) + \sum_1^\infty 2^{-(j-1)\beta}\mu(B(x_0,2^j))$$

$$\leq \mu(B(x_0,1))\,[1 + \sum_1^\infty 2^{-(j-1)\beta}D^j] < \infty.$$

We may assume without loss of generality that $U(x) > 0$ for otherwise we may apply the following argument to $U(x)+h(x)$ which satisfies (3) if U does.

We first prove that $U^*(x) < \infty$ a.e. Fix $r > 0$. If $x \in B(x_0,r)$ and χ is the characteristic function of $B(x_0,R)$, $R = 10K^5(1+2r)$, then $U^*(x) = [U\chi]^*(x)$ and since (5) shows $f^*(x) \leq C(M_0f)(x)$, it follows that $U^*(x) < \infty$ a.e. on $B(x_0,r)$ since M_0 is of weak type (1,1) while (3) shows $U\chi$ is integrable. Since r is arbitrary, $U^*(x) < \infty$ a.e.

Now $(M_\alpha f)(x)$ is bounded by

$$(15)\qquad (L_\alpha f)(x) + \sup_{r>\frac{1+d(x,x_0)}{K+1}} [\mu(B(x,r))]^{\alpha-1}\int_{B(x,r)} |f|\,d\mu$$

$$= (L_\alpha f)(x) + (Sf)(x)$$

and Sf, in view of (5) and Hölder's inequality, is bounded by a constant multiple of

$$(16)\qquad [1+\mu(B(x_0,d(x,x_0)))]^{\alpha-1}\left(\int_\Omega |f|^p h^{1-p}d\mu\right)^{1/p}\left(\int_\Omega h\,d\mu\right)^{1/p'}.$$

Combining (15) and (16), the Lemma shows that if $V(x)$ is a constant multiple of $\max\{h(x)^{1-p},U^*(x)^{1-\alpha p}\}$ then (2) is satisfied.

4. PROOF OF THEOREM 2

We prove the necessity of (4) first; indeed we shall prove a slightly stronger result, namely that (4) follows from (2) under the (weaker) hypothesis that $U \geq 0$ and $U \neq 0$ a.e. Choose

$B = B(x_0,n)$ with n sufficiently large that $\int_B U d\mu > 0$. If $x \in B$ and $r \geq n$, then $B(x_0,r) \subset B(x,2Kr)$ so if $\eta > 0$ and $f = [V+\eta]^{-1/(p-1)}$ on $B(x_0,r)$ and zero otherwise it follows that

$$(M_\alpha f)(x) \geq [\mu(B(x,2Kr))]^{\alpha-1} \int_{B(x,2Kr)} f d\mu$$

$$\geq C[\mu(B(x_0,r))]^{\alpha-1} \int_{B(x_0,r)} f d\mu$$

where we have used (5) to obtain the last inequality. Using this in (2) shows

$$\left(\int_B U d\mu\right)^{1/q} C[\mu(B(x_0,r))]^{\alpha-1} \int_{B(x_0,r)} f d\mu \leq \left(\int_\Omega f^p V d\mu\right)^{1/p}$$

$$\leq \left(\int_{B(x_0,r)} f d\mu\right)^{1/p}$$

so that

$$[\mu(B(x_0,r))]^{\alpha-1} \left(\int_{B(x_0,r)} [V+\eta]^{-1/(p-1)} d\mu\right)^{1/p'}$$

is bounded independent of η and r. This implies (4).

The main step in the proof that (4) implies (3) is to produce, for each integer $j > 0$, a U_j which is supported and positive a.e. on $B(x_0,j)$ for which

$$\left(\int_\Omega [M_\alpha f]^q U_j d\mu\right)^{1/q} \leq C_j \left(\int_\Omega |f|^p V d\mu\right)^{1/p} \tag{17}$$

for then $U(x) = \Sigma\, 2^{-j} C_j^{-q} U_j(x) > 0$ a.e. and satisfies (2).

Since $\min\{1,V(x)\}$ satisfies (4) whenever V does, we may assume without loss of generality that $V(x) \leq 1$ a.e. Fix an integer $j > 0$, put $\sigma(x) = V(x)^{-1/(p-1)}$ on $B(x_0,2Kj)$, $\sigma(x) = 0$ otherwise, and set $U_j(x) = [(M_0\sigma)(x)]^{-\beta}$ on $B(x_0,j)$, $U_j(x) = 0$ otherwise, where $\beta > q/p'$. Since (4) shows $\sigma \in L(\Omega,d\mu)$ and M_0 is of weak type (1,1), it follows that $U_j > 0$ a.e. on $B(x_0,j)$. Let χ_k denote the characteristic function of the set $E_k = \{x \in B(x_0,j): (M_0\sigma)(x) \leq 2^k\}$. Since M_α is of weak type

$(1,1/(1-\alpha))$ it follows that the sublinear operation $(Tf)(x) = (M_\alpha f\sigma)(x)\chi_k(x)$ satisfies

$$\left(\int_{\{x:(Tf)(x) > \lambda\}} d\mu\right)^{1-\alpha} \leq C\lambda^{-1}\int_\Omega |f|\ \sigma d\mu .$$

On the otherhand, with the usual interpretation if $\alpha = 0$, Hölder's inequality shows

$$(Tf)(x) = \chi_k(x)\left(\sup_{r>0} [\mu(B(x,r))]^{\alpha-1}\int_{B(x,r)} |f|\ \sigma d\mu\right)$$

$$\leq \chi_k(x)\left(\sup_{r>0} [\mu(B(x,r))]^{-1}\int_{B(x,r)} \sigma d\mu\right)^{1-\alpha}\left(\int_{B(x,r)} |f|^{1/\alpha}\ \sigma d\mu\right)^\alpha$$

$$\leq \chi_k(x)\ ((M_0\sigma)(x))^{1-\alpha}\left(\int_\Omega |f|^{1/\alpha}\ \sigma d\mu\right)^\alpha$$

$$\leq 2^{k(1-\alpha)}\left(\int_\Omega |f|^{1/\alpha}\ \sigma d\mu\right)^\alpha.$$

The Marcinkiewicz Interpolation Theorem then shows there is a constant C depending only on p and q such that

$$\left(\int_{E_k} [M_\alpha f\sigma]^q\ d\mu\right)^{1/q} \leq C\ 2^{k/p'}\left(\int_\Omega |f|^p\ \sigma d\mu\right)^{1/p}$$

or equivalently,

$$(18) \quad \left(\int_{E_k} [M_\alpha f]^q\ d\mu\right)^{1/q} \leq C\ 2^{k/p'}\left(\int_{B(x_0,2Kj)} |f|^p\ Vd\mu\right)^{1/p}.$$

Now, write $f = f_1 + f_2$ where $f_1 = f$ on $B(x_0,2Kj)$ and zero elsewhere. From (18) and the definition of E_k we have

$$(19) \int_\Omega [M_\alpha f_1]^q\ U_j d\mu = \int_{E_0} [M_\alpha f_1]^q\ U_j d\mu + \sum_1^\infty \int_{E_k\setminus E_{k-1}} [M_\alpha f_1]^q\ U_j d\mu$$

$$\leq C\left(\int_{B(x_0,2Kj)} |f|^p\ Vd\mu\right)^{q/p}\left(1 + \sum_1^\infty 2^{-\beta(k-1)+kq/p'}\right)$$

$$\leq C\left(\int_\Omega |f|^p\ Vd\mu\right)^{q/p} .$$

If $x \in B(x_0,j)$, then $(M_\alpha f_2)(x)$ is bounded by

$$\sup_{r\geq j} [\mu(B(x,r))]^{\alpha-1}\int_{B(x,r)} |f_2| d\mu$$

$$\leq \sup_{r\geq j} [\mu(B(x,r))]^{\alpha-1}\left(\int_\Omega |f|^p V d\mu\right)^{1/p}\left(\int_{B(x,r)} V^{-1/(p-1)} d\mu\right)^{1/p'}$$

$$\leq \sup_{r\geq j} [\mu(B(x_0,2Kr))]^{\alpha-1}\left(\int_\Omega |f|^p V d\mu\right)^{1/p}\left(\int_{B(x_0,2Kr)} V^{-1/(p-1)} d\mu\right)^{1/p'}$$

$$\leq C\left(\int_\Omega |f|^p V d\mu\right)^{1/p}$$

where we have used Hölder's inequality, (5) and (4). Thus

$$(20) \qquad \int_\Omega [M_\alpha f_2]^q\, U_j d\mu \;\leq\; C\mu(B(x_0,j))\left(\int_\Omega |f|^p V d\mu\right)^{q/p}.$$

Summing (19) and (20) we obtain (17) as required.

5. PROOF OF THEOREM 3

Since

$$\int_{R^n} (I_\alpha f)(x)g(x)\,dx = \int_{R^n} f(x)(I_\alpha g)(x)\,dx$$

for $f,g \geq 0$ and since $1 < q' < 1/\alpha$, $1/p' = 1/q' - \alpha$, Hölder's inequality and it's converse shows that (b) is an immediate consequence of (a). Thus we prove only (a).

Observe first that $B(0,r)$ contains the rectangle $\{x: |x_i| < (r/\sqrt{n})^{\gamma_i},\ 1\leq i\leq n\}$ and is contained in the rectangle $\{x: |x_i| < r^{\gamma_i},\ 1\leq i\leq n\}$ and since Lebesgue measure μ satisfies $\mu(B(x,r)) = \mu(B(0,r))$ it follows that

$$(21) \qquad 2^n n^{-\gamma/2}\, r^\gamma \leq \mu(B(x,r)) \leq 2^n\, r^\gamma .$$

Following an idea of Welland [9] we shall show that for $0 < \varepsilon < 1-\alpha$ there is a constant C depending on ε such that

$$(22) \qquad |(I_\alpha f)(x)| \;\leq\; C\,[(M_{\alpha-\varepsilon}f)(x)(M_{\alpha+\varepsilon}f)(x)]^{1/2}.$$

To see this, let $\delta > 0$ and observe that

$$\int_{2^{j-1}\delta \leq \rho(x-y) < 2^j\delta} [\rho(x-y)]^{\gamma(\alpha-1)} |f(y)|dy$$

$$\leq (2^{j-1}\delta)^{\gamma(\alpha-1)} \int_{B(x,2^j\delta)} |f(y)|dy$$

$$\leq 2^{\gamma(1-\alpha)}[2^j\delta]^{-(\pm\gamma\varepsilon)} 2^{-n(\alpha\pm\varepsilon-1)} (M_{\alpha\pm\varepsilon}f)(x)$$

because of (21), then choosing the + sign for $j \geq 0$, the - sign for $j < 0$ and summing these inequalities over all j yields (22) if δ satisfies

$$[2^n\delta^\gamma]^\varepsilon = [(M_{\alpha+\varepsilon}f)(x)/(M_{\alpha-\varepsilon}f)(x)]^{1/2}.$$

Suppose (7) holds. Choose $\varepsilon < \min\{\alpha, 1/p - \alpha\}$ and set $1/q_1 = 1/p - (\alpha-\varepsilon)$, $1/q_2 = 1/p - (\alpha+\varepsilon)$ so that $2q_1/q$ and $2q_2/q$ are conjugate exponents; let

$$U_1(x) = U(x)[1+\rho(x)]^{\gamma(q_1-q)/p'},$$

$$U_2(x) = U(x)[1+\rho(x)]^{\gamma(q_2-q)/p'}.$$

Then $U = (U_1^{1/q_1}U_2^{1/q_2})^{q/2}$ and

$$(23)\quad \int_{R^n} \frac{U_1(x)dx}{[1+\rho(x)]^{\gamma(1-\alpha+\varepsilon)q_1}} = \int_{R^n} \frac{U_2(x)dx}{[1+\rho(x)]^{\gamma(1-\alpha-\varepsilon)q_2}}$$

$$= \int_{R^n} \frac{U(x)dx}{[1+\rho(x)]^{\gamma(1-\alpha)q}} < \infty .$$

From (22), Hölder's inequality shows that the left side of (6) is bounded by a multiple of

$$\left(\int_{R^n} |(M_{\alpha-\varepsilon}f)(x)|^{q_1} U_1(x)dx\right)^{1/2q_1} \left(\int_{R^n} |(M_{\alpha+\varepsilon}f)(x)|^{q_2} U_2(x)dx\right)^{1/2q_2}$$

and in view of (21), (23) and Theorem 1 shows that there are

$0 \leq V_1(x), V_2(x) < \infty$ a.e. such that this is bounded by

$$\left(\int_{R^n} |f(x)|^p V_1(x)dx\right)^{1/2p}\left(\int_{R^n} |f(x)|^p V_2(x)dx\right)^{1/2p} .$$

Thus, if $V(x)$ is a suitable multiple of $\max\{V_1(x),V_2(x)\}$ then (6) is satisfied.

For the necessity of (7), observe that

$$\begin{aligned}(I_\alpha|f|)(x) &= \sup_{r>0} \int_{B(x,r)} [\rho(x-y)]^{\gamma(\alpha-1)}|f(y)|dy \\ &\geq \sup_{r>0} r^{\gamma(\alpha-1)} \int_{B(x,r)} |f(y)|dy \\ &\geq [2^{-n}n^{\gamma/2}]^{\alpha-1}(M_\alpha f)(x)\end{aligned}$$

where we have used (21) to obtain the last inequality. Thus (6) implies (2) and Theorem 1 then shows that (3), which is equivalent to (7), holds.

REFERENCES

1. H. Aimar and R. Macias, Weighted norm inequalities for the Hardy-Littlewood maximal operator on spaces of homogeneous type. Proc. Amer. Math. Soc. 91(1984), 213-216.

2. K.F. Andersen, Weighted inequalities for fractional integrals. In: Fractional Calculus, (Papers presented at a workshop held at Ross Priory, University of Strathclyde, Aug. 1984). Res. Notes in Math., No.138. Pitman, Boston, 1985, 12-25.

3. A.P. Calderón, Inequalities for the maximal function relative to a metric. Studia Math. 57(1976), 297-306.

4. L. Carleson and P. Jones, Weighted norm inequalities and a theorem of Koosis. Mittag-Leffler Inst. Rep. No.2, 1981.

5. R.R. Coifman and G. Weiss, Analyse Harmonique Non-Commutative sur Certains Espaces Homogènes. Lecture Notes in Math., Vol.242, Springer-Verlag, Berlin and New York, 1971.

6. A. Gatto and C.E. Gutiérrez, On weighted norm inequalities for the maximal function. Studia Math. 76(1983), 59-62.

7. D.S. Kurtz, Weighted norm inequalities for the Hardy-Littlewood maximal function for one parameter rectangles. Studia Math. 53(1975), 39-54.

8. J. Rubio de Francia, Boundedness of maximal functions and singular integrals in weighted L^p spaces. Proc. Amer. Math. Soc. 83(1981), 673-679.

9. G.V. Welland, Weighted norm inequalities for fractional integrals. Proc. Amer. Math. Soc. 51(1975), 143-148.

10. W.-S. Young, Weighted norm inequalities for the Hardy-Littlewood maximal function. Proc. Amer. Math. Soc. 85(1982), 24-26.

Kenneth F. Andersen, Department of Mathematics, University of Alberta, Edmonton, Alberta, Canada, T6G-2G1.

International Series of
Numerical Mathematics, Vol. 80

SOME INEQUALITIES CONCERNING CONVOLUTIONS OF KERNEL FUNCTIONS

István S. Fenyö

Abstract. We define norms for kernel-functions and prove some inequalities between the norm of convolutions of kernels and the norms of the factors. Some of them are used to give estimates of the norm of solutions of certain differential equations.

1. RELATIONS BETWEEN NORMS

It is known that convolutions of kernels of integral operators play an essential role in the theory of integral equations. The aim of the present paper is to give some estimates for certain norms of convolutions of kernel functions in terms of the norms of their factors.

Let $D \subset \mathbb{R}$ be a measurable set and p and q given reals. We denote by $H_{p,q}$ the set of kernel functions $K(s,t)$: $D \times D \to C$ with the following properties:

1) K is measurable over $D \times D$.

2) $\|K\|_{p,q} := \left(\int_D [\int_D |K(s,t)|^p ds]^{q/p} dt\right)^{1/q} < \infty.$

In what follows we simply write $\int$ instead of $\int_D$. If $p = q$ we write H_p and $\|K\|_p$ instead of $H_{p,p}$ and $\|K\|_{p,p}$, i.e.,

$$\|K\|_p = \left(\int \int |K(s,t)|^p \, ds\, dt\right)^{1/p} .$$

For the convolution of two kernels P and Q we will use the notation $(P\cdot Q)(s,t)$ or briefly $P\cdot Q$,

$$(P\cdot Q)(s,t) = P\cdot Q := \int P(s,r)Q(r,t)\, dr .$$

This paper is in final form and no version of it will be submitted for publication elsewhere.

THEOREM 1. *Let* $p \geq 1$ *and* q *its adjoint, i.e.,* $(1/p)+(1/q)=1$. *Suppose that* $P,Q \in H_{q,p} \cap H_p \cap H_q$. *Then the convolution* $P \cdot Q$ *exists and belongs to* $H_{q,p}$, *and the inequality*

$$||P\cdot Q||_{q,p} \leq ||P||_q ||Q||_p$$

holds.

Proof. By assumption $||P||_q$ and $||Q||_p$ are finite. This implies $P(s,\cdot) \in L^q(D)$ for a.e. s and $Q(\cdot,t) \in L^p(D)$ for a.e. t. Therefore, by the Hölder inequality,

$$(2) \qquad |(P\cdot Q)(s,t)| \leq \left(\int |P(s,r)|^q dr\right)^{1/q} \left(\int |Q(r,t)|^p dr\right)^{1/p} .$$

This shows the existence of $P\cdot Q$ a.e. in D. Inequality (2) can be written as follows:

$$|(P\cdot Q)(s,t)|^q \leq \left(\int |P(s,r)|^q dr\right) \left(\int |Q(r,t)|^p dr\right)^{q/p} .$$

The right hand side is integrable with respect to s, therefore

$$\int |(P\cdot Q)(s,t)|^q ds \leq ||P||_q^q \left(\int |Q(r,t)|^p dr\right)^{q/p} .$$

By a similar argument as above,

$$\int \left(\int |(P\cdot Q)(s,t)|^q ds\right)^{p/q} dt \leq ||P||_q^p ||Q||_p^p ,$$

which proves the statement.

Now we prove some inequalities of the type

$$||P\cdot Q||_p \leq k ||P||_p ||Q||_q$$

for certain p and q. In general $||P\cdot Q||_p \leq ||P||_p ||Q||_q$ does not hold. We show now that under certain conditions there exists a constant $k > 0$, called by M. Goldberg 'multiplicativity factor' [3].

THEOREM 2. *Let* $p \geq 2$ *and* $P,Q \in H_p$ *and suppose that* $0 < |D| < \infty$, *where* $|D|$ *is the measure of the set* D. *Then* $P\cdot Q$ *exists for a.e.* $s,t \in D$, *is in* H_p, *and the inequality*

$$(3) \qquad ||P\cdot Q||_p \leq |D|^{1-2/p} ||P||_p ||Q||_p$$

holds. The factor $|D|^{1-2/p}$ *is best possible*.

Proof. Denote by q the adjoint of p. From $p \geq 2$ follows $q \leq p$. By (2), Theorem 189 (p.140) and Theorem 192 (p.143) in [4], it follows that ($s,t \in D$ fixed)

$$|(P\cdot Q)(s,t)| \le \Big(\int |P(s,r)|^p dr\Big)^{1/p} \Big(\int |Q(r,t)|^q dr\Big)^{1/q}$$

$$= |D| \,\frac{\big(\int |P(s,r)|^p dr\big)^{1/p}}{|D|^{1/p}} \,\frac{\big(\int |Q(r,t)|^q dr\big)^{1/q}}{|D|^{1/p}}$$

$$\le |D| \,\frac{\big(\int |P(s,r)|^p dr\big)^{1/p}}{|D|^{1/p}} \,\frac{\big(\int |Q(r,t)|^p dr\big)^{1/p}}{|D|^{1/p}}$$

$$= |D|^{1-2/p} \Big(\int |P(s,r)|^p dr\Big)^{1/p} \Big(\int |Q(r,t)|^p dr\Big)^{1/p}$$

for a.e. $s,t \in D$. As the right hand side is integrable in $D\times D$, we can write

$$\iint |(P\cdot Q)(s,t)|^p \,ds\,dt \le |D|^{p-2} \iint |P(s,r)|^p \,ds\,dr \iint |Q(r,t)|^p \,dr\,dt$$

and so we have

$$\|P\cdot Q\|_p \le |D|^{1-2/p} \|P\|_p \|Q\|_p \, .$$

If we consider the functions $P(s,t) = 1$, $Q(s,t) = 1$ for $s,t \in D$, we get equality, which shows that the factor $|D|^{1-2/p}$ cannot be replaced by a smaller number.

Theorem 2 is the continuous analogon of Theorem 1.3 in [2].

In order to prove the next theorem we need the following

LEMMA. _Let_ $p \ge 2$ _and_ $K \in H_p$ _and_ $0 < |D| < \infty$. _Then_ $K \in H_{p,q}$, _where_ q _is the adjoint of_ p, _and the inequality_

(4) $$\|K\|_{p,q} \le |D|^{1-2/p} \|K\|_p$$

holds. The constant $|D|^{1-2/p}$ _is best possible._

Proof. Define

$$k(t) := \Big(\int |K(s,t)|^p ds\Big)^{1/p} \qquad (\text{a.e. } t \in D) ,$$

then

$$\|k\|_{L^p} = \|K\|_p < \infty$$

Since $p \ge 2$ implies $q \le p$, it follows that

$$\|K\|_{p,q} = \|k\|_{L^q} < \infty$$

This shows that $K \in H_{p,q}$. By Theorem 192 (p.143) in [4] we have

$$||K||_{p,q} = ||k||_{L^q} = |D|^{1/q} \frac{||k||_{L^q}}{|D|^{1/q}} \le |D|^{1/q} \frac{||k||_{L^p}}{|D|^{1/p}}$$

$$= |D|^{1/q-1/p} ||K||_p = |D|^{1-2/p} ||K||_p ,$$

as it was stated.

If we apply (4) for $K(s,t) = 1$, $s,t \in D$, we get an equality, i.e. $|D|^{1-2/p}$ cannot be replaced by a smaller constant.

THEOREM 3. _Let_ $1 < p < 2$ _and_ q _the adjoint of_ p. _Suppose_ $P,Q \in H_{q,p}$, _then_ $P \cdot Q$ _exists_ _a.e._ _in_ $D \times D$ _and is in_ H_p. _The following estimate_

(5) $$||P \cdot Q||_p \le |D|^{1-2/q} \min \{||P||_p ||Q||_q, ||P||_q ||Q||_p\}$$

holds. _The constant_ $|D|^{1-2/q}$ _cannot be improved_.

Proof. By the Hölder inequality (Theorem 189 in [4]) we have

$$|(P \cdot Q)(s,t)|^p \le \left(\int |P(s,r)|^p dr\right) \left(\int |Q(r,t)|^q dr\right)^{p/q} .$$

The right hand side exists as $P \in H_{q,p}$, this implies $P \in H_p$ as $p < q$. From this follows the existence of the convolution. The integrals on the right hand side of the preceding inequality with respect to s resp. to t exist, therefore

$$\iint |P \cdot Q|^p \, dt \, ds \le \left(\iint |P(s,r)|^p \, dr \, ds\right) \left(\int \left(\int |Q(r,t)|^q dr\right)^{p/q} dt\right)$$

or

(6) $$||P \cdot Q||_p \le ||P||_p ||Q||_{q,p} .$$

In the same way we get

(7) $$||P \cdot Q||_p \le ||Q||_p ||P^*||_{q,p} ,$$

where P^* denotes the adjoint of P. As $q \ge 2$ we have by the lemma

$$||Q||_{q,p} \le |D|^{1-2/q} ||Q||_q ,$$

and therefore (6) implies

$$||P \cdot Q||_p \le |D|^{1-2/q} ||P||_p ||Q||_q .$$

If we apply now the lemma to P^*, we get by (7) and the Fubini theorem

$$\|P\cdot Q\|_p \le |D|^{1-2/q}\|Q\|_p\|P^*\|_q = |D|^{1-2/q}\|Q\|_p\|P\|_q ,$$

which completes the proof.

Theorem 3 is the continuous analogon of Satz VIII in A. Ostrowski [5].

2. APPLICATIONS

Let us first apply the definitions of $\|\cdot\|_{p,q}$ and $\|\cdot\|_p$ to the case where $K(s,t)$ does not depend on t, i.e. $K(s,t)=k(s)$ with $k\in L^p$. Then we see immediately that

$$(8) \qquad \|k\|_{p,q} = |D|^{1/q}\|k\|_{L^p} ,$$

where q is an arbitrary real ($\neq 0$), and if $q=p$, we have

$$(9) \qquad \|k\|_p = |D|^{1/p}\|k\|_{L^p} .$$

We suppose here that $|D|$ is finite.

Consider now a linear integral transformation of the form

$$(\mathcal{K}f)(s) = \int_D K(s,r)f(r)\,dr ,$$

which can be considered as a convolution. Then by (8) resp. (9) we have

$$(8') \qquad \|\int K(s,r)f(r)dr\|_{p,q} = |D|^{1/q}\|\mathcal{K}f\|_{L^p}$$

resp.

$$(9') \qquad \|\int K(s,r)f(r)dr\|_p = |D|^{1/p}\|\mathcal{K}f\|_{L^p}$$

under the assumption that $\mathcal{K}f\in L^p$ for some $p>1$.

It is well known that the n-th indefinite integral (n is not necessarily an integer) of f can be written as the following convolution,

$$(I^n f)(s) = \frac{1}{\Gamma(n)}\int_0^s (s-r)^{n-1}f(r)\,dr .$$

Here we have the convolution of $f(r)$ (not depending on t) with the kernel

$$K(s,r) = \begin{cases} \dfrac{(s-r)^{n-1}}{\Gamma(n)} & \text{for } s \geq r \\ 0 & \text{elsewhere .} \end{cases} \tag{10}$$

Now let D be the interval (0,1), $f \in L^p$, where $p > 1$. Denote by q the adjoint to p and let $n > 1/p$. Then

PROPOSITION 1. *The following inequality is valid:*

$$\|I^n f\|_{L^q} \leq \frac{1}{[(q(n-1)+1)(q(n-1)+2)]^{1/q}\Gamma(n)} \|f\|_{L^p} .$$

Proof. Using Theorem 1 and (8') we have

$$\|I^n f\|_{L^q} \leq \|K(s,r)\|_q \|f\|_{L^p} .$$

Considering expression (10), a simple calculation yields the proposition. Since $n > 1/p$, we have $q(n-1)+1 > 0$.

Let again $D = (0,1)$ and $p > 1$, $n > 1/p$, $f \in L^p$.

PROPOSITION 2. *The following estimate holds:*

$$\|I^n f\|_{L^p} \leq \frac{1}{[(p(n-1)+1)(p(n-1)+2)]^{1/p}\Gamma(n)} \|f\|_{L^p} . \tag{11}$$

Proof. Apply Theorem 2 for $p \geq 2$ and Theorem 3 for $1 < p < 2$ using the form (10) of the kernel.

Let us consider now an ordinary linear differential expression

$$(Lx)(t) = \sum_{j=0}^{n} P_j(t)x^{(j)} \qquad \left(x^{(j)} = \frac{d^j x}{dx^j},\ j = 0,1,2,\ldots,n\right),$$

where $P_n(t) \not\equiv 0$, $t \in D = [a,b]$ (a finite interval) and $P_j \in C^n[a,b]$. Consider further the boundary functionals

$$R_k x = \sum_{j=0}^{n-1} [\alpha_{jk} x^{(j)}(a) + \beta_{kj} x^{(j)}(b)] \qquad (k = 1,2,\ldots,n) ,$$

where α_{jk} and β_{jk} are given constants.

Let us suppose that the differential operator L with the homogeneous boundary conditions $R_k x = 0$ $(k = 1,2,\ldots,n)$ has a Green kernel (conditions for it see e.g. [1], Theorem 2, p.421) G(s,t) which is a continuous function defined in $[a,b]\times[a,b]$. It is well known that for a given function $f\in L^p$ $(p>1)$ the differential problem

$$Lx = f \quad \text{with} \quad R_k x = 0 \quad (k=1,2,\dots,n) \tag{12}$$

has a unique solution of the form

$$x(s) = \int_a^b G(s,t)f(t)dt\,, \qquad s\in[a,b]\,. \tag{13}$$

If we apply Theorems 3 and 2 to (13) considering (8'), (9') and (9), we get

PROPOSITION 3. *If* $1<p<2$ *and* q *is the conjugate of* p, *then for the* L^p*-norm of the solution of* (12) *the following estimate holds*:

$$\|x\|_{L^p} \le (b-a)^{1-2/q}\|G\|_q\|f\|_{L^p} \;; \tag{14}$$

if $p\ge 2$, *then the inequality*

$$\|x\|_p \le (b-a)^{1-2/p}\|G\|_p\|f\|_{L^p} \tag{15}$$

is valid.

Similar estimates can be derived for solutions of partial differential equations if the Green function exists.

EXAMPLE. Consider the differential equation

$$x'' = f\,, \qquad x'(0) = 0\,, \qquad x(1) = 0\,.$$

The Green function of this problem is (see e.g. [1], p.425)

$$G(s,t) = \begin{cases} 1-s & \text{if} \quad 0\le t\le s\le 1 \\ 1-t & \text{if} \quad 0\le s\le t\le 1\,. \end{cases}$$

If we now assume that $x\colon[0,1]\to\mathbb{R}$ satisfies $x''\in L^p$, $p\ge 2$, and $x'(0)=x(1)=0$, then a simple calculation shows

$$\|x\|_{L^p} \le \frac{2^{1/p}}{[(p+1)(p+2)]^{1/p}}\,\|x''\|_{L^p}\,. \tag{16}$$

If we apply (11) with $n=2$ and $p>1$ to $I^2f = x$, we get

$$\|x\|_{L^p} \leq \frac{1}{[(p+1)(p+2)]^{1/p}} \|x''\|_{L^p} . \tag{17}$$

The constant in (17) is smaller than in (16). But we have to consider that the inequality (16) is valid for all functions in $C^2(0,1)$ for which $x'(0) = 0$, $x(1) = 0$, while the relation (17) refers to functions in $C^2(0,1)$ for which $x(0) = x'(0) = 0$ holds. In the example $x(t) = t^2$ and $p = 1$, equality holds in (17).

REFERENCES

1. S. Fenyö und W. Stolle, Theorie und Praxis der linearen Integralgleichungen, Bd. 4. Birkhäuser Verlag, Basel - Boston - Stuttgart, 1984.

2. M. Goldberg and E.G. Straus, Multiplicity of l_p norms for matrices. Linear Algebra Appl. 52/53 (1983), 351-360.

3. M. Goldberg, Mixed multiplicativity and l_p norms for matrices. Linear Algebra 73 (1986), 123-131.

4. G.H. Hardy, J.E. Littlewood and G. Pólya, Inequalities. Cambridge Univ. Press, Cambridge, 2nd edition, 1952.

5. A. Ostrowski, Über Normen von Matrizen. Math. Z. 63 (1955), 1-18.

István S. Fenyö, Mathematical Research Institute of the Hungarian Academy of Sciences, H-1053 Budapest, Realtanoda u. 13-15, Hungary

International Series of
Numerical Mathematics, Vol. 80

INEQUALITIES FOR SOME SPECIAL FUNCTIONS AND THEIR ZEROS

Carla Giordano and Andrea Laforgia

Abstract. We establish inequalities for the Bessel functions $J_\nu(x)$ of the first kind, by means of the arithmetic geometric mean inequality and the infinite product formula for $J_\nu(x)$. A concavity property is also obtained for the positive zeros $j_{\nu k}$ $(k=1,2,\dots)$ of $J_\nu(x)$ using a lower bound for the second derivative of $j_{\nu k}$ recently established in [3]. Finally we show a monotonicity property of the zeros of Legendre polynomials. This property is proved as a consequence of the classical Sturm comparison theorem.

1. INTRODUCTION

In the present paper we deal with some methods which can be used to obtain inequalities for some Special Functions and their zeros. In particular we consider the case of the Bessel function $J_\nu(x)$ of the first kind. Our principal tools are the infinite product formula for $J_\nu(x)$ and the arithmetic-geometric mean inequality. The results obtained here can be extended to many other Special Functions.

It is also a purpose of the paper to obtain a concavity property of the positive zeros $j_{\nu k}$ $(k=1,2,\dots)$ of $J_\nu(x)$. The result is proved as a consequence of a Lemma established in [3].

Finally an inequality of Turán-type for the zeros

This paper is in final form and no version of it will be submitted for publication elsewhere.

$x_{n,k}$ $(k=1,2,...,n)$ of the Legendre polynomials $P_n(x)$ is proved. The proof of this result is given by using the Sturm comparison theorem in a form due to Szegö [15, p. 19] .

2. BESSEL FUNCTION OF THE FIRST KIND

In a private communication D. Kershaw observed that the classical result [13, p. 285]

$$\Gamma(x_1)\,\Gamma(x_2)\cdots\Gamma(x_n) \geq [\Gamma(x)]^n, \quad x_i > 0, \quad i=1,2,\cdots,n$$

where $x = (x_1+x_2+...+x_n)/n$, can be proved as a consequence of the infinite product formula [7, p. 1]

$$z\,\Gamma(z) = \prod_{n=1}^{\infty} [(1+1/n)^z\,(1+z/n)^{-1}] \quad , \; z > 0$$

and the arithmetic-geometric mean inequality.

We show here that Kershaw's observation is also useful in the case of the Bessel function $J_\nu(x)$ of the first kind. Namely we prove the following result.

THEOREM 2.1 *For* $\nu > -1$ *let* $J_\nu(x)$ *be the Bessel function of the first kind. Then*

$$(2.1) \qquad J_\nu(x_1)\,J_\nu(x_2)\cdots J_\nu(x_n) \leq [J_\nu(x)]^n$$

where

$$x = [(x_1^2+x_2^2+\cdots+x_n^2)/n]^{1/2}, \quad 0 < x_i < j_{\nu 1}, \qquad i=1,2,...,n \quad ,$$

$j_{\nu 1}$ *being the first positive zero of* $J_\nu(x)$.

Proof. The infinite product formula [1, p. 370; 9.5.10] states that for arbitrary z and $\nu \neq -1, -2, ...$

(2.2) $$J_\nu(z) = \frac{(z/2)^\nu}{\Gamma(\nu+1)} \prod_{k=1}^{\infty} \left(1 - \frac{z^2}{j_{\nu k}^2}\right).$$

Using this n times we find

(2.3) $$J_\nu(x_1)\, J_\nu(x_2) \cdots J_\nu(x_n) = \left(\frac{x_1 x_2 \cdots x_n}{2^n}\right)^\nu \frac{1}{[\Gamma(\nu+1)]^n} \cdot$$

$$\cdot \prod_{k=1}^{\infty} \left(1 - \frac{x_1^2}{j_{\nu k}^2}\right)\left(1 - \frac{x_2^2}{j_{\nu k}^2}\right) \cdots \left(1 - \frac{x_n^2}{j_{\nu k}^2}\right).$$

Since $0 < x_i < j_{\nu 1} (i=1,2,...,n)$, every term of the product above is positive. Therefore applying the arithmetic-geometric mean inequality, we obtain

$$\left(1 - \frac{x_1^2}{j_{\nu k}^2}\right)\left(1 - \frac{x_2^2}{j_{\nu k}^2}\right) \cdots \left(1 - \frac{x_n^2}{j_{\nu k}^2}\right) \le \left(1 - \frac{x^2}{j_{\nu k}^2}\right)^n.$$

Moreover

$$x_1 x_2 \cdots x_n \le \left(\frac{x_1^2+x_2^2+\cdots+x_n^2}{n}\right)^{n/2}.$$

Using these inequalities in the right-hand side of (2.3), we find

$$J_\nu(x_1)\, J_\nu(x_2) \cdots J_\nu(x_n) \le \left[\frac{(x/2)^\nu}{\Gamma(\nu+1)} \prod_{k=1}^{\infty} \left(1 - \frac{x^2}{j_{\nu k}^2}\right)\right]^n$$

and in view of (2.2), this reduces to

$$J_\nu(x_1)\, J_\nu(x_2) \cdots J_\nu(x_n) \le [J_\nu(x)]^n,\ x = [(x_1^2 + x_2^2 + \cdots + x_n^2)/n]^{1/2}.$$

This completes the proof of Theorem 2.1.

REMARK. The equality in (2.1) occurs if and only if the arithmetic and geometric means coincide, that is when $x_1=x_2=\dots=x_n$.

When ν is half an odd integer, the function $J_\nu(x)$ reduces to a combination of trigonometric functions. Therefore Theorem 2.1 gives inequalities for trigonometric functions.

One of the possible consequences of Theorem 2.1 is the following.

COROLLARY 2.1 *For* $0 < x_i < \pi$ $(i=1,2,\dots,n)$ *and* $x = [(x_1^2 + x_2^2 + \dots + x_n^2)/n]^{1/2}$

$$\sin x_1 \sin x_2 \cdots \sin x_n \le \sin^n x. \tag{2.4}$$

Proof. The conclusion of Corollary 2.1 follows from Theorem 2.1 by setting $\nu=1/2$ and

$$J_{1/2}(x) = \sqrt{\frac{2}{\pi x}}\ \sin x$$

to obtain

$$\frac{\sin x_1 \sin x_2 \cdots \sin x_n}{\sqrt{x_1 x_2 \dots x_n}} \le \left(\frac{\sin x}{\sqrt{x}}\right)^n , \tag{2.5}$$

$$0 < x_i < \pi, \qquad i=1,2,\cdots n .$$

In order to complete the proof of Corollary 2.1, we need only the result

$$\sqrt{x_1 x_2 \dots x_n} \le x^{n/2}$$

which of course, follows from the arithmetic-geometric mean inequality.

The inequality (2.4) can be complemented by the one that we

obtain using the following Jensen's inequality [13, p. 12]

$$\begin{aligned} &\log f(x_1) + \log f(x_2) + \cdots + \log f(x_n) \geq n \log f(x) \ , \\ &x = (x_1 + x_2 + \cdots + x_n)/n, \quad x_i > 0, \quad i=1,2,\cdots n. \end{aligned} \tag{2.6}$$

Indeed the function $1/\sin x$ is log-convex, for $0 < x < \pi$, so we can apply (2.6), leading to

$$\begin{aligned} &\sin x_1 \sin x_2 \cdots \sin x_n \leq \sin^n \overline{x}, \\ &0 < x_i < \pi, \quad i=1,2,\cdots,n \end{aligned} \tag{2.7}$$

where

$$\overline{x} = (x_1 + x_2 + \cdots + x_n)/n \ ;$$

equality in (2.7) occurs if and only if $x_1=x_2=\ldots=x_n$. Now $\sin^n\overline{x} \leq \sin^n x$, for $0 < x_i < \pi/2$, but when $\pi/2 < x_i < \pi$ this inequality is reversed. Therefore the result of Corollary 2.1 can be included in the more informative inequalities

$$\sin x_1 \sin x_2 \cdots \sin x_n \leq \sin^n \overline{x} \leq \sin^n x, \quad 0 < x_i < \pi/2$$

and

$$\sin x_1 \sin x_2 \cdots \sin x_n \leq \sin^n x \leq \sin^n \overline{x}\, , \quad \pi/2 < x_i < \pi,$$

where $i=1,2,\ldots,n$.

Inequality (2.7) admits the following geometrical interpretation. Let $x_1,x_2,\ldots,x_n$ be the angles of a convex polygon, then

$$\sin x_1 \sin x_2 \cdots \sin x_n \leq \left(\sin \frac{2\pi}{n}\right)^n ,$$

with $x_1+x_2+\ldots+x_n = (n-2)\pi$. For $n=3$ (2.7) gives that in a triangle whose angles are indicated by x_1, x_2, x_3, the inequality

$$\sin\frac{x_1}{2}\sin\frac{x_2}{2}\sin\frac{x_3}{2}\le\frac{1}{8}$$

holds. This result has been already established by Ehret in [2] using a more sophisticated analysis.

Further inequalities can be obtained using (2.5) as follows.

1. For $n=2$, $x_1=\pi/2$ and $x_2=t$

$$\sqrt{\pi t}\,\sin^2\sqrt{\frac{t^2+\pi^2/4}{2}}-\sqrt{t^2+\pi^2/4}\,\sin t\ge 0,\quad 0<t<\pi$$

2. For $n=2$, $x_1=\pi/4$ and $x_2=t$

$$\sqrt{\pi t}\,\sin^2\sqrt{\frac{t^2+\pi^2/16}{2}}-\sqrt{t^2+\pi^2/16}\,\sin t\ge 0,\quad 0<t<\pi$$

3. For $n=2$, $\sqrt{\frac{x_1^2+x_2^2}{2}}=\frac{\pi}{2}$ and $x_2=t$

$$\sin t\,\sin\sqrt{\pi^2/2-t^2}\le\frac{2}{\pi}\,t^{1/2}\sqrt[4]{\pi^2/2-t^2},\qquad 0<t<\frac{\pi}{\sqrt{2}}\,.$$

These inequalities become equalities when $t=\pi/2$, $\pi/4$, $\pi/\sqrt{2}$, respectively.

New trigonometric inequalities can be proved as consequence of Theorem 2.1 for $\nu=k+1/2$ where $k=0,\pm 1,\pm 2,\ldots$. In fact for these values of ν the Bessel function $J_\nu(x)$ reduces to a combination of trigonometric functions with an overall factor $x^{-1/2}$. The expressions of $J_{k+1/2}(x)$ can be obtained from the recurrence relation [9, p. 16]

$$J_{\nu+1}(z)=\frac{2\nu}{z}J_\nu(z)-J_{\nu-1}(z)\,.$$

3. ZEROS OF BESSEL FUNCTIONS

For $\nu \geq 0$ let $j_{\nu k}$ be the k-th positive zero of the Bessel function $J_\nu(x)$ of the first kind.

Recently several monotonicity, concavity and convexity properties of $j_{\nu k}$ as a function of ν have been investigated. The work [5] should be consulted for further information and complete references. In that paper the reader can also find results on the zeros of the general Bessel function

$$C_\nu(x) = J_\nu(x)\cos\alpha - Y_\nu(x)\sin\alpha, \qquad 0 \leq \alpha < \pi$$

where $Y_\nu(x)$ is the Bessel function of the second kind.

Here we present a new concavity result on $j_{\nu k}$. This will be proved as a consequence of the following Lemma [3].

LEMMA 3.1 *For* $\nu \geq 0$ *and* $k=1,2,\ldots$ *the function* $j=j_{\nu k}$ *satisfies the inequality*

$$(\nu+j)\,j'' > \frac{\nu j'^2}{j} - j'$$

where $' = \dfrac{d}{d\nu}$.

THEOREM 3.1 *For* $\nu \geq 0$ *and* $k=1,2,\ldots$ *let* $j_{\nu k}$ *be the k-th positive zero of the Bessel function of the first kind. Then the function*

$$f(\nu) = \frac{\nu}{j_{\nu k}}$$

is concave as a function of ν.

Proof. We have to show that

$$f''(\nu) = \frac{-\nu j^2 j'' - 2jj'(j-\nu j')}{j^4} < 0$$

or

$$-\nu j j'' - 2 j j' + 2 \nu j'^2 < 0$$

which can be written in the following way

$$j'' > \frac{2 j'^2}{j} - \frac{2 j'}{\nu} .$$

By Lemma 3.1 it is sufficient to show that

$$\frac{\nu j'^2}{(\nu+j) j} - \frac{j'}{\nu+j} > \frac{2j'^2}{j} - \frac{2j'}{\nu}$$

or equivalently

$$(j-\nu j')(\nu+2j) > 0.$$

In order to complete the proof of Theorem 3.1 we need only the inequality $j - \nu j' > 0$. But this is true because j/ν decreases with respect to ν [12]. The proof of Theorem 3.1 is complete.

REMARK. We observe that using results on the concavity (convexity) of functions of $j_{\nu k}$, it is possible to obtain inequalities for these zeros. For example using the concavity of $j_{\nu k}$ as a function of ν, various linear inequalities have been established in [11]. Many other consequences of the concavity properties of the zeros of cylinder functions have been studied in [8].

4. INEQUALITIES FOR THE ZEROS OF THE LEGENDRE POLYNOMIALS

P. Turán established [16] the following inequality

$$(4.1)\quad \begin{vmatrix} P_n(x) & P_{n+1}(x) \\ P_{n+1}(x) & P_{n+2}(x) \end{vmatrix} < 0 , \quad -1 < x < 1, \quad n=0,1,...$$

where $P_n(x)$ is the Legendre polynomial. Inequalities similar to (4.1) have been proved by many authors for the Bessel functions, other classes of orthogonal polynomials etc.... Recently À.Elbert and A. Laforgia have established inequalities of Turán-type for the zeros of classical orthogonal polynomials [4], [6]. One of the results proved is the inequality

$$\begin{vmatrix} x^{(\lambda)}_{n+1,k-1} & x^{(\lambda)}_{n,k-1} \\ x^{(\lambda)}_{n+1,k} & x^{(\lambda)}_{n,k} \end{vmatrix} < 0 \quad , \qquad k = 1,2,\dots,\left[\frac{n}{2}\right], -\frac{1}{2} < \lambda \le \frac{3}{2}$$

where $x^{(\lambda)}_{n,k}$ is the k-th positive zero, in decreasing order, of the ultraspherical polynomial $P^{(\lambda)}_n(x)$.

In this section we continue our investigations on the inequalities of Turán-type involving the n zeros $x_{n,k}$, $k=1,2,\dots,n$ of Legendre polynomials $P_n(x)$.

Solving the problem 4 in [10, p. 571], C.C. Grosjean proved the inequality

$$(1-x_{n,k-1})\cdot(1-x_{n,k+1}) < (1-x_{n,k})^2$$

where $x_{n,k}$ denotes the k-th zero of $P_n(x)$ in increasing order. This result shows, among other things, that, for fixed n, the ratio $(1-x_{n,k-1})/(1-x_{n,k})$ increases with respect to k. Now we want to investigate the behaviour of a similar quotient when k is fixed and n varies. In order to make this, we shall need the following classical result [15, p. 19].

LEMMA (Sturm comparison theorem in Szegö's form). *Let the functions y and Y be nontrivial solutions of the differential equations*

$$y''+f(x)y = 0 \quad , \quad Y''+F(x)Y = 0$$

and let them have consecutive zeros at $x_1,x_2,\dots,x_m$ *and*

$X_1, X_2, \ldots, X_m$ *respectively on an interval* (a,b). *Suppose that* f *and* F *are continuous, that*

$$f(x) < F(x) \quad , \quad a < x < x_m$$

and that

$$\lim_{x\to a^+} [\, y'(x)\, Y(x) - y(x)\, Y'(x)\,] = 0 \, .$$

Then

$$X_k < x_k \quad , \quad k=1,2,\ldots,m \, .$$

THEOREM 4.1 *For* $k=1,2,\ldots,n$ *let* $x_{n,k}$ *the k-th zero in increasing order, of the Legendre polynomial* $P_n(x)$. *Then*

$$\begin{vmatrix} 1-x_{n+m,k-1} & 1-x_{n,k-1} \\ 1-x_{n+m,k} & 1-x_{n,k} \end{vmatrix} < 0 \quad , \qquad m=1,2\ldots$$

Proof. The function $u_n(x) = [x(2-x)]^{1/2} P_n(1-x)$, $0 \le x \le 2$ is a solution of the differential equation

$$U'' + f_n(x)\, U = 0$$

where

$$f_n(x) = \frac{n(n+1)}{x(2-x)} + \frac{1}{x^2(2-x)^2} \quad .$$

Therefore the function $u_n\,[(1-x_{n,k})x]$ satisfies the differential equation

$$V'' + g_n(x)\, V = 0 \tag{4.2}$$

where $g_n(x) = (1-x_{n,k})^2 f_n\,[(1-x_{n,k})x]$. Besides (4.2) we consider the differential equation

$$Z'' + g_{n+m}(x)\, Z = 0 \ , \quad \forall\, m \in \{1,2\ldots\}$$

satisfied by $u_{n+m}[(1-x_{n+m,k})x]$. Since $x_{n+m,k} < x_{n,k}$, hence $g_n(x) < g_{n+m}(x)$. Moreover $u_n[(1-x_{n,k})x]$ and $u_{n+m}[(1-x_{n+m,k})x]$ have a common zero at x=1 and it is easy to check that any condition of Lemma above is satisfied. Thus we get that the next zero of $u_{n+m}[(1-x_{n+m,k})x]$ occurs before the next zero of $u_n[(1-x_{n,k})x]$. This gives

$$(4.3) \qquad \frac{1-x_{n+m,k-1}}{1-x_{n+m,k}} < \frac{1-x_{n,k-1}}{1-x_{n,k}} \ , \quad m=1,2\ldots \ , \quad k=2,3,\ldots n$$

which completes the proof of Theorem 4.1.

The previous result shows that, for fixed $\underline{k}$, the ratio $(1-x_{n,k-1})/(1-x_{n,k})$ decreases as $\underline{n}$ increases. Inequality (4.3) enables us to obtain lower and upper bounds for this ratio. These bounds are stringent for large values of $\underline{n}$. For example from (4.3) we get

$$1.0014 = \frac{1-x_{64,1}}{1-x_{64,2}} < \frac{1-x_{63,1}}{1-x_{63,2}} < \frac{1-x_{62,1}}{1-x_{62,2}} = 1.0016$$

where the values of the zeros are taken from [14].

REFERENCES

1. M. Abramowitz and I.A. Stegun, Editors, Handbook of Mathematical Functions. Appl. Math. Series No. 55, National Bureau of Standards, Washington, D.C., 1964.
2. H. Ehret, An approach to trigonometric inequalities. Math. Mag. 43 (1970), 254-257.
3. Á. Elbert and A. Laforgia, Further results on the zeros of Bessel functions. Analysis. 5 (1985), 71-86.

4. Á. Elbert and A. Laforgia, Monotonicity results on the zeros of generalized Laguerre polynomials. J. Approx. Theory. To appear.

5. Á. Elbert and A. Laforgia, Some consequences of a lower bound for the second derivative of the zeros of Bessel functions. J. Math. Anal. Appl. To appear.

6. Á. Elbert and A. Laforgia, Some monotonicity properties for the zeros of ultraspherical polynomials. Acta Math. Hung. To appear.

7. A. Erdélyi, W. Magnus, F. Oberhettinger and F.G. Tricomi, Higher Transcendental Functions. Vol. 1, McGraw-Hill Book Company, Inc., New York, Toronto and London, 1953.

8. C. Giordano and A. Laforgia, Elementary approximations for zeros of Bessel functions. J. Comp. Appl. Math. 9 (1983), 221-228.

9. A. Gray and G.B. Mathews, A Treatise on Bessel Functions. Mac-Millan and Co., London, 2nd Edition, 1952.

10. C.C. Grosjean, Solution of problem 4. Lect. Notes Math. 1171, Polynômes Orthogonaux et Applications. Springer-Verlag, Berlin and New York, 1984.

11. A. Laforgia and M.E. Muldoon, Inequalities and approximations for zeros of Bessel Functions of small order. SIAM J. Math. An. 14 (1983), 383-388.

12. E. Makai, On zeros of Bessel Functions. Univ. Beograd Publ. Elektrotehn. Fak. Ser. Mat. Fiz. 602-603 (1978), 109-110.

13. D.S. Mitrinović, Analytic Inequalities. Springer-Verlag, New York, Heidelberg and Berlin, 1970.

14. A.H. Stroud and Don Secrest, Gaussian Quadrature Formulas. Prentice-Hall, Inc., Englewood Cliffs, 1966.

15. G. Szegö, Orthogonal Polynomials. Amer. Math. Soc. Colloquium Publications 23, Amer. Math. Soc. RI, 4th Edition, 1975.

16. P. Turán, On the zeros of the polynomials of Legendre. Casopis pro Pestování Mat. a Fys. 75 (1950), 113-122.

Carla Giordano, Dipartimento di Matematica, Università di Torino, Via Carlo Alberto, 10 - 10123 Torino - Italia

Andrea Laforgia, Dipartimento di Matematica, Università di Torino, Via Carlo Alberto, 10 - 10123 Torino - Italia

International Series of
Numerical Mathematics, Vol. 80

UNIQUENESS INEQUALITY AND BEST HARMONIC L^1-APPROXIMATION

Werner Haussmann and Lothar Rogge

Abstract. If v_1 and v_2 are two best L^1-approximants to $f \in L^1(X)$ from a vector subspace $V \subset L^1(X)$, then $(f-v_1)(f-v_2) \geq 0$ a.e. on X, where $(X, \mathcal{A}, \mu)$ is a given measure space. This simple inequality helps to derive the uniqueness of a best harmonic L^1-approximant to a given subharmonic function under weak assumptions. In addition, an existence theorem for best harmonic L^1-approximants is given.

1. INTRODUCTION

Let $(X, \mathcal{A}, \mu)$ be a measure space, and $L^1(X) := L^1(X, \mathcal{A}, \mu)$ the vector space of all real valued integrable functions. Given a (not necessarily finite-dimensional) subspace $V \subset L^1(X)$, then we consider the approximation problem to determine a $v^* \in V$ such that

$$\| f - v^* \|_1 \leq \| f - v \|_1 \qquad \text{for all } v \in V,$$

where $\| . \|_1$ is the corresponding L^1-norm.

Besides the case of a finite-dimensional V, in general, a best approximant v^* does not need to exist. Also, even if we identify functions g and $\overline{g}$ with $g = \overline{g}$ a.e., uniqueness will not be guaranteed.

This paper is in final form and no version of it will be submitted for publication elsewhere.

Indeed, there is the famous theorem due to Krein [7] and Phelps [9], see also Kripke-Rivlin [8]:

If $(X,\mathcal{A},\mu)$ *is atom-free and* V *is a finite-dimensional subspace of* $L^1(X)$ *, then there is an* $f \in L^1(X)$ *which possesses infinitely many best approximants out of* V.

In order to get uniqueness results, the following *uniqueness inequality* is very useful:

Suppose $v_1, v_2 \in V$ *are best approximants to* $f \in L^1(X)$, *then*

$$\text{(UI)} \qquad (f - v_1)(f - v_2) \geq 0 \qquad \text{a.e. in } X.$$

This result is well-known, in particular in several special cases (see e. g. Rice [10]). Its short and simple proof runs along the following line:

Since v_1 and v_2 are best approximants to f, this is also true for $v := \frac{1}{2}(v_1 + v_2)$, thus

$$\int (|f - v| - \tfrac{1}{2}|f - v_1| - \tfrac{1}{2}|f - v_2|)d\mu = 0.$$

The integrand is non-positive, hence

$$\tfrac{1}{2}|(f - v_1) + (f - v_2)| = |f - v| = \tfrac{1}{2}(|f - v_1| + |f - v_2|)$$

a.e. in X, thus $(f - v_1)(f - v_2) \geq 0$ a.e. in X. □

We shall employ (UI) in Section 2 in order to show unicity of a best approximant in the case of harmonic approximation to subharmonic functions. Section 3 will be devoted to the problem of existence of a best harmonic approximant; the proof is based on the compactness principle of harmonic functions.

2. UNIQUENESS OF BEST HARMONIC APPROXIMATION

Now we consider harmonic L^1-approximation to functions $f \in L^1(B)$, where $B := \{x \in \mathbb{R}^n \mid |x| < 1\}$ denotes the open unit ball. The vector space of approximating functions will be $V := H(B) \cap L^1(B)$, where $H(B)$ is the space of harmonic functions on B. If we denote $B_o := \{x \in \mathbb{R}^n \mid |x| < 1/\sqrt[n]{2}\}$, then the following result holds true:

<u>Let</u> $f \in L^1(B)$, <u>and</u> $h^* \in H(B) \cap L^1(B)$, <u>and suppose that</u>

(α) $f - h^* \leq 0$ <u>a.e. in</u> B_o,

(β) $f - h^* \geq 0$ <u>a.e. in</u> $B \setminus B_o$.

<u>Then</u> h^* <u>is a best</u> <u>L^1-approximant to</u> f <u>from</u> $H(B) \cap L^1(B)$.

Indeed, let $E := B_o \cap Z(f - h^*)$ and $F := (B \setminus B_o) \cap Z(f - h^*)$, where $Z(f - h^*)$ denotes the set of zeros of $f - h^*$ in B. Then

$$\alpha(t) := \begin{cases} -1 & \text{for } x \in E \\ 1 & \text{for } x \in F \end{cases}$$

is measurable on $Z(f - h^*)$, and $|\alpha(t)| \leq 1$. By the mean value equation for harmonic functions, we have for all elements $h \in H(B) \cap L^1(B)$

$$\int_{Z(f-h^*)} \alpha h \, d\mu + \int_{B \setminus Z(f-h^*)} h \operatorname{sign}(f-h^*) \, d\mu = \int_{B \setminus B_o} h \, d\mu - \int_{B_o} h \, d\mu = 0.$$

According to a well-known characterization theorem for best L^1-approximants (cf. Singer [11]), this implies that h^* is a best L^1-approximant to f from $H(B) \cap L^1(B)$. □

Without further assumptions, h^* will by no means be unique. Therefore, in the sequel we shall consider subharmonic functions

$f \in L^1(B)$ which satisfy the following conditions (see e. g. the book by Hayman-Kennedy [5]):

(a) $-\infty \leq f < \infty$ in B,

(b) f is upper semi-continuous in B,

(c) if x_o is any point of B, then there exist arbitrary small positive values of r such that

$$f(x_o) \leq \frac{1}{\lambda(\partial B(x_o,r))} \int_{\partial B(x_o,r)} f(x)\, d\sigma(x)$$

where $d\sigma(x)$ denotes the surface area on the sphere $\partial B(x_o,r)$ of radius r centred at x_o (*mean value inequality*).

The set of all subharmonic functions f on B will be denoted by $S(B)$.

Now we state our

THEOREM 1. *Let* $f \in S(B) \cap L^1(B)$, *and suppose that* f *is continuous in all points of* ∂B_o. *Furthermore, let the function* $h^* \in H(B) \cap L^1(B)$ *solve the Dirichlet problem*

(i) $h^*_{|\partial B_o} = f_{|\partial B_o}$,

and suppose that

(ii) $f - h^* \geq 0$ *a.e. on* $B \setminus B_o$, *and*

(iii) $\lambda(Z(f - h^*)) \leq \frac{1}{2}\lambda(B)$.

Then h^* *is the unique best* L^1*-approximant to* f *from the space* $H(B) \cap L^1(B)$.

Proof. Since (i) together with the subharmonicity of f yields $f - h^* \leq 0$ on B_o, this combined with (ii) implies that h^* is a best harmonic L^1-approximant (according to the above mentioned sufficient condition for a best L^1-approximant). So we only have to prove the unicity of h*.

Suppose, h^* and also $h' \in H(B) \cap L^1(B)$ are best approximants to f. We distinguish two cases:

(1) Let $f - h^* < 0$ in B_o. Then, by (UI), $f - h' \leq 0$ a.e. in B_o. The continuity of f in all points of ∂B_o implies $f - h' \leq 0$ on ∂B_o, hence, by the subharmonicity of $f - h'$, we can only have either

$$\alpha) \quad f - h' = 0 \quad \text{on } B_o, \text{ or}$$

$$\beta) \quad f - h' < 0 \quad \text{on } B_o.$$

In the first case, $h' = h^*$ on B_o, thus on B, by the harmonicity.

Thus let $f - h' < 0$ on B_o. Assume, there exists an $x_o \in \partial B_o$ such that $(f - h')(x_o) < 0$. Then, by continuity, there is a ball $B(x_o,r)$, such that $(f - h')|_{B(x_o,r)} < 0$. This yields

$$\lambda(\{x \in B \mid (f - h')(x) < 0\}) > \frac{1}{2}\lambda(B),$$

which contradicts the characterization theorem of Singer [11]. Also, $(f - h')(x_1) > 0$ for $x_1 \in \partial B_o$ is impossible, hence $f - h' = 0$. This implies $h' = h^*$ on B by the maximum principle and the harmonicity of h' and h^*.

(2) Now, let $f - h^* = 0$ in one point of B_o and thus in all of B_o, and, by continuity, on $\overline{B}_o$. By (iii) we have $f - h^* > 0$ a.e. on $B \setminus B_o$. Therefore, (UI) implies

$$f - h' \geq 0 \quad \text{a.e. on } B \setminus B_o,$$

and, in particular, using the continuity in ∂B_o,

$$f - h' \geq 0 \quad \text{on } \partial B_o.$$

This means $h^* \geq h'$ on ∂B_o. If we have equality, then the uniqueness is settled.

So we assume $h^* > h'$ for at least one point $x_2 \in \partial B_o$. Then, by the minimum principle,

$$h^* > h' \qquad \text{on} \quad B_o,$$

hence

$$f - h' > f - h^* = 0 \qquad \text{on} \quad B_o.$$

But we have $(f - h')(x_2) > 0$ with $x_2 \in \partial B_o$. By the continuity condition on ∂B_o, there exists an $s > 0$ such that for the ball $B(x_2,s)$ we have

$$(f - h')_{|B(x_2,s)} > 0.$$

This implies

$$\lambda(\{x \in B \mid (f - h')(x) > 0\}) > \frac{1}{2}\lambda(B),$$

which contradicts the characterization theorem for best L^1-approximants (see Singer [11]). Hence $h' = h^*$ on ∂B_o which implies $h' = h^*$ on B. □

REMARK. (i) The condition "f continuous in all points of ∂B_o" in Theorem 1 cannot be dropped. This can be seen by the function

$$f(x) = \begin{cases} 0 & \text{for} \quad 0 \le |x| < 1/\sqrt[n]{2}, \\ 1 + m(|x| - 1/\sqrt[n]{2}) & \text{for} \quad 1/\sqrt[n]{2} \le |x| < 1 \end{cases}$$

with a sufficiently large (positive) $m \in \mathbb{R}$ (such that f satisfies the mean value inequality in the points of ∂B_o). This function f possesses infinitely many best harmonic L^1-approximants, namely all constant functions $h_\alpha(x) = \alpha$ $(0 \le \alpha \le 1)$.

(ii) A special case of Theorem 1 was considered in Goldstein-Haussmann-Rogge [3], where f was supposed to be in $S(B) \cap C(\overline{B})$.

3. EXISTENCE OF BEST HARMONIC APPROXIMANTS

Now we prove the existence theorem for best harmonic L^1-approximants.

THEOREM 2. *Let* $D \subset \mathbb{R}^n$ *be an open set, and* $f : D \to \mathbb{R}$ *Lebesgue integrable. Then there exists a best* L^1*-approximant to* f *from* $H(D) \cap L^1(D)$, *the harmonic and integrable functions on* D.

This theorem is based on the following

LEMMA. *Let* $D \subset \mathbb{R}^n$ *be open, and* $h_n \in H(D)$ *be a locally uniformly bounded sequence. Then there exists an* $h_0 \in H(D)$ *and a subsequence* $(h_n)_{n \in \mathbb{N}_1}$ *with* $\mathbb{N}_1 \subset \mathbb{N}$ *such that*

$$h_n \to h_0 \qquad \text{for} \quad n \in \mathbb{N}_1,\ n \to \infty,$$

where the convergence is pointwise on D.

Proof. By assumption, for every $x \in D$ there exists a $\rho(x) > 0$ such that the ball $B(x,\rho(x)) \subset D$ and h_n, $n \in \mathbb{N}$, is uniformly bounded on $B(x,\rho(x))$. By the Lindelöf property of $\mathbb{R}^n$, there exists a sequence $(x_k)_{k \in \mathbb{N}}$ such that

$$D = \bigcup_{k \in \mathbb{N}} B(x_k,\rho(x_k)).$$

Now we apply the compactness principle for harmonic functions (see e. g. Wermer [12, p. 160]): There exists a subsequence $(h_{n,1})$ which converges pointwise on $B(x_1,\rho(x_1))$ to a harmonic function on $B(x_1,\rho(x_1))$. For $2 \le k \in \mathbb{N}$, let $(h_{n,k})$ be a subsequence of $(h_{n,k-1})$ which converges pointwise on

$$\bigcup_{\lambda=1}^{k} B(x_\lambda,\rho(x_\lambda))$$

to a harmonic function. The diagonal sequence $(h_{n,n})$ converges pointwise on $D = \cup_{k \in \mathbb{N}} B(x_k,\rho(x_k))$ to a function $h_0 \in H(D)$. □

Proof of Theorem 2. Let $\alpha := \inf\{\|f - h\|_1 \mid h \in H(D) \cap L^1(D)\}$. Then there exists a sequence $h_n \in H(D) \cap L^1(D)$ such that

$$\|f - h_n\|_1 \to \alpha \qquad \text{for} \quad n \to \infty .$$

We show that $(h_n)_{n\in\mathbb{N}}$ is locally uniformly bounded. Indeed, let $x \in D$ and $\rho > 0$ such that $B(x,\rho) \subset D$. Choose $y \in B(x,\frac{1}{2}\rho)$, then, by the mean value property

$$|h_n(y)| \le c_1 \cdot \int_{B(y,\frac{1}{2}\rho)} |h_n| \, d\lambda$$

$$\le c_1 \Big(\int_D |f| \, d\lambda + \int_D |h_n - f| \, d\lambda \Big) \le c_2$$

for $n \in \mathbb{N}$, where $c_1 = \dfrac{1}{\lambda(B(y,\frac{1}{2}\rho))}$.

Since the hypotheses of the Lemma are satisfied, there exists an $h_o \in H(D)$ and a subsequence $\mathbb{N}_1 \subset \mathbb{N}$, such that

$$h_n \to h_o \qquad \text{for} \quad n \in \mathbb{N}_1,\ n \to \infty,$$

where the convergence is pointwise on D. From this we have

$$|f - h_o| = \lim_{n\in\mathbb{N}_1} |f - h_n| ,$$

thus by Fatou's lemma

$$\int_D |f - h_o| \, d\lambda \le \liminf_{n\in\mathbb{N}_1} \int_D |f - h_n| \, d\lambda \le \alpha < \infty .$$

Hence $|f - h_o| \in L^1(D)$, thus $h_o \in L^1(D)$. By the definition of α, h_o is a best harmonic L^1-approximant to f from $H(D) \cap L^1(D)$. This concludes the proof. □

4. CONCLUDING REMARKS

In the previous papers [2,3] the problem of best harmonic L^1-approximation was considered in the case when the function f to be approximated was in $S(B) \cap C(\overline{B})$. We would also like to mention some papers which deal with harmonic approximation with respect to the sup-norm: Burchard [1], Hayman-Kershaw-Lyons [6] as well as [4].

REFERENCES

1. H.G. Burchard, Best uniform harmonic approximation. In: "Approximation Theory II" (G.G. Lorentz, C.K. Chui and L.L. Schumaker, Eds.), 309-314. Academic Press, New York and London, 1976.

2. M. Goldstein, W. Haussmann and K. Jetter, Best harmonic L^1 approximation to subharmonic functions. J. London Math. Soc. (2) 30 (1984), 257-264.

3. M. Goldstein, W. Haussmann and L. Rogge, On the mean value property of harmonic functions and best harmonic L^1-approximation. To appear.

4. W. Haussmann and K. Zeller, Best approximation by harmonic functions. In "Approximation Theory V" (C.K. Chui and L.L. Schumaker, Eds.). Academic Press, New York and London, to appear.

5. W.K. Hayman and P.B. Kennedy, Subharmonic Functions. Academic Press, London, 1976.

6. W.K. Hayman, D. Kershaw and T. J. Lyons, The best harmonic approximant to a continuous function. In: "Anniversary Volume on Approximation Theory and Functional Analysis" (P.L. Butzer, R.L. Stens and B.Sz.-Nagy, Eds.), Internat. Ser. Numer. Math. 65, 317-327. Birkhäuser-Verlag, Basel, Boston and Stuttgart, 1984.

7. M.G. Krein and N.I. Achieser, On some problems in the theory of moments. Charkov 1938; Transl. Math. Monographs, Vol. 2, Amer. Math. Soc., Providence, R. I., 1962.

8. B.R. Kripke and T.J. Rivlin, Approximation in the metric of $L^1(X,\mu)$. Trans. Amer. Math. Soc. 119 (1965), 101-122.

9. R.R. Phelps, Uniqueness of Hahn-Banach extensions and unique best approximation. Trans. Amer. Math. Soc. 95 (1960), 238-255.

10. J.R. Rice, The Approximation of Functions, Vol. I. Addison-Wesley, Reading, Mass., 1964.

11. I. Singer, Best Approximation in Normed Linear Spaces by Elements of Linear Subspaces. Springer-Verlag, Berlin, 1970.

12. J. Wermer, Potential Theory, Lecture Notes Math. 408, Springer-Verlag, Berlin, Heidelberg and New York, 1981.

W. Haussmann and L. Rogge, Department of Mathematics, University of Duisburg, Lotharstr. 65, 4100 Duisburg, West Germany.

International Series of
Numerical Mathematics, Vol. 80

ON THE STRUCTURE OF (s,t)-CONVEX FUNCTIONS

Norbert Kuhn

Abstract. A function $f:I\to[-\infty,\infty[$ ($I\subset\mathbb{R}$ interval) is said to be (s,t)-convex (for fixed $s,t\in]0,1[$) iff one has $f(su+(1-s)v)\leqq tf(u)+(1-t)f(v)$ for all $u,v\in I$. We prove that such a function is necessarily (t,t)-convex for all rational $t\in]0,1[$. Furthermore we show that a function which fulfills the (s,t)-convexity inequality in a weakened sense is closely related to a uniquely determined (s,t)-convex function.

1. INTRODUCTION

In the following let $I\subset\mathbb{R}$ be a non-trivial interval. For $s,t\in]0,1[$, a function $f:I\to[-\infty,\infty[$ is called (s,t)-convex iff one has $f(su+(1-s)v)\leqq tf(u)+(1-t)f(v)$ for all $u,v\in I$. And f is called t-convex iff f is (t,t)-convex. Furthermore we define $K'(f)$ to be the set of all pairs $(s,t)\in]0,1[^2$ such that f is (s,t)-convex and $K(f):=\{t\in]0,1[: f \text{ is } t\text{-convex}\}$. The structure of t-convex functions was examined in [3] (note that K(f) was defined there to contain 0 and 1).

In this paper we discuss the structure of (s,t)-convex functions. Our results are based on two different techniques. One is to reduce results on (s,t)-convex functions to results on "(s,t)-affine" functions via a certain theorem of Hahn-Banach-type. Using this we prove for example that the condition $K'(f)\neq\emptyset$ already implies that $K(f)\neq\emptyset$. The other technique was used by M.Kuczma [2] in a somewhat simpler situation. We examine functions which fulfill the (s,t)-convexity inequality in a weakened sense and we prove that such a function is closely related to a

This paper is in final form and no version of it will be submitted for publication elsewhere.

uniquely determined (s,t)-convex function.

2. THE STRUCTURE OF $K'(f)$

First we present a result of Hahn-Banach-type. It is a special case of Rodé's powerful Hahn-Banach-theorem [5], but its proof is much simpler than the proof of Rodé's result.

For $s,t\in]0,1[$, a function $f:I\to[-\infty,\infty[$ is called (s,t)-affine iff one has $f(su+(1-s)v)=tf(u)+(1-t)f(v)$ for all $u,v\in I$. Furthermore we define $A'(f)$ to be the set of all pairs $(s,t)\in]0,1[^2$ such that f is (s,t)-affine.

THEOREM 1. _For_ $f:I\to[-\infty,\infty[$ _and_ $a\in I$ _there exists_ $\varphi:I\to[-\infty,\infty[$ _with_ $\varphi\leqq f$, $\varphi(a)=f(a)$ _and_ $K'(f)\subset A'(\varphi)$.

Proof. If $f(a)=-\infty$ then the function $\varphi:=-\infty$ has the desired properties. Therefore we may assume $f(a)\neq-\infty$. Let M be the set of all functions $\Phi:I\to[-\infty,\infty[$ such that $\Phi\leqq f$, $\Phi(a)=f(a)$ and $K'(f)\subset K'(\Phi)$. Then $f\in M$. And it is easy to see that for each totally ordered subset $\emptyset\neq L\subset M$ the pointwise defined function $\inf\{\Phi\colon \Phi\in L\}$ belongs to M. Therefore Zorn's lemma implies the existence of a minimal $\varphi\in M$. We fix $(s,t)\in K'(f)$ and have to show that φ is (s,t)-concave, that means

$$\varphi(su+(1-s)v)\geqq t\varphi(u)+(1-t)\varphi(v)\quad \forall u,v\in I.$$

For $u\in[\varphi\neq-\infty]$ define $\varphi_u:I\to[-\infty,\infty[$ by $\varphi_u(x):=\varphi(su+(1-s)x)+\varphi(sx+(1-s)u)-\varphi(u)$ for all $x\in I$. Then an elementary calculation shows

(1) $K'(\varphi)\subset K'(\varphi_u)$ and (2) $\varphi_u\leqq\varphi$.

And we claim

(3) $\varphi_u(a)=f(a)$.

This is clear in the case $u=a$. Thus the minimality of φ combined with (1) and (2) implies that $\varphi=\varphi_a$. Now for arbitrary $u\in[\varphi\neq-\infty]$ it follows that

$$\varphi_u(a)=\varphi_a(u)+\varphi(a)-\varphi(u)=\varphi(a)=f(a).$$

This proves the assertion (3).

In order to prove the (s,t)-concavity we fix $u,v\in I$ and may assume that $\varphi(u),\varphi(v)\neq-\infty$. Again the minimality of φ combined with (1),(2) and (3) implies that $\varphi_u=\varphi$. It follows that

$$\begin{aligned}\varphi(u)+\varphi(v)&=\varphi(u)+\varphi_u(v)\\&=\varphi(su+(1-s)v)+\varphi(sv+(1-s)u)\\&\leqq\varphi(su+(1-s)v)+t\varphi(v)+(1-t)\varphi(u)\end{aligned}$$

and therefore

$$\varphi(su+(1-s)v)\geqq t\varphi(u)+(1-t)\varphi(v).$$

Thus the proof is finished.

Of course the above proof also works in even more general situations. For example: without changing anything one may replace I by a convex subset of a real vector space.

LEMMA 2. _For a function_ $f:I\to[-\infty,\infty[$ _we have_:

i) (s,t), $(\sigma,\tau)\in A'(f) \Rightarrow (s\sigma,t\tau)\in A'(f)$

ii) _If_ $A'(f)\neq\emptyset$ _and_ $f(a)=-\infty$ _for some_ $a\in I$ _then_ $f(x)\neq-\infty$ _in at most one point of_ I _which then must be a boundary point of_ I, _and consequently_ $A'(f)=]0,1[^2$.

iii) _Assume_ $(s,t),(\sigma,\tau)\in A'(f)$ _with_ $s<\sigma$ _and_ $t<\tau$. _Then_ $(s/\sigma,t/\tau)\in A'(f)$.

Proof. i) and ii): An easy calculation shows that i) is true. In particular we have $(0,0)\in\overline{A'(f)}$. From this it follows first that $f=-\infty$ on int I and then the full assertion.

iii) Let $(s,t),(\sigma,\tau)\in A'(f)$ with $s<\sigma$ and $t<\tau$. Because of ii) we may assume $f(x)\neq-\infty$ for all $x\in I$. For $u,v\in I$ we have the identity

$$\sigma\left(\frac{s}{\sigma}u+\left(1-\frac{s}{\sigma}\right)v\right)+(1-\sigma)v=su+(1-s)v$$

and hence

$$\tau f(\frac{s}{\sigma}u+(1-\frac{s}{\sigma})v)+(1-\tau)f(v)=tf(u)+(1-t)f(v).$$

It follows that

$$f(\frac{s}{\sigma}u+(1-\frac{s}{\sigma})v)=\frac{t}{\tau}f(u)+(1-\frac{t}{\tau})f(v).$$

Now we can prove the main result of this section.

THEOREM 3. <u>For a function</u> $f:I\to[-\infty,\infty[$ <u>we have</u>:

i) $(s,t),(\sigma,\tau)\in K'(f) \Rightarrow (s\sigma,t\tau)\in K'(f)$

ii) <u>If</u> $(s,t),(\sigma,\tau)\in K'(f)$ <u>with</u> $s<\sigma$ <u>and</u> $t<\tau$ <u>then</u> $(s/\sigma,t/\tau)$, $(\sigma-s,\tau-t)\in K'(f)$.

iii) <u>If</u> $(s,t),(\sigma,\tau)\in K'(f)$ <u>with</u> $s+\sigma,t+\tau<1$ <u>then</u> $(s+\sigma,t+\tau)\in K'(f)$.

iv) $K'(f)\neq\emptyset \Rightarrow t\in K(f)$ <u>for all rational</u> $t\in]0,1[$

<u>Proof.</u> i) is easy.

ii) Assume $(s,t),(\sigma,\tau)\in K'(f)$ with $s<\sigma$, $t<\tau$ and fix $u,v\in I$. From Theorem 1 we obtain $\varphi:I\to[-\infty,\infty[$ such that $\varphi\leqq f$, $\varphi(\frac{s}{\sigma}u+(1-\frac{s}{\sigma})v)=f(\frac{s}{\sigma}u+(1-\frac{s}{\sigma})v)$ and $K'(f)\subset A'(\varphi)$. And in view of Lemma 2 iii) we have $(s/\sigma,t/\tau)\in A'(\varphi)$. By combining the above it follows that

$$f(\frac{s}{\sigma}u+(1-\frac{s}{\sigma})v)\leqq\frac{t}{\tau}f(u)+(1-\frac{t}{\tau})f(v),$$

hence $(s/\sigma,t/\tau)\in K'(f)$. Furthermore

$$(\sigma-s,\tau-t)=(\sigma(1-\frac{s}{\sigma}),\tau(1-\frac{t}{\tau}))\in K'(f)$$

because of $(\sigma,\tau),(1-\frac{s}{\sigma},1-\frac{t}{\tau})\in K'(f)$ and i).

iii) From $s+\sigma$, $t+\tau<1$ we have $s<1-\sigma$, $t<1-\tau$ and therefore

$$(1-(s+\sigma),1-(t+\tau))=((1-\sigma)-s,(1-\tau)-t)\in K'(f),$$

hence $(s+\sigma,t+\tau)\in K'(f)$.

iv) Fix $(s,t)\in K'(f)$. In view of i) we may assume $s,t<\frac{1}{2}$. By iii) it follows that $(2s,2t)\in K'(f)$, and by ii) we obtain

$$(\frac{1}{2},\frac{1}{2})=(\frac{s}{2s},\frac{t}{2t})\in K'(f).$$

Thus we have $(t,t) \in K'(f)$ for all rational $t \in]0,1[$.

In [3] we proved that the condition $K(f) \neq \emptyset$ implies that $K(f) = [K(f)] \cap]0,1[$ where $[K(f)]$ denotes the subfield of the reals generated by $K(f)$. For the set $K'(f)$ we do not have such a plain result. By reducing to the affine case one can prove for a function $f: I \to [-\infty, \infty[$ which is non-constant on int I :

$$\left.\begin{array}{c}(s,t),(\sigma,\tau) \in K'(f)\\ s=\sigma \text{ or } t=\tau\end{array}\right\} \Rightarrow (s,t)=(\sigma,\tau)$$

In particular we have

$$s \in K(f) \Rightarrow (s,t) \notin K'(f) \quad \forall t \in]0,1[\setminus \{s\}.$$

Thus a function $f: I \to [-\infty, \infty[$ which is (s,t)-convex for some $s \neq t$ and s rational must be constant on int I . On the other hand the author does not know of any example of a function which is (s,t)-convex for some $s \neq t$ and non-constant on int I . And the following corollary in connection with the above property shows that such a function would be very irregular.

COROLLARY 4. Assume $f: I \to [-\infty, \infty[$ and $(s,t) \in K'(f)$ with $s \neq t$. Then $\overline{K'(f)} = [0,1]^2$.

Proof. First observe the rule

$$(s,t),(\sigma,\tau),(\gamma,\delta) \in K'(f) \Rightarrow (s\sigma+(1-s)\gamma, t\tau+(1-t)\delta) \in K'(f).$$

This combined with the fact $(1/2,1/2) \in K'(f)$ gives

$$\tfrac{1}{2}K'(f) + \tfrac{1}{2}K'(f) \subset K'(f)$$

and therefore $\overline{K'(f)} = \overline{\text{co}\, K'(f)}$. Furthermore we know $(0,0),(1,1) \in \overline{K'(f)}$. Consequently we have only to show that $(0,1),(1,0) \in \overline{K'(f)}$. Of course we may assume $s<t$. Define

$$k_n := \max\{k \in \mathbb{N} : kt^n < 1\} \quad \forall n \in \mathbb{N}.$$

From $k_n t^n < 1 \leq k_n t^n + t^n$ for all $n \in \mathbb{N}$ it follows that $k_n t^n \to 1$ and $k_n s^n = (\frac{s}{t})^n k_n t^n \to 0$. And by Theorem 3 we have $(k_n s^n, k_n t^n) \in K'(f)$ for all $n \in \mathbb{N}$. Thus we obtain $(0,1) \in \overline{K'(f)}$ and in view of this $(1,0) \in \overline{K'(f)}$, too.

3. A UNIQUENESS RESULT

First we prove that t-convex functions are continuous in a certain sense. This generalizes a known result of Bernstein and Doetsch [1].

THEOREM 5. *Let* $f: I \to \mathbb{R}$ *with* $K(f) \neq \emptyset$. *Assume* $a, b \in \operatorname{int} I$ *with* $a < b$. *Then* f *restricted to* $|a,b| := \{ta + (1-t)b :\ t \in K(f)\}$ *is bounded and Lipschitz-continuous*.

Proof. 1) For $t \in K(f)$ we have $f(ta+(1-t)b) \leq \max\{f(a), f(b)\}$. And from $\frac{1}{2}(ta+(1-t)b) + \frac{1}{2}((1-t)a+tb) = \frac{1}{2}a + \frac{1}{2}b$ combined with the fact $\frac{1}{2} \in K(f)$ it follows that

$$f(\tfrac{1}{2}a+\tfrac{1}{2}b) \leq \tfrac{1}{2}f(ta+(1-t)b) + \tfrac{1}{2}\max\{f(a), f(b)\}$$

and therefore

$$f(ta+(1-t)b) \geq 2f(\tfrac{1}{2}a+\tfrac{1}{2}b) - \max\{f(a), f(b)\}.$$

Thus f is bounded on $|a,b|$.

2) Using the "field properties" of $K(f)$ we obtain $c, d \in \operatorname{int} I$ such that $c<a$, $b<d$ and $|a,b| \subset |c,d|$. Let $\varepsilon := \min\{d-b, a-c\} > 0$. Then an obvious compactness argument shows that we have only to prove the following:

(1) Let $u, v \in |a,b|$ with $0 < v-u \leq \varepsilon$. Then f restricted to $|u,v|$ is Lipschitz-continuous.

Let us fix such points u, v. By 1) there exists $M>0$ with $|f| \leq M$ on $|c,d|$. We claim:

(2) $|f(su+(1-s)v) - f(tu+(1-t)v)| \leq 2M|s-t| \quad \forall s, t \in K(f)$

Fix $s, t \in K(f)$ and without loss of generality $s>t$. Let $x := su+(1-s)v$

and $y:=tu+(1-t)v$. Then $x=y-(s-t)(v-u)$. Furthermore

$$z:=y-(v-u)=u-t(v-u)\in|c,d|.$$

In fact, for $\alpha,\beta\in K(f)$ with $u=\alpha c+(1-\alpha)d$ and $v=\beta c+(1-\beta)d$ we have $z=(2\alpha-\beta)c+(1-(2\alpha-\beta))d$ and that is why $2\alpha-\beta\in]0,1[$ (since $c<z<d$). It follows that $2\alpha-\beta\in[K(f)]\cap]0,1[=K(f)$.
Now we obtain

$$\begin{aligned} f(x)-f(y) &= f(y-(s-t)(v-u))-f(y) \\ &= f((s-t)z+(1-(s-t))y)-f(y) \\ &\leqq (s-t)(f(z)-f(y)) \\ &\leqq 2M(s-t). \end{aligned}$$

And an analogous argument gives

$$f(y)-f(x)\leqq 2M(s-t).$$

Consequently the proof is finished.

The following results and proofs are motivated by those of M.Kuczma [2].

Denote by $\mathcal{I}$ the set of non-void open intervals $I\subset\mathbb{R}$. And let $\mathcal{M}$ be an arbitrary class of subsets of $\mathbb{R}$ with the following properties.

I	$\alpha>0$, $M\in\mathcal{M}$	$\Rightarrow$	$M\in\mathcal{M}$
II	$x\in\mathbb{R}$, $M\in\mathcal{M}$	$\Rightarrow$	$x+M\in\mathcal{M}$
III	$M_1,M_2,\cdots\in\mathcal{M}$	$\Rightarrow$	$\cup M_1\in\mathcal{M}$
IV	$I\in\mathcal{I}$, $M\in\mathcal{M}$	$\Rightarrow$	$I\setminus M\neq\emptyset$

EXAMPLES. i) $\mathcal{M}:=\{\emptyset\}$

ii) $\mathcal{M}:=\{M\subset R$: M has Lebesgue measure zero$\}$

iii) $\mathcal{M}:=\{M\subset R$: M is of the first category$\}$

For $I\in\mathcal{I}$, $x\in I$ and $0<t<1$ define

$$I(t,x):=\{u\in I:\ x\in tu+(1-t)I\}=I\cap(\tfrac{1}{t}x-\tfrac{1-t}{t}I).$$

Then $I(t,x)$ is an open interval with $x\in I(t,x)$.

THEOREM 6. Let I be an open interval and $f:I\to\mathbb{R}$ t-convex for some $t\in]0,1[$. Then

$$f(x)=\inf_{u\in I(t,x)\setminus M}\left(tf(u)+(1-t)f\left(\frac{x-tu}{1-t}\right)\right)\quad \forall x\in I, M\in\mathcal{M}.$$

Proof. Fix $x\in I$ and $M\in\mathcal{M}$. By the conditions I, II and III

$$N:=\bigcup_{l=0}^{\infty} t^{-l}(M-(1-t^{l})x)\in\mathcal{M}.$$

And M is a subset of N. Condition IV implies that

$$(I(t,x)\cap]-\infty,x[)\cap N\neq\emptyset.$$

Thus there exists $y\in I(t,x)\setminus N$ with $y<x$. We have

$$u_l:=(1-t^l)x+t^l y\in I(t,x)\setminus M\quad \forall l\in\mathbb{N},$$

furthermore $u_l\to x$ and $v_l:=\frac{1}{1-t}(x-tu_l)\to x$. We choose $z\in I(t,x)$ such that $x<z$ and $x\in|y,z|:=\{sy+(1-s)z:\ s\in K(f)\}$. Then $u_l,v_l\in|y,z|$ for almost all $l\in\mathbb{N}$. And Theorem 5 implies that

$$\begin{aligned} f(x)&=\lim(tf(u_l)+(1-t)f(v_l))\\ &\geqq \inf_{u\in I(t,x)\setminus M}\left(tf(u)+(1-t)f\left(\frac{x-tu}{1-t}\right)\right). \end{aligned}$$

Clearly the opposite is true, too. This proves the assertion.

COROLLARY 7. Let I be an open interval, $M\in\mathcal{M}$ and $f,g:I\to[-\infty,\infty[$ such that $g\leqq f$ on $I\setminus M$. Assume $K'(f),K'(g)\neq\emptyset$. Then $g\leqq f$ on I.

Proof. If $g(x)=-\infty$ for some $x\in I$ then $g=-\infty$. In the other case the assertion follows from Theorem 6 combined with Theorem 3.

4. WEAKENED (s,t)-CONVEXITY

Again let $\mathcal{I}$ denote the set of non-void open intervals $I\subset\mathbb{R}$. And let $\mathcal{M}$ be a class of subsets of $\mathbb{R}$ with the following properties:

I $\alpha>0,\ M\in\mathcal{M}\ \Rightarrow\ \alpha M\in\mathcal{M}$

II $x\in\mathbb{R},\ M\in\mathcal{M}\ \Rightarrow\ x+M\in\mathcal{M}$

III $M_1, M_2, \ldots \in \mathcal{M} \Rightarrow \bigcup M_1 \in \mathcal{M}$

IV $I \in \mathcal{I} \Rightarrow I \notin \mathcal{M}$

V $M \in \mathcal{M},\ S \subset M \Rightarrow S \in \mathcal{M}$

Note that IV and V imply the condition IV in section 3.

Furthermore let $\mathcal{N}$ be a class of subsets of $\mathbb{R}^2$ such that

VI $N_1, N_2 \in \mathcal{N} \Rightarrow N_1 \cup N_2 \in \mathcal{N}$

VII $I \in \mathcal{I} \Rightarrow I \times I \notin \mathcal{N}$

VIII $N \in \mathcal{N},\ P \subset N \Rightarrow P \in \mathcal{N}$

are fulfilled. Additionally the classes $\mathcal{M}$ and $\mathcal{N}$ have to be compatible in the following way:

IX $M \in \mathcal{M} \Rightarrow M \times \mathbb{R}, \mathbb{R} \times M \in \mathcal{N}$

X $M \in \mathcal{M},\ 0<t<1 \Rightarrow \{(x,y) \in \mathbb{R}^2 : tx+(1-t)y \in M\} \in \mathcal{N}$

XI For $a,b,c,d \in \mathbb{R}$ with $ad-bc \neq 0$ and $N \in \mathcal{N}$ there exists some $M \in \mathcal{M}$ such that $\{x \in \mathbb{R} : (ax+by, cx+dy) \in N\} \in \mathcal{M}$ for all $y \in \mathbb{R} \setminus M$.

EXAMPLES. i) $\mathcal{M} := \{\emptyset\}$ and $\mathcal{N} := \{\emptyset\}$

ii) $\mathcal{M} := \{M \subset \mathbb{R} : M$ has Lebesgue measure zero$\}$
$\mathcal{N} := \{N \subset \mathbb{R}^2 : N$ has Lebesgue measure zero$\}$

iii) $\mathcal{M} := \{M \subset \mathbb{R} : M$ is of the first category$\}$
$\mathcal{N} := \{N \subset \mathbb{R}^2 : N$ is of the first category$\}$

Of course Example i) is a trivial one. And in Examples ii) and iii) the conditions I-IX are well-known. Furthermore condition X is easy to verify (in the case of Example iii) see Theorem 15.4 in [4]). For the proof of condition XI note that in view of $ad-bc \neq 0$ the linear mapping $\mathbb{R}^2 \to \mathbb{R}^2, (x,y) \to (ax+by, cx+dy)$ is invertible. Thus XI follows from the Fubini-theorem or in case of Example iii) from the Kuratowski-Ulam-theorem (see [4]).

Let $f: I \to [-\infty, \infty[$ and $s,t \in]0,1[$. We say f is (s,t)-convex mod $\mathcal{N}$ iff there exists some $N \in \mathcal{N}$ such that

$$f(sx+(1-s)y)\leq tf(x)+(1-t)f(y) \quad \forall (x,y)\in(I\times I)\setminus N.$$

And our purpose is to prove that for such a function f there exist some $M\in\mathcal{M}$ and a uniquely determined (s,t)-convex function $g:I\to[-\infty,\infty[$ such that g=f on $I\setminus M$. In the case of Example ii) and for $s=t=\frac{1}{2}$ this was proven by M.Kuczma [2].

In the following let I be an open interval. For $x\in I$ and $0<t<1$ let $I(t,x)$ defined as in section 3.

THEOREM 8. *Let* $f:I\to[-\infty,\infty[$ *be* (s,t)-*convex mod* $\mathcal{N}$ *for some pair* $(s,t)\in]0,1[^2$. *Then there exists some* $M'\in\mathcal{M}$ *such that*

$$f(x)=\sup_{M\in\mathcal{M}}\ \inf_{u\in I(s,x)\setminus M}\left(tf(u)+(1-t)f\left(\frac{x-su}{1-s}\right)\right) \quad \forall x\in I\setminus M'.$$

Proof. Let $P:=\{(x,y)\in I^2:\ f(sx+(1-s)y)>tf(x)+(1-t)f(y)\}$. Then $P\in\mathcal{N}$ by condition VIII. Therefore the conditions V and XI imply the existence of sets $M_1,M_2\in\mathcal{M}$ such that

$$S_x:=\{u\in I(s,x):\ (u,x)\in P\}\in\mathcal{M} \qquad \forall x\in I\setminus M_1,$$
$$T_x:=\{u\in I(s,x):\ (u,\tfrac{x-su}{1-s})\in P\}\in\mathcal{M} \ \forall x\in I\setminus M_2.$$

It follows that $M':=M_1\cup M_2\in\mathcal{M}$ and $S_x,T_x\in\mathcal{M}$ for all $x\in I\setminus M'$. Fix $x\in I\setminus M'$. For arbitrary $u\in I(s,x)\setminus T_x$ we have $u\notin T_x$, hence $(u,\frac{x-su}{1-s})\notin P$ and therefore

$$f(x)=f\left(su+(1-s)\frac{x-su}{1-s}\right)\leq tf(u)+(1-t)f\left(\frac{x-su}{1-s}\right).$$

Consequently we have

$$f(x)\leq \inf_{u\in I(s,x)\setminus T_x}\left(tf(u)+(1-t)f\left(\frac{x-su}{1-s}\right)\right),$$

and because of $T_x\in\mathcal{M}$ this is

$$\leq\sup_{M\in\mathcal{M}}\ \inf_{u\in I(s,x)\setminus M}\left(tf(u)+(1-t)f\left(\frac{x-su}{1-s}\right)\right).$$

In order to prove the other direction fix $M\in\mathcal{M}$. Then

$$N:=\bigcup_{l=0}^{\infty} s^{-l}((S_x\cup M)-(1-s^l)x)\in \mathcal{M}$$

by I,II and III. Choose $u_0\in I(s,x)\setminus N$ such that $v_0:=\frac{1}{1-s}(x-su_0)\in I(s,x)$, too. This is possible in view of IV, V and the fact that $v_0\to x$ if $u_0\to x$. Define $u_l:=(1-s^l)x+s^l u_0\in I(s,x)\setminus(S_x\cup M)$ and $v_l:=(1-s^l)x+s^l v_0\in I(s,x)$ ($l=1,2,\ldots$). Because of $u_{l+1}=(1-s)x+su_l$ and $u_l\notin S_x$ we have $f(u_{l+1})\leq(1-t)f(x)+tf(u_l)$. In the case of $f(x)=-\infty$ it follows that $f(u_l)=-\infty$ for all $l\in\mathbb{N}$ and therefore $\limsup f(u_l)\leq f(x)$. And in the case of $f(x)\in\mathbb{R}$ we obtain $f(u_{l+1})-f(x)\leq t(f(u_l)-f(x))$ for all $l\in\mathbb{N}$ and therefore $\limsup f(u_l)\leq f(x)$, too. Analogously one shows $\limsup f(v_l)\leq f(x)$. In view of $x=su_l+(1-s)v_l$ for all $l\in\mathbb{N}$ it follows that $\limsup(tf(u_l)+(1-t)f(v_l))\leq f(x)$. Thus

$$\inf_{u\in I(s,x)\setminus M}\left(tf(u)+(1-t)f\left(\frac{x-su}{1-s}\right)\right)\leq f(x).$$

This proves the assertion.

THEOREM 9. <u>Assume that</u> $f:I\to[-\infty,\infty[$ <u>is (s,t)-convex mod</u> $\mathcal{N}$ <u>for some pair</u> $(s,t)\in]0,1[^2$ <u>and that</u>

$$(*)\qquad f(x)=\sup_{M\in\mathcal{M}}\ \inf_{u\in I(s,x)\setminus M}\left(tf(u)+(1-t)f\left(\frac{x-su}{1-s}\right)\right)\quad \forall x\in I.$$

<u>Then</u> f <u>is (s,t)-convex</u>.

<u>Proof.</u> 1) Because of (*) for each $x\in I$ there exist sets $M_1(x),M_2(x),\ldots\in\mathcal{M}$ such that

$$f(x)\leq tf(u)+(1-t)f\left(\frac{x-su}{1-s}\right)+\frac{1}{l}\quad \forall u\in I(s,x)\setminus M_l(x)\quad (l=1,2,\ldots).$$

For $M(x):=\bigcup_{l=1}^{\infty} M_l(x)\in\mathcal{M}$ it follows that

$$f(x)\leq tf(u)+(1-t)f\left(\frac{x-su}{1-s}\right)\quad \forall u\in I(s,x)\setminus M(x).$$

2) Since f is (s,t)-convex mod $\mathcal{N}$ condition VIII implies that

$$P:=\{(x,y)\in I^2:\ f(sx+(1-s)y)>tf(x)+(1-t)f(y)\}\in\mathcal{N}.$$

Therefore by V and XI there exists $M\in\mathcal{M}$ such that

$$S_x := \{y\in I:\ (x,y)\in P\}\in\mathcal{M}\quad \forall y\in I\setminus M.$$

3) Define $M(\lambda,x):=\{u\in I(s,x):\ tf(u)+(1-t)f(\frac{x-su}{1-s})<\lambda\}$ $\forall x\in I, \lambda\in\mathbb{R}$. Then

$$M(\lambda,x)\notin\mathcal{M}\quad \forall x\in I \text{ and } \lambda\in\mathbb{R} \text{ with } f(x)<\lambda.$$

In fact, for such x and λ the relation $M(\lambda,x)\in\mathcal{M}$ combined with (*) would imply that

$$f(x)\geq \inf_{u\in I(s,x)\setminus M(\lambda,x)} (tf(u)+(1-t)f(\frac{x-su}{1-s}))\geq\lambda>f(x).$$

4) Fix $x,y\in I$ and $\lambda,\mu\in\mathbb{R}$ such that $f(x)<\lambda$ and $f(y)<\mu$. Note $M\cup(\frac{1}{s}x-\frac{1-s}{s}M)\in\mathcal{M}$. In view of $M(\lambda,x)\notin\mathcal{M}$ and condition V there exists $u\in M(\lambda,x)\setminus(M\cup(\frac{1}{s}x-\frac{1-s}{s}M))$. We have $S_u, S_w\in\mathcal{M}$ where $w:=\frac{x-su}{1-s}$. Note

$$N:=S_u\cup(\frac{1}{s}y-\frac{1-s}{s}S_w)\cup(\frac{1}{1-s}M(sx+(1-s)y)-\frac{1}{1-s}u)\in\mathcal{M}$$

and $M(\mu,y)\notin\mathcal{M}$. Again by condition V there exists $v\in M(\mu,y)\setminus N$. From $u\in I\setminus M$ and $v\notin S_u$ it follows that $(u,v)\notin P$. Analogously we obtain $(\frac{x-su}{1-s},\frac{y-sv}{1-s})\notin P$. Thus

$$f(su+(1-s)v)\leq tf(u)+(1-t)f(v),$$

$$f(s\frac{x-su}{1-s}+(1-s)\frac{y-sv}{1-s})\leq tf(\frac{x-su}{1-s})+(1-t)f(\frac{y-sv}{1-s}).$$

From $v\notin\frac{1}{1-s}M(sx+(1-s)y)-\frac{s}{1-s}u$ it follows that $su+(1-s)v\notin M(sx+(1-s)y)$. On the other hand we have

$$su+(1-s)v\in sI(s,x)+(1-s)I(s,y)\subset I(s,sx+(1-s)y).$$

Therefore it follows that

$$f(sx+(1-s)y)\leq tf(su+(1-s)v)+(1-t)f(\frac{(sx+(1-s)y)-s(su+(1-s)v)}{1-s})$$

$$=tf(su+(1-s)v)+(1-t)f(s\frac{x-su}{1-s}+(1-s)\frac{y-sv}{1-s}).$$

Now we obtain

$$f(sx+(1-s)y)\leq t(tf(u)+(1-t)f(v))+(1-t)(tf(\frac{x-su}{1-s})+(1-t)f(\frac{y-sv}{1-s}))$$

$$=t(tf(u)+(1-t)f(\frac{x-su}{1-s}))+(1-t)(tf(v)+(1-t)f(\frac{y-sv}{1-s}))$$

$$<t\lambda+(1-t)\mu.$$

Thus the proof is finished.

THEOREM 10. *Let* $f,g:I\to[-\infty,\infty[$ *and* $M\in\mathcal{M}$ *such that* $f=g$ *on* $I\setminus M$. *Assume that* f *is* (s,t)*-convex mod* $\mathcal{N}$ *for some pair* $(s,t)\in]0,1[^2$. *Then* g *is* (s,t)*-convex mod* $\mathcal{N}$, *too*.

Proof. The conditions VI, VIII and IX imply that $N_0:=(I\times M)\cup(M\times I)\in\mathcal{N}$. Furthermore $N_1:=\{(x,y)\in I^2: sx+(1-s)y\in M\}\in\mathcal{N}$ by VIII and X. And by assumption there exists some $N_2\in\mathcal{N}$ such that

$$f(sx+(1-s)y)\leq tf(x)+(1-t)f(y)\quad\forall(x,y)\in I^2\setminus N_2.$$

By VI we have $N:=N_0\cup N_1\cup N_2\in\mathcal{N}$. It follows that

$$g(sx+(1-s)y)=f(sx+(1-s)y)$$
$$\leq tf(x)+(1-t)f(y)$$
$$=tg(x)+(1-t)g(y)$$

for all $(x,y)\in I^2\setminus N$. Thus g is (s,t)-convex mod $\mathcal{N}$.

For a function $f:I\to[-\infty,\infty[$ define $K'_{\mathcal{N}}(f):=\{(s,t)\in]0,1[^2: f$ is (s,t)-convex mod $\mathcal{N}\}$. Now we can state the main result of this section. It reads as follows.

THEOREM 11. *Let* $f:I\to[-\infty,\infty[$ *with* $K'_{\mathcal{N}}(f)\neq\emptyset$. *Then exist* $M'\in\mathcal{M}$ *and a uniquely determined function* $g:I\to[-\infty,\infty[$ *such that* $g=f$ *on* $I\setminus M'$ *and* $K'(g)=K'_{\mathcal{N}}(f)$.

Proof. 1) Fix $(s,t)\in K'_{\mathcal{N}}(f)$ and define

$$g(x):=\sup_{M\in\mathcal{M}}\ \inf_{u\in I(s,x)\setminus M}\left(tf(u)+(1-t)f(\frac{x-su}{1-s})\right)\quad\forall x\in I.$$

An easy argumentation shows $g(x)<\infty$ for all $x\in I$. Thus g defines a function $I\to[-\infty,\infty[$. By Theorem 8 there exists $M'\in\mathcal{M}$ such that g=f on $I\setminus M'$. It follows that

$$g(x)=\sup_{M\in\mathcal{M}}\ \inf_{u\in I(s,x)\setminus M}\left(tg(u)+(1-t)g\left(\frac{x-su}{1-s}\right)\right)\quad\forall x\in I.$$

Furthermore g is (s,t)-convex mod $\mathcal{N}$ by Theorem 10. Thus Theorem 9 implies that g is (s,t)-convex.

2) Fix $(\sigma,\tau)\in K'(g)$. In view of g=f on $I\setminus M'$ Theorem 10 implies that $(\sigma,\tau)\in K'_{\mathcal{N}}(f)$.

3) Fix $(\sigma,\tau)\in K'_{\mathcal{N}}(f)$. By 1) we obtain some $M''\in\mathcal{M}$ and some (σ,τ)-convex function $h:I\to[-\infty,\infty[$ such that h=f on $I\setminus M''$. It follows that h=g on $I\setminus(M'\cup M'')$. In view of $M'\cup M''\in\mathcal{M}$ Corollary 7 implies that h=g on I. Thus $(\sigma,\tau)\in K'(g)$. And at the same time we have proved that g is uniquely determined by the condition g=f on $I\setminus M'$.

REFERENCES

1. F.Bernstein, G.Doetsch, Zur Theorie der konvexen Funktionen. Math.Ann. 76(1915), 514-526.

2. M.Kuczma, Almost convex functions. Colloq.Math. 21(1970), 279-284.

3. N.Kuhn, A note on t-convex functions. General Inequalities 4 (Oberwolfach 1983), 269-276, Internat.Ser.Numer.Math. 71, Birkhäuser, Basel Stuttgart, 1984.

4. J.C.Oxtoby, Maß und Kategory. Springer, Berlin-Heidelberg-New York, 1971.

5. G.Rodé, Eine abstrakte Version des Satzes von Hahn-Banach. Arch.Math. 31(1978), 474-481.

Norbert Kuhn, Fachbereich Mathematik, Universität des Saarlandes, D-6600 Saarbrücken, Fed.Rep.Germany

Added in Proof. Recently König has given a simple proof of the full Rodé result and even of a more general result (to appear in Aequationes Math.).

International Series of
Numerical Mathematics, Vol. 80

MOMENTS OF CONVEX AND MONOTONE FUNCTIONS

Peter Schöpf

Abstract. If $f:C \to R$ is a nonnegative, monotone decreasing function from the n-dimensional cube $C := [0,1]^n$ into the real numbers with $\int_C f > 0$, then one can find two convex functions $g_1 \leq f \leq g_2$ such that for $k > 0$ the following inequalities hold.

$$\prod_{i=2}^{n+1} \left(\frac{k+i}{i}\right)^{1/k} M_k(g_1) \geq M_k(f) \geq (n+1)^{-(k+n+1)/k} M_k(g_2) ,$$

where $M_k(f) := \left(\frac{1}{k+1} \frac{\int_C f^{k+1}}{\int_C f}\right)^{1/k}$ is called the k-moment of f.

The factor at $M_k(g_1)$ is best possible.

INTRODUCTION

In [1] one can find the following theorem.

THEOREM. Let $f:[0,1] \to R$ be nonnegative and monotone. Then there exist two convex functions $g_1, g_2: [0,1] \to R$, $g_1 \leq f \leq g_2$ with $\left(\frac{k+2}{2}\right)^{1/k} M_k(g_1) \geq M_k(f) \geq 2^{-(k+2)/k} M_k(g_2)$. Furthermore, the constant $((k+2)/2)^{1/k}$ is best possible and the convex function g_2 can be chosen to be independent of k. [1])

---

[1]) The constant at $M_k(g_2)$ is not printed correctly in [1], p.136.

This paper is in final form and no version of it will be submitted for publication elsewhere.

In this paper we present a generalisation of this theorem to the n-dimensional space. Our main theorem is the following.

THEOREM. *Let* $C := [0,1]^n$, $f: C \to R$ *nonnegative and monotone decreasing in the following sense.* $f(x_1+h_1,\dots,x_n+h_n) \le \le f(x_1,\dots,x_n)$ *for every vector* $(h_1,\dots,h_n)$ *with all* $h_i \ge 0$. *Let further* $\int_C f > 0$ *and* $k > 0$. *Then there exist two convex functions* $g_1, g_2: C \to R$ *with* $0 \le g_1 \le f \le g_2$ *and*

$$(1) \quad \prod_{i=2}^{n+1} \left(\frac{k+i}{i}\right)^{1/k} M_k(g_1) \ge M_k(f) \ge (n+1)^{-(k+n+1)/k} M_k(g_2).$$

Furthermore, the factor at $M_k(g_1)$ *is best possible and the convex function* g_2 *can be chosen to be independent of* k.

Proof. Part 1: Construction of a convex, piecewise linear minorant function for f.
The monotonicity of f on the compact cube C and $k > 0$ imply monotonicity and boundedness for f and f^{k+1}. Thus f and f^{k+1} are Lebesgue integrable. Now let $x := (x_1,\dots,x_n)$, $|\mathbf{x}| := = (\sum_{i=1}^{n} x_i^2)^{1/2}$ and $h(\delta) := \inf \{f(x) \mid |\mathbf{x}| < \delta,\ x \in C\}$ for $\delta > 0$.
Then $H := \lim_{\delta \to 0} h(\delta)$ exists because h is monotone and $0 \le h(\delta) \le \le f(0)$. f monotone decreasing and $\int_C f > 0$ implies $H > 0$.
Case 1: There exists a point $a = (a_1,\dots,a_n) \in \overset{o}{C}$ (= interior of the cube) with $f(a) = H$. Then

$$g_{a,H}(x) := \max \{0, H(1 - \sum_{i=1}^{n} x_i/a_i)\}$$

is a convex, piecewise linear function with $g_{a,H}(x) \le f(x)$ for all $x \in C$. *Case 2:* For every $x \in \overset{o}{C}$ we have $f(x) < H$. Thus for every $b \in]0,H]$ we can find a $\delta > 0$ with $h(\delta) > b$. For given b we always can choose an $a \in \overset{o}{C}$ with $|a| < \delta$, and then we define

$$g_{a,b}(x) := \max \{0,\ b(1 - \sum_{i=1}^{n} x_i/a_i)\}.$$

If every $x_i < a_i$, then $f(x) \geq f(a) \geq h(\delta) > b$ and therefore $g_{a,b}(x) = b(1 - \sum_{i=1}^{n} x_i/a_i) < h(\delta) \leq f(x)$. If one of the $x_i \geq a_i$, then we have $g_{a,b}(x) = 0 \leq f(x)$. This implies $g_{a,b} \leq f$. The k-moment of $g_{a,b}$ $(0 < a_i < 1,\ 0 < b)$ is obtained after an easy calculation as follows

$$(2) \qquad M_k(g_{a,b}) = b\left(\prod_{i=1}^{n+1} \frac{i}{k+i}\right)^{1/k}$$

and does not depend on a.

Part 2: Choice of a,b in such a way that $g_1 := g_{a,b}$ fulfills the left inequality in (1). For every $x \in \overset{o}{C}$ we have $f(x) \leq H$ and therefore $\int_C f^{k+1} \leq \int_C H^k f = H^k \int_C f$ or with the definition of the k-moment $M_k(f) \leq (k+1)^{-1/k} H$. Case 1: $M_k(f) = (k+1)^{-1/k} H$, i.e. $f(a) = H$ for some $a \in \overset{o}{C}$. In this situation we choose the minorant $g_{a,H}$ as described in case 1 of part 1. Case 2: $M_k(f) < (k+1)^{-1/k} H$. We take $b \in]0,H[$ such that $M_k(f) < (k+1)^{-1/k} b$, and then we choose a minorant $g_{a,b}$ as in case 2 of part 1. With the aid of (2) it is easily checked that in both cases the left inequality in (1) is fulfilled.

Part 3: The constant factor at $M_k(g_1)$ is best possible.

REMARK. The nonnegative, monotone decreasing functions

$$f(x) := \begin{cases} 1 & \text{for } 0 \leq x \leq 1/2 \\ 0 & \text{for } 1/2 < x \leq 1 \end{cases}, \quad g(x) := \begin{cases} 1 & \text{for } 0 \leq x \leq 1/2 \\ 2 - 2x & \text{for } 1/2 < x \leq 1 \end{cases}$$

show that we can have $f \leq g$ and $M_k(f) > M_k(g)$. But now we will prove that for every convex minorant function g of a function $g_{a,b}$ we always get $M_k(g) \leq M_k(g_{a,b})$. For this purpose we need the following generalisation of the geometric lemma of [2].

LEMMA 1. Let $F,G:C \to R$ monotone decreasing functions (according to our definition) and let $A \cup B = C$ with $A \cap B = \emptyset$. Assume further that there is a real number s with

(i) $s \leq F(x) \leq G(x)$ for $x \in A$

(ii) $G(x) \leq F(x) \leq s$ for $x \in B$

(iii) $\int_C F = \int_C G$.

This implies for every $k \geq 0$ the validity of $\int_C F^{k+1} \leq \int_C G^{k+1}$.

Proof. For every $x \in A$ the set $\{x' \mid x' \leq x\}$ is contained in A because F is monotone decreasing. This implies that A,B are Lebesgue measurable sets and that the integrals

$\int_A \int_{F(x)}^{G(x)} y^k \, dy \, dx$, $\int_B \int_{G(x)}^{F(x)} y^k \, dy \, dx$ exist. The remainder of the proof is the same as in [3].

We also need the following "invariance" lemma.

LEMMA 2. Let $Q := \{y \in C \mid |y| = 1\}$, $u:Q \to R$ a positive, continuous function with $u(y) \leq 1$ and $b > 0$. Then the function

$$h(x) := \begin{cases} \max \{0, b(1 - |x|/u(\frac{x}{|x|}))\} & \text{for } x \in C \setminus \{0\} \\ b & \text{for } x = 0 \end{cases}$$

is continuous and bounded with $M_k(h) = b\left(\prod_{i=1}^{n+1} \frac{i}{k+i}\right)^{1/k}$ and $M_k(h)$ is independent of u.

Proof. We have

$\int_C h^{k+1}(x) \, dx = \int_Q \int_0^{u(y)} h^{k+1}(ry) \, r^{n-1} \, dr \, dy$, Where dy is the areal element on the unit sphere. Substitution for h on the right side and n-fold integration by parts yields

$b^{k+1} \prod_{i=2}^{n+1} \left(\frac{i}{k+i}\right) \cdot \frac{1}{n+1} \int_Q \frac{u^n(y)}{n} \, dy$. With the definition of the k-moment we immediatedly get from this

$$M_k(h) = b\left(\prod_{i=1}^{n+1} \frac{i}{k+i}\right)^{1/k} .$$

Let us now continue with the proof of part 3.
We consider the function $f:C \to R$ with

$$f(x) := \begin{cases} 1 & \text{if every } x_i \neq 1 \\ 0 & \text{if some } x_i = 1 , \end{cases}$$

which is monotone decreasing (according to our definition) and we claim that the convex function $g_1 := g_{a,b}$ with $a_1 = \ldots = a_n = 1$, $b = 1$ has the greatest possible k-moment, namely

$$M_k(g_1) = \prod_{i=2}^{n+1} \left(\frac{k+i}{i}\right)^{-1/k} M_k(f) \quad (\text{with } M_k(f) = (k+1)^{-1/k})$$

under all nonnegative, convex minorant functions of f. Let g be a further nonnegative, convex minorant function of f. Then it is easy to see that $0 \leq g \leq g_1$, i.e. g_1 is the greatest convex minorant function of f. g is bounded and convex on C and therefore we can modify g on the boundary of C such that the new g is continuous and convex on C. In order to have a welldefined k-moment, we have to assume $\int_C g > 0$. This implies $S := \max g(C) = g(0) > 0$ and for every $y \in Q$ the convex function $\lambda \mapsto g(\lambda y)$ is strictly monotone decreasing from $S = g(0)$ down to zero and then identical with zero. Now we can define the following function

$$u(s,y) := \inf \{\lambda \mid g(\lambda y) = s\}$$

from $[0,S] \times Q$ into R. It is easily proven from the convexity of **g and** $0 \leq g \leq g_1$ that u is continuous and fulfills $g(u(s,y)y) = s$. (Special $u(S,y) = 0$). With this u we introduce for every $s \in [0,S[$ the cone-functions

$$h_s(x) := \begin{cases} \max \{0, S - (S-s)|x|/u(s,x/|x|)\} & \text{for } x \in C \setminus \{0\} \\ S & \text{for } x = 0 , \end{cases}$$

which are continuous on C and therefore integrable with

$$\int_C h_s(x)\, dx = \frac{1}{(n+1)n} \int_Q \left[\frac{Su(s,y)}{S - s}\right]^n dy.$$

We will show now that $s \mapsto \int_C h_s$ (with $S \mapsto \lim_{s\to S} \int_C h_s$) is continuous on [0,S]. From the continuity of u on the compact cube C we get
$$\lim_{s\to s'} \int_C h_s = \frac{1}{(n+1)n} \int_Q \frac{Su(s',y)}{S-s'}^n dy = \int_C h_{s'} \text{ for every } s' \neq S.$$
The following inequalities for $s' \geq s$
$$S = g(0) \geq g(u(s',y)y) \geq g(u(s,y)\mathbf{y})$$
and the convexity of g show that
$$s' \leq S - (S-s)\frac{u(s',y)}{u(s,y)}$$
and this implies
$$\frac{u(s',y)}{S-s'} \leq \frac{u(s,y)}{S-s} .$$
Therefore $\lim_{s\to S} \int_C h_s$ **exists** because $s \mapsto \int_C h_s$ is monotone decreasing and ≥ 0. The function $s \mapsto \int_C h_s$ attains every value between its maximum and minimum. Because of the definition of h_s we have $g \leq h_0$ and therefore $\int_C g \leq \int_C h_0$. The function u is continuous in s and uniformly with respect to y. Therefore we can find for every $\varepsilon > 0$ a $\delta > 0$ such that $S - s < \delta$ implies $|u(s,y)-u(0,y)| = |u(s,y)| < \varepsilon$. From the definition of h_s we conclude that $h_s(\mathbf{x}) \leq S$ for $|x| \leq \varepsilon$ and $h_s(x) \leq g(x)$ for $|x| \geq \varepsilon$. This leads for $S - s < \delta$ to $\int_C h_s \leq \int_{|x|\leq\varepsilon} S + \int_{|x|>\varepsilon} g \leq \varepsilon^n S + \int_C g$ and in the sequel to $\lim_{s\to S} \int_C h_s \leq \int_C g$. This implies the existence of a number $s \in [0,S]$ with $\int_C h_s = \int_C g$. The functions g and h_s satisfy the requirements of lemma 1 and therefore we get $\int_C h_s^{k+1} \geq \int_C g^{k+1}$, which gives us (with the use of lemma 2) the desired inequality
$$M_k(g) \leq M_k(h_s) = S\left(\prod_{i=1}^{n+1} \frac{i}{k+i}\right)^{1/k} \leq \left(\prod_{i=1}^{n+1} \frac{i}{k+i}\right)^{1/k} = M_k(g_1) .$$

Part 4: Proof of the right inequality in (1).
If $P := \{t \in R \mid t \geq 0\}$, we define for every pair (a,b) with $a = (a_1,\ldots,a_n) \in \overset{\circ}{P}{}^n \cap C$ and $0 < b < f(a)$ the functions

$$g_{a,b}(x) := \max \{0,\ b(1 - \sum_{i=1}^{n} x_i/a_i)\} \text{ for } x \in P^n$$

$$G_{a,b}(x) := \begin{cases} (n+1)g_{a,b}(\frac{x}{n+1}) & \text{for } x \in \overset{\circ}{P}{}^n \\ (n+1)f(0) & \text{for } x \in P^n \setminus \overset{\circ}{P}{}^n. \end{cases}$$

The existence of such pairs (a,b) was proven in part 1. Since $g_{a,b}$ and $G_{a,b}$ are convex, **we have the same** for the following functions

$$g(x) := \sup \{g_{a,b}(x) \mid a \in \overset{\circ}{P}{}^n \cap C,\ 0 < b \leq f(a)\}$$

$$G(x) := \sup \{G_{a,b}(x) \mid a \in \overset{\circ}{P}{}^n \cap C,\ 0 < b \leq f(a)\}.$$

We see that $0 \leq g \leq f \leq G$ on C. Because integration over C respectively P^n ignores the values of the functions on the corresponding boundaries we have

$$\int_{P^n} G^{k+1} = \int_{\overset{\circ}{P}{}^n} G^{k+1} = \int_{\overset{\circ}{P}{}^n} (n+1)^{k+1}\ g^{k+1}(\tfrac{x}{n+1})\ dx = (n+1)^{n+k+1} \int_{P^n} g^{k+1}$$

and we can derive the right side of inequality (1) with $g_2 := G|_C$ in the same way as in [1].

$$(M_k(g_2))^k \int_C g_2 = (k+1)^{-1} \int_C g_2^{k+1} \leq (k+1)^{-1} \int_{P^n} G^{k+1} =$$

$$= (n+1)^{n+k+1} (k+1)^{-1} \int_{P^n} g^{k+1} \leq (n+1)^{n+k+1} (k+1)^{-1} \int_C f^{k+1} =$$

$$= (n+1)^{n+k+1} (M_k(f))^k \int_C f \leq (n+1)^{n+k+1} (M_k(f))^k \int_C g_2.$$

REFERENCES

1. T. Nishiura and F. Schnitzer, Moments of convex and monotone functions. Monatsh. Math. 76 (1972), 135-137.

2. T. Nishiura and F. Schnitzer, A proof of an inequality of H. Thunsdorff. Publications de la faculte de electrotechnique de l universite a Belgrade, Serie: Mathematiques et physique, No.357-No.380 (1971), 1-2.

3. F. Schnitzer and P. Schöpf, Verschärfung der Integralungleichung für das Potenzmittel von Funktionen mit sternförmigem Epigraphen. Arch. Math. 41 (1983), 459-463.

Peter Schöpf, Institut für Mathematik, Universität Graz
A - 8010 Graz, Hans-Sachs-Gasse 3, Österreich.

International Series of
Numerical Mathematics, Vol. 80

AN EVEN ORDER SEARCH PROBLEM

Roger J. Wallace

Abstract. How might simple real zeros of real valued continuous k-th derivatives $g^{(k)}$ be efficiently estimated, by using only values of g and points in dom(g)? A standard approach to these questions entails successively choosing a (prescribed) total of $n>k$ points to be the abscissae for sequences of k-th divided differences, whose signs are then used to locate the zeros. Of central importance is the particular rule (or strategy) by which these n points are selected. In this paper, it is shown how analysis of a maximal solution of a Booth inequality determines the most efficient strategy for k=14.

1. INTRODUCTION

Let k be a given non-negative integer and let g denote a real valued function possessing a continuous k-th derivative $g^{(k)}$. How might simple real zeros of $g^{(k)}$ be efficiently approximated, by using only values of g and points in the domain of g? A standard approach to this question entails successively choosing a (prescribed) total of $n>k$ points to be the abscissae for sequences of k-th divided differences. The signs of the differences are then used to locate the zeros; see Isaacson and Keller [6] (k=0), Kiefer [7], [8] (k=1,2), Johnson (cited in Bellman [1]) (k=1), Booth [2], [3], [4], [5] (k=0,3,4,5,6), Wallace [9] (k=4,8,10).

Of central importance is the particular rule (or strategy) by which these n points are selected. Some strategies estimate the zeros of $g^{(k)}$ more efficiently than others, so previous workers have sought the most efficient strategy $\bar{S}_k(n)$, for given k. To date, $\bar{S}_k(n)$ has been exhibited for k=0,1,2,3,4,5,6 only. In this paper, $\bar{S}_{14}(n)$ is determined, by analysis of a particular maximal solution of the Booth inequality (see Section 3)

This paper is in final form and no version of it will be submitted for publication elsewhere.

$$L(n+15) \le \begin{cases} L(n) + L(n+14) \\ L(n+1) + L(n+13) \\ \vdots \\ L(n+6) + L(n+8) \\ 2L(n+7) \end{cases}, \quad n\ge 0.$$

The structure of $\bar{S}_{14}(n)$ is found to be similar to that of $\bar{S}_6(n)$, $\bar{S}_2(n)$ and $\bar{S}_0(n)$.

2. PRELIMINARIES AND NOTATION

The objective is to estimate a simple real zero of the continuous k-th derivative of a real valued function defined on a known open interval. This problem can be reduced to that of approximating the unique real zero ξ_k of the continuous k-th derivative $f_k^{(k)}$ of a function f_k, where

(a) f_k is defined on $(0,1)$;

(b) $f_k^{(k)}(x)<0$ if $x\varepsilon(0,\xi_k)$;
$f_k^{(k)}(x)>0$ if $x\varepsilon(\xi_k,1)$.

ξ_k is to be estimated by using only values of f_k and points in $(0,1)$. The algorithm to be employed to locate ξ_k is now described.

REMARK 1: A STANDARD ALGORITHM FOR APPROXIMATING ξ_k; SEE BOOTH [2]

(a) Prescribe an $n>k$. Next, select $x_1=x_1(k)$, $x_2=x_2(k)$, $x_3=x_3(k),\ldots,x_{k+1}=x_{k+1}(k)$ $(0<x_1<x_2<x_3<\ldots<x_{k+1}<1)$, and evaluate the ordinates $f_k(x_1),f_k(x_2),f_k(x_3),\ldots,f_k(x_{k+1})$;

(b) Use these abscissae and ordinates to compute the k-th divided difference $D^{(k)}[f_k;x_1,x_2,x_3,\ldots,x_{k+1}]$, defined recursively by $D^{(0)}[f_k;x_1] = f_k(x_1)$ and

$$D^{(k)}[f_k;x_1,x_2,x_3,\ldots,x_{k+1}]$$

$$= \frac{D^{(k-1)}[f_k;x_1,x_2,x_3,\ldots,x_k]-D^{(k-1)}[f_k;x_2,x_3,x_4,\ldots,x_{k+1}]}{x_1-x_k}, \quad k>0;$$

(c) Use standard mean value theorems to conclude that, if $D^{(k)}[f_k;x_1,x_2,x_3,\ldots,x_{k+1}]\geq 0$ then $\xi_k\varepsilon(0,x_{k+1}]$; if $D^{(k)}[f_k;x_1,x_2,x_3,\ldots,x_{k+1}]<0$ then $\xi_k\varepsilon(x_1, 1)$;

(d) Select a (k+2)-th abscissa x_{k+2} from the appropriate sub-interval $((0,x_{k+1})$ or $(x_1,1))$;

(e) Evaluate $f_k(x_{k+2})$;

(f) Repeat the process until n distinct evaluations of f_k have been made.

DEFINITION. A (k-th order) strategy $S_k=S_k(n)$ is the particular rule by which the n(k-th order) abscissae $x_1,x_2,x_3,\ldots,x_{k+1},x_{k+2},x_{k+3},\ldots,x_n$ of Remark 1 are chosen.

For a given k, there exist many S_k's. It will prove expedient to exhibit some.

EXAMPLE 2a: k=0. A standard 0-th order strategy is the so-called Bisection strategy B; see Isaacson and Keller [6]:

(a),(b) Prescribe an n>0. Next, select $x_1=\frac{1}{2}$, the midpoint of (0,1), and evaluate $f_0(\frac{1}{2})$;

(c) If $f_0(\frac{1}{2})$ is positive (negative) then conclude that $\xi_0\varepsilon(0,\frac{1}{2})((\frac{1}{2},1))$;

(d) Select $x_2=\frac{1}{4}(\frac{3}{4})$, the midpoint of $(0,\frac{1}{2})((\frac{1}{2},1))$;

(e) Evaluate $f_0(\frac{1}{4})(f_0(\frac{3}{4}))$;

(f) Repeat the process until n distinct evaluations of f_0 have been made.

After n distinct evaluations of f_0 under B, the error in estimating ξ_0 is $(\frac{1}{2})^n$.

EXAMPLE 2b: $k=0$, $S_0=B$, $f_0(x)=3x-2$. Prescribe $n=3$. $f_0(x_1)=f_0(\frac{1}{2})$ is negative; hence $\xi_0\varepsilon(\frac{1}{2},1)$, so select $x_2=\frac{3}{4}$, the midpoint of $(\frac{1}{2},1)$. $f_0(x_2)=f_0(\frac{3}{4})$ is positive; hence $\xi_0\varepsilon(\frac{1}{2},\frac{3}{4})$, so select $x_3=\frac{5}{8}$, the midpoint of $(\frac{1}{2},\frac{3}{4})$. Finally, $f_0(x_3)=f_0(\frac{5}{8})$ is negative; hence $\xi_0\varepsilon(\frac{5}{8},\frac{3}{4})$. Note that $(\frac{5}{8},\frac{3}{4})$ is of length $(\frac{1}{2})^3$.

EXAMPLE 3a: k=0. Another 0-th order strategy is the so-called Trisection strategy T:

(a),(b) Prescribe an $n>0$. Select $x_1=\frac{1}{3}$, and evaluate $f_0(\frac{1}{3})$;

(c) If $f_0(\frac{1}{3})$ is positive (negative) then conclude that $\xi_0\varepsilon(0,\frac{1}{3})((\frac{1}{3},1))$;

(d) Select $x_2=\frac{1}{9}(\frac{5}{9})$;

(e) Evaluate $f_0(\frac{1}{9})(f_0(\frac{5}{9}))$;

(f) Repeat the process until n distinct evaluations of f_0 have been made.

After n distinct evaluations of f_0 under T, the <u>maximum possible error</u> in estimating ξ_0 is $(\frac{2}{3})^n$.

EXAMPLE 3b: $k=0$, $S_0=T$, $f_0(x)=3x-2$, $n=3$ (compare with Example 2b). $f_0(x_1)=f_0(\frac{1}{3})$ is negative; hence $\xi_0\varepsilon(\frac{1}{3},1)$, so select $x_2=\frac{5}{9}$. $f_0(\frac{5}{9})$ is negative; hence $\xi_0\varepsilon(\frac{5}{9},1)$, so select $x_3=\frac{19}{27}$. Finally, $f_0(x_3)=f_0(\frac{19}{27})$ is positive; hence $\xi_0\varepsilon(\frac{5}{9},\frac{19}{27})$. Note that $(\frac{5}{9},\frac{19}{27})$ is of length $\frac{4}{27}<(\frac{2}{3})^3$. In contrast, if $f_0(x)=9x-7$, then, after 3 evaluations of f_0, T confines ξ_0 to $(\frac{19}{27},1)$, an interval of length $(\frac{2}{3})^3$.

REMARK 4. Selection of 0-th order abscissae need not be based solely on fixed ratios such as $\frac{1}{2}$ or $\frac{1}{3}$. Other strategies exist for approximating ξ_0, notably <u>stochastic</u> ones that employ a given probability density function χ to select the abscissae in a random fashion. Denote such a strategy by R_χ. Booth [3] has shown that, for any such χ, the maximum possible <u>expected</u> error, associated with n distinct evaluations of f_0 under R_χ, exceeds $(\frac{1}{2})^n$.

Before exhibiting an important example of an S_1, a useful notation will be introduced.

REMARK 5. The notation (see Booth [4])

$$[a;x_1,x_2,x_3,\ldots,x_{k+1};b]$$

will be used to signify that (i) ξ_k is known to lie in $[a,b]$, and (ii) f_k has been evaluated at distinct $x_1,x_2,x_3,\ldots,x_{k+1}\varepsilon(a,b)$. $[a;x_1,x_2,x_3,\ldots,x_{k+1};b]$ is termed a (k-th order) <u>configuration</u>.

EXAMPLE. Recall Remark 1 with k=1. At the end of step (a), the (1-st order) configuration is $[0;x_1,x_2;1]$. By the end of step (e), it is either $[0;x_3,x_1;x_2]$ or $[x_1;x_2,x_3;1]$, according as $\xi_1 \varepsilon (0,x_2]$ or $\xi_1 \varepsilon (x_1,1)$.

REMARK 6. Two (k-th order) configurations $[a;x_1,x_2,x_3,\ldots,x_{k+1};b]$ and $[c;y_1,y_2,y_3,\ldots,y_{k+1};d]$ are said to be <u>equivalent</u> if

$$(x_i-a):(b-a)=(y_i-c):(d-c) \quad \text{for } \underline{\text{all}} \text{ } i \text{ in } 1\le i\le k+1.$$

EXAMPLE. The (1st-order) configurations $[0;\frac{1}{4},\frac{1}{2};1]$, $[0;\frac{1}{8},\frac{1}{4};\frac{1}{2}]$ and $[1;5,9;17]$ are all equivalent, since $\frac{1}{4}:1=\frac{1}{8}:\frac{1}{2}=4:16$ and $\frac{1}{2}:1=\frac{1}{4}:\frac{1}{2}=8:16$. But, $[0;\frac{1}{4},\frac{1}{2};1]$ and $[0;\frac{1}{6},\frac{1}{4};\frac{1}{2}]$ are not equivalent, since $\frac{1}{4}:1\neq\frac{1}{6}:\frac{1}{2}$.

EXAMPLE 7a: k=1. As with the problem concerning ξ_0, there exist many strategies for selecting those abscissae $x_1=x_1(1)$, $x_2=x_2(1)$, $x_3=x_3(1),\ldots,x_n=x_n(1)$ that are to be used to locate ξ_1. One such strategy, formulated by Kiefer [7], is based on the classic Fibonacci sequence $\{F(n)\},n\ge0$, defined by

$$F(0)=F(1)=1;$$
$$F(n+2)=F(n)+F(n+1),\ n\ge0:$$

"(a) Initiate this so-called <u>Fibonacci strategy</u> F by prescribing an $n\ge3$. Next, choose

$$x_1=\frac{F(n-2)}{F(n)} \qquad \text{and} \qquad x_2=\frac{F(n-1)}{F(n)},$$

and evaluate $f_1(\frac{F(n-2)}{F(n)})$ and $f_1(\frac{F(n-1)}{F(n)})$ (Note that the initial configuration (recall Remark 5) is, therefore,

$$A_0 = [0;\ \frac{F(n-2)}{F(n)},\ \frac{F(n-1)}{F(n)};\ 1]).$$

(b) Compute the 1-st divided difference (recall Remark 1(b) with k=1)

$$D^{(1)}[f_1;\frac{F(n-2)}{F(n)},\frac{F(n-1)}{F(n)}] \quad \left(= \frac{f_1(\frac{F(n-2)}{F(n)}) - f_1(\frac{F(n-1)}{F(n)})}{(\frac{F(n-2)}{F(n)}) - (\frac{F(n-1)}{F(n)})} \right).$$

(c) Thus, conclude that,

if $D^{(1)}[f_1; \frac{F(n-2)}{F(n)}, \frac{F(n-1)}{F(n)}] \geq 0$ then $\xi_1 \varepsilon (0, \frac{F(n-1)}{F(n)}]$;

if $D^{(1)}[f_1; \frac{F(n-2)}{F(n)}, \frac{F(n-1)}{F(n)}] < 0$ then $\xi_1 \varepsilon (\frac{F(n-2)}{F(n)}, 1)$.

(d) <u>Case (i)</u>: $\xi_1 \varepsilon (0, \frac{F(n-1)}{F(n)}]$

Here, choose

$$x_3 = \frac{F(n-3)}{F(n)} \quad (< \frac{F(n-2)}{F(n)}).$$

Therefore, the configuration is

$$A_1 = [0; \frac{F(n-3)}{F(n)}, \frac{F(n-2)}{F(n)}; \frac{F(n-1)}{F(n)}].$$

<u>Case (ii)</u>: $\xi_1 \varepsilon (\frac{F(n-2)}{F(n)}, 1)$

Here, choose

$$x_3 = \frac{2F(n-2)}{F(n)} \quad (> \frac{F(n-1)}{F(n)}).$$

Therefore, the configuration is

$$[\frac{F(n-2)}{F(n)}; \frac{F(n-1)}{F(n)}, \frac{2F(n-2)}{F(n)}; 1],$$

equivalent (recall Remark 6) to

$$[0; \frac{F(n-1)-F(n-2)}{F(n)}, \frac{F(n-2)}{F(n)}; \frac{F(n)-F(n-2)}{F(n)}],$$

which equals A_1.

So, in both cases, appropriate choice of $x_3 (\frac{F(n-3)}{F(n)}$ or $\frac{2F(n-2)}{F(n)})$ yields a configuration equal to, or equivalent to,

(2.1) $$A_1 = [0; \frac{F(n-3)}{F(n)}, \frac{F(n-2)}{F(n)}; \frac{F(n-1)}{F(n)}].$$

But,

$$\frac{F(n-3)}{F(n)} : \frac{F(n-1)}{F(n)} = \frac{F(n-3)}{F(n-1)} : 1 \quad \text{and} \quad \frac{F(n-2)}{F(n)} : \frac{F(n-1)}{F(n)} = \frac{F(n-2)}{F(n-1)} : 1.$$

Hence, A_1 is equivalent to

$$[0; \frac{F(n-3)}{F(n-1)}, \frac{F(n-2)}{F(n-1)}; 1],$$

the initial configuration A_0 with n replaced by n-1. Therefore, the process described above can be repeated. Appropriate choice of x_4 now yields a configuration equal to, or equivalent to (recall (2.1)),

$$[0; \frac{F(n-4)}{F(n)}, \frac{F(n-3)}{F(n)}; \frac{F(n-2)}{F(n)}];$$

whereupon appropriate choice of x_5 yields a configuration equal to, or equivalent to,

$$[0;\ \frac{F(n-5)}{F(n)}, \frac{F(n-4)}{F(n)}; \frac{F(n-3)}{F(n)}\ ];$$

...; appropriate choice of x_{n-1} yields a configuration equal to, or equivalent to,

$$[0;\ \frac{F(1)}{F(n)}, \frac{F(2)}{F(n)}; \frac{F(3)}{F(n)}]\ (=[0;\ \frac{1}{F(n)}, \frac{2}{F(n)}; \frac{3}{F(n)}\ ]).$$

Form the associated 1-st divided difference, and so confine ξ_1 to a sub-interval of length $\frac{2}{F(n)}$. This has required n-1 evaluations of f_1; at $x_1, x_2, x_3, \ldots, x_{n-1}$. Finally choose x_n arbitrarily close to the remaining abscissa within the sub-interval, and form the associated 1-st divided difference. Thus, conclude that ξ_1 is now confined to an interval of length equal to, or arbitrarily close to, $\frac{1}{F(n)}$. Note that it has required n distinct evaluations of f_1 to reach this conclusion."

EXAMPLE 7b: k=1, $S_1=F$, $f_1(x)=2x^2-3x+2$. Prescribe n=4. Next, choose $x_1=\frac{F(2)}{F(4)}=\frac{2}{5}$, $x_2=\frac{F(3)}{F(4)}=\frac{3}{5}$, and evaluate $f_1(\frac{2}{5})$, $f_1(\frac{3}{5})$. The initial configuration is, therefore, $[0;\ \frac{2}{5},\frac{3}{5};1]$. Compute $D^{(1)}[f_1;\frac{2}{5},\frac{3}{5}]=(f_1(\frac{2}{5})-f_1(\frac{3}{5}))/(\frac{2}{5}-\frac{3}{5})$. It is negative; hence $\xi_1\varepsilon(\frac{2}{5},1)$, so select $x_3=\frac{2F(2)}{F(4)}=\frac{4}{5}$ (Note that the resultant configuration $[\frac{2}{5};\frac{3}{5},\frac{4}{5};1]$ is equivalent to $[0;\frac{1}{5},\frac{2}{5};\frac{3}{5}]=[0;\ \frac{1}{F(4)},\frac{2}{F(4)};\ \frac{3}{F(4)}]$). Form $D^{(1)}\ [f_1;\frac{3}{5},\frac{4}{5}]$. It is negative; hence $\xi_1\varepsilon(\frac{3}{5},1)$, a sub-interval of length $\frac{2}{5}=\frac{2}{F(4)}$. This has required 3 evaluations of f_1; at $x_1=\frac{2}{5}$, $x_2=\frac{3}{5}$, $x_3=\frac{4}{5}$. Finally, choose $x_4=(\frac{4+\epsilon}{5})$, $\epsilon>0$ arbitrarily small, and compute $D^{(1)}[f_1;\frac{4}{5},\frac{4+\epsilon}{5}]$. It is positive; hence $\xi_1\varepsilon(\frac{3}{5},\frac{4+\epsilon}{5}]$, an interval of length arbitrarily close to $\frac{1}{5}=\frac{1}{F(4)}$. Note that it has required 4 distinct evaluations of f_1 to reach this conclusion.

Examples 7, 3, 2 and Remark 4 illustrate the following remark.

REMARK. In estimating ξ_k, by (at most) n distinct evaluations of f_k under a particular $S_k=S_k(n)$, there is an associated maximum possible error. Denote this maximum error by $E_n^{(k)}(S_k)$.

EXAMPLE.

$E_n^{(0)}(B)=(\frac{1}{2})^n$; $E_n^{(0)}(T)=(\frac{2}{3})^n$; $E_n^{(0)}(R_\chi)>(\frac{1}{2})^n$ for any χ;

$E_n^{(1)}(F)=\frac{1}{F(n)}$.

DEFINITION. Denote $\frac{1}{E_n^{(k)}(S_k)}$ by $L_{S_k}(n)$.

DEFINITION. One (k-th order) strategy S_k is said to be <u>more efficient</u> than another (k-th order) strategy S_k' if

$$E_n^{(k)}(S_k) < E_n^{(k)}(S_k');$$

(or, equivalently, if

$$L_{S_k}(n) \quad > L_{S_k'}(n) \text{).}$$

EXAMPLE: k=0. B is more efficient than T or any R_χ.

DEFINITION. The <u>most efficient</u> (k-th order) strategy is that strategy $\bar{S}_k$ for which

$$E_n^{(k)}(\bar{S}_k) = \inf_{S_k} E_n^{(k)}(S_k);$$

(that is, for which

$$L_{\bar{S}_k}(n) \quad = \sup_{S_k} L_{S_k}(n) \text{).}$$

Denote $L_{\bar{S}_k}(n)$ by $L_k(n)$.

REMARK. Thus, for any (k-th order) strategy S_k,

(2.2) $$L_{S_k}(n) \leq L_k(n), \quad n\geq 0.$$

EXAMPLE 8.

$$(2^n=)\ L_B(n) \leq L_0(n), \quad n\geq 0;$$

$$(F(n)=)\ L_F(n) \leq L_1(n), \quad n\geq 0.$$

THEOREM 9 (CLASSIC; SEE, FOR EXAMPLE, ISAACSON AND KELLER [6])

$\bar{S}_0=B$ <u>and</u> $L_0(n)=2^n$, $n\geq 0$.

THEOREM 10 (KIEFER [7])

$\bar{S}_1=F$ <u>and</u> $L_1(n)=F(n)$, $n\geq 0$.

REMARK. Later, S. Johnson (cited in Bellman [1]) produced an alternative proof of Theorem 10. Essentially, he proved that $L_1(n)$ satisfies

(2.3a) $$L_1(0)=L_1(1)=1 ;$$

(2.3b) $$L_1(n+2)\le L_1(n)+L_1(n+1), \quad n\ge 0 .$$

This means that

$$L_1(n)\le F(n), \qquad n\ge 0.$$

However (recall Example 8),

$$F(n)\le L_1(n), \quad n\ge 0;$$

thus,

$$L_1(n)=F(n), \qquad n\ge 0.$$

THEOREM 11 (KIEFER [8]). $L_2(3)=2$; and, for $t\ge 2$,

$$L_2(2t)=2^{t-1}; \quad L_2(2t+1)=3.2^{t-2}.$$

3. BOOTH'S INEQUALITY FOR $L_k(n)$

REMARK. In [2], Booth extended Johnson's result (2.3) by proving that $L_k(n)$ satisfies

(3.1a) $$L_k(0)=L_k(1) = L_k(2) = \dots = L_k(k) = 1;$$

(3.1b) $$L_k(n+k+1) \le \begin{cases} L_k(n) + L_k(n+k) \\ L_k(n+1) + L_k(n+k-1) \\ L_k(n+2) + L_k(n+k-2) \\ \vdots \\ L_k(n+[\frac{k}{2}]) + L_k(n+k-[\frac{k}{2}]) \end{cases}, \qquad n\ge 0.$$

Result (3.1) provides no explicit information on the value of any $L_k(n)$, but it does imply that

(3.2) $$L_k(n)\le U_k(n), \quad n\ge 0,$$

where $U_k(n)$ is the n-th term of the sequence $U_k= \{U_k(n)\}$, $n\ge 0$, that is defined by

(3.3a) $$U_k(0)=U_k(1) = U_k(2) = \dots = U_k(k) = 1;$$

(3.3b) $$U_k(n+k+1)=\min \begin{cases} U_k(n) + U_k(n+k) \\ U_k(n+1) + U_k(n+k-1) \\ U_k(n+2) + U_k(n+k-2) \\ \vdots \\ U_k(n+[\frac{k}{2}]) + U_k(n+k-[\frac{k}{2}]) \end{cases}, \qquad n\ge 0.$$

For a given k, the value of any $U_k(n)$ can be obtained recursively from (3.3), thereby giving an upper bound on the value of the corresponding $L_k(n)$. But the greater importance of inequality (3.2) lies in the fact that, for small k at least, certain properties of $U_k(n)$ can be determined, and then used to establish $\bar{S}_k$. This procedure will be illustrated by recounting how Booth [2] established $\bar{S}_3$.

First, it was recalled that $U_3(n)$ is given by

$$U_3(0) = U_3(1) = U_3(2) = U_3(3) = 1;$$

$$(3.4) \qquad U_3(n+4) = \min \begin{cases} U_3(n) + U_3(n+3) \\ U_3(n+1) + U_3(n+2) \end{cases}, \quad n\geq 0.$$

It was then established that the lower sum in (3.4) dominates throughout; that is, for $n\geq 0$, $U_3(n)$ satisfies the single recurrence

$$(3.5) \qquad U_3(n+4) = U_3(n+1) + U_3(n+2).$$

A particular (3-rd order) strategy S_3^* was then exhibited. The structure of S_3^* is based on (3.5) and on a subsidiary recurrence and inequality $(U_3(n+4)-U_3(n+3)=U_3(n-1), U_3(n+4)-U_3(n)<U_3(n+3))$ that $U_3(n)$ satisfies as a consequence of satisfying (3.5). It was shown that applying S_3^* to any f_3 guarantees to confine the unique zero $\xi_3 \varepsilon (0,1)$ of $f_3^{(3)}$ to an interval of length $\frac{1}{U_3(n)}$ or $\frac{1+\varepsilon}{U_3(n)}$. This means that $L_{S_3^*}(n)=U_3(n)$; hence (recall (2.2)), $U_3(n)\leq L_3(n)$, $n\geq 0$. However (recall (3.2)), $L_3(n)\leq U_3(n)$, $n\geq 0$. Thus,

$$(3.6) \qquad L_3(n) = U_3(n), \quad n\geq 0; \quad \text{and so } \bar{S}_3 = S_3^*.$$

REMARK: k=4. In [4], Booth proved that

$$(3.7) \qquad L_4(n) = U_4(n), \quad n\geq 0,$$

and gave an implicit description of $\bar{S}_4$. This was achieved as follows.

First, it was recalled that $U_4(n)$ is given by

$$(3.8a) \qquad U_4(0) = U_4(1) = U_4(2) = U_4(3) = U_4(4) = 1;$$

$$(3.8b) \qquad U_4(n+5) = \min \begin{cases} U_4(n) + U_4(n+4) \\ U_4(n+1) + U_4(n+3) \\ 2U_4(n+2) \end{cases}, \quad n\geq 0.$$

A particular (4-th order) strategy S_4^* was then exhibited. The structure of S_4^* is based on (3.8) and on <u>periodic</u> recurrences and inequalities (such as $U_4(n+5)=U_4(n+3)+U_4(n+1)$, if $n\equiv 1$

(mod 3), $U_4(n+5) \le U_4(n+4)+U_4(n-3)$, if $n \equiv 2 \pmod 3$) that $U_4(n)$ satisfies as a consequence of satisfying (3.8). It was shown that applying S_4^* to any f_4 guarantees to confine the unique zero $\xi_4 \varepsilon (0,1)$ of $f_4^{(4)}$ to an interval of length $\frac{1}{U_4(n)}$ or $\frac{1+\varepsilon}{U_4(n)}$. This means that $L_{S_4^*}(n)=U_4(n)$; hence, $U_4(n) \le L_4(n)$, $n \ge 0$. But $L_4(n) \le U_4(n)$, $n \ge 0$; thus,

$$L_4(n)=U_4(n), \quad n \ge 0; \quad \text{and so} \quad \bar{S}_4=S_4^*.$$

REMARK. Some of the abscissae of $\bar{S}_4$ are chosen in a direct manner, similar to that by which the abscissae of $\bar{S}_3$ are chosen. However, some of the later points are selected by more involved criteria. For example, one of $\bar{S}_4$'s abscissae is

$$(3.9) \qquad \min \begin{cases} \dfrac{U_4(n-6) + U_4(n-12)}{U_4(n)} \\[2ex] \dfrac{U_4(n-5)}{U_4(n)} \end{cases}, \text{ when } n \equiv 2 \pmod 3.$$

In [4], Booth established that

$$L_4(3t+1) = 2^{t-1} \quad , t \ge 2.$$

Later, a complete closed form expression for $L_4(n)$, $n \ge 0$, was given by Wallace [9]:

$$(3.10) \qquad L_4(5)=2; \text{ and } \begin{cases} L_4(3t) &= \frac{5}{12} \cdot 2^t - \frac{1}{6}(-1)^t + \frac{1}{2} \\ L_4(3t+1) &= 2^{t-1} \\ L_4(3t+2) &= \frac{2}{3} \cdot 2^t - \frac{1}{6}(-1)^t + \frac{1}{2} \end{cases}, \quad t \ge 2.$$

Formula (3.10) now permits a numerical value to be attached to all the abscissae of $\bar{S}_4$, particularly ones such as (3.9).

REMARK: $k=5,6$. In [4], Booth announced that

$$(3.11) \qquad L_6(n) = U_6(n), \quad n \ge 0.$$

Later, in [5], Booth proved that

$$(3.12) \qquad L_5(n) < U_5(n), \quad \text{for all } n \text{ sufficiently large}.$$

Also (recall (3.3)) with $k=0$), $U_0(0)=1$; and $U_0(n+1)=\min\{U_0(n)+U_0(n)\} = 2U_0(n)$, $n \ge 0$. So, $U_0(n)=2^n$, $n \ge 0$; hence (recall Theorem 9),

$$(3.13) \qquad L_0(n) = U_0(n), \quad n \ge 0.$$

Moreover, $U_1(0)=U_1(1)=1$; and $U_1(n+2)=\min\{U_1(n)+U_1(n+1)\} = U_1(n)+U_1(n+1)$, $n \ge 0$. So, $U_1(n)=F(n)$, $n \ge 0$; hence,

$$(3.14) \qquad L_1(n) = U_1(n), \quad n \ge 0.$$

REMARK 12. $L_2(n)=U_2(n), n\geq 0$; by virtue of (3.3) and Theorem 11.

REMARK: $k>6$. It appears that, for $k>6$, no closed expression for $L_k(n)$, $n\geq 0$, has been exhibited. However, recall (3.13), (3.14), Remark 12, (3.6), (3.7), (3.12), (3.11). Thus, observe that, for $n\geq 0$,

$$L_k(n) = U_k(n),\ k=0,1,2,3,4,6;$$

whereas, for all n sufficiently large,

$$L_5(n) < U_5(n).$$

These two observations suggest the open question:

"Does $L_k(n) = U_k(n)$, $n\geq 0$, for all even k?"

With a view to answering this question, Wallace [10] determined a closed form expression for $U_{2p}(n)$, p a fixed non-negative integer. As a consequence, $U_{2p}(n)$ satisfies a set of periodic recurrences and inequalities, like those associated with $U_4(n)$. However, unlike with $U_4(n)$, it is not immediately apparent how this set, or indeed the expression for $U_{2p}(n)$ itself, can be utilized to establish $\bar{S}_{2p}$. Such is the case, for example, when $2p=8,10,12$. One particular exception, though, is when $2p=14$.

4. DETERMINATION OF $\bar{S}_{14}$

Set $r=3$ in Theorem 5.3 of Wallace [10], and obtain

$$(4.1)\quad \begin{cases} U_{14}(15)=U_{14}(16)=U_{14}(17)=U_{14}(18)=U_{14}(19)=U_{14}(20)=U_{14}(21) \\ \quad =U_{14}(22)=2; \\ U_{14}(23)=U_{14}(24)=U_{14}(25)=U_{14}(26)=3; \\ \quad U_{14}(27)=U_{14}(28)=U_{14}(29)=U_{14}(30)=4; \\ U_{14}(31)=U_{14}(32)=5;\ U_{14}(33)=U_{14}(34)=6; \\ \quad U_{14}(35)=U_{14}(36)=7;\ U_{14}(37)=U_{14}(38)=8; \\ U_{14}(8t+39+\ell)=(9+\ell)2^t,\ t\geq 0,\ 0\leq\ell<8. \end{cases}$$

A (14-th order) strategy S^*_{14}, based on (4.1), will now be exhibited. It will transpire that $S^*_{14}=\bar{S}_{14}$.

"(a) Initiate S^*_{14} by prescribing an $n\geq 54$. It will prove necessary to distinguish between the 8 cases $n\equiv\ell \pmod 8$, $0\leq\ell<8$:

Case (i): $n\equiv 6 \pmod 8$

On (0,1), choose the 15 distinct abscissae

$$x_1 = \frac{U_{14}(n-32)}{U_{14}(n)}\ ;\ x_2 = \frac{U_{14}(n-24)}{U_{14}(n)};\ x_3 = \frac{U_{14}(n-20)}{U_{14}(n)}\ ;\ x_4 = \frac{U_{14}(n-16)}{U_{14}(n)};$$

$$x_5 = \frac{U_{14}(n-14)}{U_{14}(n)}\ ;\ x_6 = \frac{U_{14}(n-12)}{U_{14}(n)};\ x_7 = \frac{U_{14}(n-10)}{U_{14}(n)}\ ;$$

$$x_{8+\ell} = \frac{U_{14}(n-8+\ell)}{U_{14}(n)}\ ,\ 0\le\ell<8;$$

and evaluate $f_{14}(\frac{U_{14}(n-32)}{U_{14}(n)})$, $f_{14}(\frac{U_{14}(n-24)}{U_{14}(n)})$, ..., $f_{14}(\frac{U_{14}(n-1)}{U_{14}(n)})$.

Denote this (original) configuration (recall Remark 5)

$$[0;\ \frac{U_{14}(n-32)}{U_{14}(n)},\ \frac{U_{14}(n-24)}{U_{14}(n)},\ \frac{U_{14}(n-20)}{U_{14}(n)},\ \frac{U_{14}(n-16)}{U_{14}(n)},\ \frac{U_{14}(n-14)}{U_{14}(n)},$$
$$\frac{U_{14}(n-12)}{U_{14}(n)},\ \frac{U_{14}(n-10)}{U_{14}(n)},\ \frac{U_{14}(n-8)}{U_{14}(n)},\ \frac{U_{14}(n-7)}{U_{14}(n)},\ \frac{U_{14}(n-6)}{U_{14}(n)},\ \frac{U_{14}(n-5)}{U_{14}(n)},$$
$$\frac{U_{14}(n-4)}{U_{14}(n)},\ \frac{U_{14}(n-3)}{U_{14}(n)},\ \frac{U_{14}(n-2)}{U_{14}(n)},\ \frac{U_{14}(n-1)}{U_{14}(n)};\ 1]$$

by C_0. Compute $D^{(14)}[f_{14};\ \frac{U_{14}(n-32)}{U_{14}(n)},\ \frac{U_{14}(n-24)}{U_{14}(n)},\ldots,\ \frac{U_{14}(n-1)}{U_{14}(n)}]$ and, according as it is non-negative or negative, conclude that

$\xi_{14}\varepsilon(0,\ \frac{U_{14}(n-1)}{U_{14}(n)}]$ or $\xi_{14}\varepsilon(\frac{U_{14}(n-32)}{U_{14}(n)},\ 1)$.

In the former instance, choose $x_{16} = \frac{U_{14}(n-9)}{U_{14}(n)}$ in $(0,\ \frac{U_{14}(n-1)}{U_{14}(n)})$; and denote the resultant configuration

$$[0;\ \frac{U_{14}(n-32)}{U_{14}(n)},\ \frac{U_{14}(n-24)}{U_{14}(n)},\ \frac{U_{14}(n-20)}{U_{14}(n)},\ \frac{U_{14}(n-16)}{U_{14}(n)},\ \frac{U_{14}(n-14)}{U_{14}(n)},$$
$$\frac{U_{14}(n-12)}{U_{14}(n)},\ \frac{U_{14}(n-10)}{U_{14}(n)},\ \frac{U_{14}(n-9)}{U_{14}(n)},\ \frac{U_{14}(n-8)}{U_{14}(n)},\ \frac{U_{14}(n-7)}{U_{14}(n)},\ \frac{U_{14}(n-6)}{U_{14}(n)},$$
$$\frac{U_{14}(n-5)}{U_{14}(n)},\ \frac{U_{14}(n-4)}{U_{14}(n)},\ \frac{U_{14}(n-3)}{U_{14}(n)},\ \frac{U_{14}(n-2)}{U_{14}(n)};\ \frac{U_{14}(n-1)}{U_{14}(n)}\ ]$$

by C_1. Note that extra configuration points are to be written in boldface type. In the latter instance, select $\frac{U_{14}(n-9)+U_{14}(n-32)}{U_{14}(n)}$ in $(\frac{U_{14}(n-32)}{U_{14}(n)}, 1)$; and denote the resultant configuration

$$[\frac{U_{14}(n-32)}{U_{14}(n)}; \frac{U_{14}(n-24)}{U_{14}(n)}, \frac{U_{14}(n-20)}{U_{14}(n)}, \frac{U_{14}(n-16)}{U_{14}(n)}, \frac{U_{14}(n-14)}{U_{14}(n)}, \frac{U_{14}(n-12)}{U_{14}(n)}, \frac{U_{14}(n-10)}{U_{14}(n)}, \frac{U_{14}(n-8)}{U_{14}(n)}, \mathbf{\frac{U_{14}(n-9)+U_{14}(n-32)}{U_{14}(n)}}, \frac{U_{14}(n-7)}{U_{14}(n)}, \frac{U_{14}(n-6)}{U_{14}(n)}, \frac{U_{14}(n-5)}{U_{14}(n)}, \frac{U_{14}(n-4)}{U_{14}(n)}, \frac{U_{14}(n-3)}{U_{14}(n)}, \frac{U_{14}(n-2)}{U_{14}(n)}, \frac{U_{14}(n-1)}{U_{14}(n)}; 1]$$

by C_1'. Note that $\frac{U_{14}(n-8)}{U_{14}(n)} < \frac{U_{14}(n-9)+U_{14}(n-32)}{U_{14}(n)} < \frac{U_{14}(n-7)}{U_{14}(n)}$ when $n \equiv 6 \pmod 8$, by virtue of (4.1). C_1' need not be pursued further, as it is equivalent (recall Remark 6) to C_1. This is because (4.1) implies that, when $n \equiv 6 \pmod 8$,

$$(4.2)\quad \begin{cases} U_{14}(n-24)-U_{14}(n-32)=U_{14}(n-32); \; U_{14}(n-20)-U_{14}(n-32)=U_{14}(n-24); \\ U_{14}(n-16)-U_{14}(n-32)=U_{14}(n-20); \; U_{14}(n-14)-U_{14}(n-32)=U_{14}(n-16); \\ U_{14}(n-12)-U_{14}(n-32)=U_{14}(n-14); \; U_{14}(n-10)-U_{14}(n-32)=U_{14}(n-12); \\ U_{14}(n-8)-U_{14}(n-32)=U_{14}(n-10); \; U_{14}(n-7+\ell)-U_{14}(n-32)=U_{14}(n-8+\ell), \; 0\le\ell<8 . \end{cases}$$

Next, recall C_1. Compute $D^{(14)}[f_{14}; \frac{U_{14}(n-32)}{U_{14}(n)}, \frac{U_{14}(n-24)}{U_{14}(n)}, \ldots, \frac{U_{14}(n-10)}{U_{14}(n)}, \mathbf{\frac{U_{14}(n-9)}{U_{14}(n)}}, \frac{U_{14}(n-8)}{U_{14}(n)}, \ldots, \frac{U_{14}(n-2)}{U_{14}(n)}]$ and, according as it is non-negative or negative, conclude that

$$\xi_{14}\varepsilon(0, \frac{U_{14}(n-2)}{U_{14}(n)}] \quad \text{or} \quad \xi_{14}\varepsilon(\frac{U_{14}(n-32)}{U_{14}(n)}, \frac{U_{14}(n-1)}{U_{14}(n)}) .$$

In the former instance, choose $x_{17} = \frac{U_{14}(n-11)}{U_{14}(n)}$ in $(0, \frac{U_{14}(n-2)}{U_{14}(n)})$; and denote the resultant configuration

$$[0; \frac{U_{14}(n-32)}{U_{14}(n)}, \frac{U_{14}(n-24)}{U_{14}(n)}, \frac{U_{14}(n-20)}{U_{14}(n)}, \frac{U_{14}(n-16)}{U_{14}(n)}, \frac{U_{14}(n-14)}{U_{14}(n)}, \frac{U_{14}(n-12)}{U_{14}(n)}, \frac{U_{14}(n-11)}{U_{14}(n)}, \frac{U_{14}(n-10)}{U_{14}(n)}, \frac{U_{14}(n-9)}{U_{14}(n)}, \frac{U_{14}(n-8)}{U_{14}(n)}, \frac{U_{14}(n-7)}{U_{14}(n)}, \frac{U_{14}(n-6)}{U_{14}(n)}, \frac{U_{14}(n-5)}{U_{14}(n)}, \frac{U_{14}(n-4)}{U_{14}(n)}, \frac{U_{14}(n-3)}{U_{14}(n)}; \frac{U_{14}(n-2)}{U_{14}(n)}]$$

by C_2. In the latter instance, select again $x_{17} = \frac{U_{14}(n-9)+U_{14}(n-32)}{U_{14}(n)}$ (in $(\frac{U_{14}(n-32)}{U_{14}(n)}, \frac{U_{14}(n-1)}{U_{14}(n)})$); and denote the resultant configuration

$$[\frac{U_{14}(n-32)}{U_{14}(n)}; \frac{U_{14}(n-24)}{U_{14}(n)}, \frac{U_{14}(n-20)}{U_{14}(n)}, \frac{U_{14}(n-16)}{U_{14}(n)}, \frac{U_{14}(n-14)}{U_{14}(n)}, \frac{U_{14}(n-12)}{U_{14}(n)}, \frac{U_{14}(n-10)}{U_{14}(n)}, \frac{U_{14}(n-9)}{U_{14}(n)}, \frac{U_{14}(n-8)}{U_{14}(n)}, \frac{U_{14}(n-9)+U_{14}(n-32)}{U_{14}(n)}, \frac{U_{14}(n-7)}{U_{14}(n)}, \frac{U_{14}(n-6)}{U_{14}(n)}, \frac{U_{14}(n-5)}{U_{14}(n)}, \frac{U_{14}(n-4)}{U_{14}(n)}, \frac{U_{14}(n-3)}{U_{14}(n)}, \frac{U_{14}(n-2)}{U_{14}(n)}; \frac{U_{14}(n-1)}{U_{14}(n)}]$$

by C_2'. C_2' need not be pursued further, as it is equivalent to C_2. This follows by using all the identities (4.2) except for the last ($\ell=7$); and also by employing the result

$$U_{14}(n-9)-U_{14}(n-32) = U_{14}(n-11), \text{ if } n\equiv 6 \pmod 8,$$

valid because of (4.1). Next, recall C_2, and continue the process. Appropriate choice of 6 further new abscissae $(x_{18}, x_{19}, x_{20}, x_{21}, x_{22}, x_{23})$ now yields a configuration equal to, or equivalent to,

$$C_8 = [0; \frac{U_{14}(n-40)}{U_{14}(n)}, \frac{U_{14}(n-32)}{U_{14}(n)}, \frac{U_{14}(n-28)}{U_{14}(n)}, \frac{U_{14}(n-24)}{U_{14}(n)}, \frac{U_{14}(n-22)}{U_{14}(n)},$$
$$\frac{U_{14}(n-20)}{U_{14}(n)}, \frac{U_{14}(n-18)}{U_{14}(n)}, \frac{U_{14}(n-16)}{U_{14}(n)}, \frac{U_{14}(n-15)}{U_{14}(n)}, \frac{U_{14}(n-14)}{U_{14}(n)},$$
$$\frac{U_{14}(n-13)}{U_{14}(n)}, \frac{U_{14}(n-12)}{U_{14}(n)}, \frac{U_{14}(n-11)}{U_{14}(n)}, \frac{U_{14}(n-10)}{U_{14}(n)}, \frac{U_{14}(n-9)}{U_{14}(n)};$$
$$\frac{U_{14}(n-8)}{U_{14}(n)}].$$

But it is evident that C_8 is equivalent to

$$[0; \frac{U_{14}(n-40)}{U_{14}(n-8)}, \frac{U_{14}(n-32)}{U_{14}(n-8)}, \frac{U_{14}(n-28)}{U_{14}(n-8)}, \frac{U_{14}(n-24)}{U_{14}(n-8)}, \frac{U_{14}(n-22)}{U_{14}(n-8)},$$
$$\frac{U_{14}(n-20)}{U_{14}(n-8)}, \frac{U_{14}(n-18)}{U_{14}(n-8)}, \frac{U_{14}(n-16)}{U_{14}(n-8)}, \frac{U_{14}(n-15)}{U_{14}(n-8)}, \frac{U_{14}(n-14)}{U_{14}(n-8)},$$
$$\frac{U_{14}(n-13)}{U_{14}(n-8)}, \frac{U_{14}(n-12)}{U_{14}(n-8)}, \frac{U_{14}(n-11)}{U_{14}(n-8)}, \frac{U_{14}(n-10)}{U_{14}(n-8)}, \frac{U_{14}(n-9)}{U_{14}(n-8}; 1],$$

the original configuration C_0 with n replaced by n-8. Thus, because n is periodic, with period 8, the process described above can be repeated. Eventually, after a total of $(\frac{n-46}{8})$ repititions, there is left a configuration equal to, or equivalent to (recall C_8),

$$[0; \frac{U_{14}(14)}{U_{14}(n)}, \frac{U_{14}(22)}{U_{14}(n)}, \frac{U_{14}(26)}{U_{14}(n)}, \frac{U_{14}(30)}{U_{14}(n)}, \frac{U_{14}(32)}{U_{14}(n)}, \frac{U_{14}(34)}{U_{14}(n)},$$
$$\frac{U_{14}(36)}{U_{14}(n)}, \frac{U_{14}(38)}{U_{14}(n)}, \frac{U_{14}(39)}{U_{14}(n)}, \frac{U_{14}(40)}{U_{14}(n)}, \frac{U_{14}(41)}{U_{14}(n)}, \frac{U_{14}(42)}{U_{14}(n)},$$
$$\frac{U_{14}(43)}{U_{14}(n)}, \frac{U_{14}(44)}{U_{14}(n)}, \frac{U_{14}(45)}{U_{14}(n)}; \frac{U_{14}(46)}{U_{14}(n)}],$$

which (because of (4.1)) is equal to

$$[0; \frac{1}{U_{14}(n)}, \frac{2}{U_{14}(n)}, \frac{3}{U_{14}(n)}, \frac{4}{U_{14}(n)}, \frac{5}{U_{14}(n)}, \frac{6}{U_{14}(n)}, \frac{7}{U_{14}(n)},$$
$$\frac{8}{U_{14}(n)}, \frac{9}{U_{14}(n)}, \frac{10}{U_{14}(n)}, \frac{11}{U_{14}(n)}, \frac{12}{U_{14}(n)}, \frac{13}{U_{14}(n)}, \frac{14}{U_{14}(n)},$$
$$\frac{15}{U_{14}(n)}; \frac{16}{U_{14}(n)}].$$

Form the associated 14-th divided difference, and so confine ξ_{14} to a sub-interval of length $\frac{15}{U_{14}(n)}$. Thereupon, at most 31 (not at most 15) more evaluations of f_{14} finally confine ξ_{14} to an interval of length $\frac{1}{U_{14}(n)}$ or $\frac{1+\varepsilon}{U_{14}(n)}$. Note that it has required (at most) $15+(n-46)+31=n$ evaluations of f_{14} to reach this conclusion. This means that $L_{S^*_{14}}(n)=U_{14}(n)$; hence (recall (2.2)),

(4.3i) $$U_{14}(n) \leq L_{14}(n), \quad \text{if } n\equiv 6 \pmod 8, \quad n\geq 0.$$

Case (ii): $n\equiv 5 \pmod 8$

Here, choose the initial configuration to be that configuration on (0,1) which is equivalent to C_1 with n replaced by n+1; namely,

$$[0;\ \frac{U_{14}(n-31)}{U_{14}(n)}, \frac{U_{14}(n-23)}{U_{14}(n)}, \frac{U_{14}(n-19)}{U_{14}(n)}, \frac{U_{14}(n-15)}{U_{14}(n)}, \frac{U_{14}(n-13)}{U_{14}(n)},$$
$$\frac{U_{14}(n-11)}{U_{14}(n)}, \frac{U_{14}(n-9)}{U_{14}(n)}, \frac{U_{14}(n-8)}{U_{14}(n)}, \frac{U_{14}(n-7)}{U_{14}(n)}, \frac{U_{14}(n-6)}{U_{14}(n)}, \frac{U_{14}(n-5)}{U_{14}(n)},$$
$$\frac{U_{14}(n-4)}{U_{14}(n)}, \frac{U_{14}(n-3)}{U_{14}(n)}, \frac{U_{14}(n-2)}{U_{14}(n)}, \frac{U_{14}(n-1)}{U_{14}(n)}\ ;\ 1\ ]\ .$$

Now, proceed as before. Thus, conclude that

(4.3ii) $$U_{14}(n) \leq L_{14}(n), \quad \text{if } n\equiv 5 \pmod 8, \quad n\geq 0.$$

Case (iii): $n\equiv \ell \pmod 8$, $-1\leq \ell<5$

Here, choose the initial configuration to be that configuration on (0,1) which is equivalent to $C_{6-\ell}$ with n replaced by $n+6-\ell$. Then proceed as before. Thus, conclude that, for $-1\leq\ell<5$,

(4.3iii) $$U_{14}(n) \leq L_{14}(n), \quad \text{if } n\equiv \ell \pmod 8, \quad n\geq 0.$$

It follows, from (4.3iii), (4.3ii), (4.3i), that

$$U_{14}(n) \leq L_{14}(n), \quad \text{for all } n\geq 0.$$

However, (recall (3.2)),

$$L_{14}(n) \leq U_{14}(n), \quad \text{for all } n\geq 0.$$

Thus,

$$L_{14}(n) = U_{14}(n), \quad \text{and so } \bar{S}_{14}=S^*_{14}."$$

REMARK 13. Formula (4.1) elucidates how to use $\bar{S}_{14}$. For (4.1) implies that

$$C_0 = [0; \frac{1}{16}, \frac{2}{16}, \frac{3}{16}, \frac{4}{16}, \frac{5}{16}, \frac{6}{16}, \frac{7}{16}, \frac{8}{16}, \frac{9}{16}, \frac{10}{16}, \frac{11}{16}, \frac{12}{16}, \frac{13}{16}, \frac{14}{16}, \frac{15}{16}; 1] ;$$

$$C_1 = [0; \frac{1}{16}, \frac{2}{16}, \frac{3}{16}, \frac{4}{16}, \frac{5}{16}, \frac{6}{16}, \frac{7}{16}, \frac{15}{32}, \frac{8}{16}, \frac{9}{16}, \frac{10}{16}, \frac{11}{16}, \frac{12}{16}, \frac{13}{16}, \frac{14}{16}; \frac{15}{16}] ;$$

$$(\text{equivalent to } C_1^* = [0; \frac{1}{15}, \frac{2}{15}, \frac{3}{15}, \frac{4}{15}, \frac{5}{15}, \frac{6}{15}, \frac{7}{15}, \frac{15}{30}, \frac{8}{15}, \frac{9}{15}, \frac{10}{15}, \frac{11}{15}, \frac{12}{15}, \frac{13}{15}, \frac{14}{15}; 1])$$

$$C_2 = [0; \frac{1}{16}, \frac{2}{16}, \frac{3}{16}, \frac{4}{16}, \frac{5}{16}, \frac{6}{16}, \frac{13}{32}, \frac{7}{16}, \frac{15}{32}, \frac{8}{16}, \frac{9}{16}, \frac{10}{16}, \frac{11}{16}, \frac{12}{16}, \frac{13}{16}; \frac{14}{16}] ;$$

$$(\text{equivalent to } C_2^* = [0; \frac{1}{14}, \frac{2}{14}, \frac{3}{14}, \frac{4}{14}, \frac{5}{14}, \frac{6}{14}, \frac{13}{28}, \frac{7}{14}, \frac{15}{28}, \frac{8}{14}, \frac{9}{14}, \frac{10}{14}, \frac{11}{14}, \frac{12}{14}, \frac{13}{14}; 1])$$

$$C_3 = [0; \frac{1}{16}, \frac{2}{16}, \frac{3}{16}, \frac{4}{16}, \frac{5}{16}, \frac{11}{32}, \frac{6}{16}, \frac{13}{32}, \frac{7}{16}, \frac{15}{32}, \frac{8}{16}, \frac{9}{16}, \frac{10}{16}, \frac{11}{16}, \frac{12}{16}; \frac{13}{16}] ;$$

$$(\text{equivalent to } C_3^* = [0; \frac{1}{13}, \frac{2}{13}, \frac{3}{13}, \frac{4}{13}, \frac{5}{13}, \frac{11}{26}, \frac{6}{13}, \frac{13}{26}, \frac{7}{13}, \frac{15}{26}, \frac{8}{13}, \frac{9}{13}, \frac{10}{13}, \frac{11}{13}, \frac{12}{13}; 1])$$

$$C_4 = [0; \frac{1}{16}, \frac{2}{16}, \frac{3}{16}, \frac{4}{16}, \frac{9}{32}, \frac{5}{16}, \frac{11}{32}, \frac{6}{16}, \frac{13}{32}, \frac{7}{16}, \frac{15}{32}, \frac{8}{16}, \frac{9}{16}, \frac{10}{16}, \frac{11}{16}; \frac{12}{16}] ;$$

$$(\text{equivalent to } C_4^* = [0; \frac{1}{12}, \frac{2}{12}, \frac{3}{12}, \frac{4}{12}, \frac{9}{24}, \frac{5}{12}, \frac{11}{24}, \frac{6}{12}, \frac{13}{24}, \frac{7}{12}, \frac{15}{24}, \frac{8}{12}, \frac{9}{12}, \frac{10}{12}, \frac{11}{12}; 1])$$

$$C_5 = [0; \frac{1}{16}, \frac{2}{16}, \frac{3}{16}, \frac{7}{32}, \frac{4}{16}, \frac{9}{32}, \frac{5}{16}, \frac{11}{32}, \frac{6}{16}, \frac{13}{32}, \frac{7}{16}, \frac{15}{32}, \frac{8}{16}, \frac{9}{16}, \frac{10}{16}; \frac{11}{16}] ;$$

(equivalent to $C_5^* = [0; \frac{1}{11}, \frac{2}{11}, \frac{3}{11}, \frac{7}{22}, \frac{4}{11}, \frac{9}{22}, \frac{5}{11}, \frac{11}{22}, \frac{6}{11}, \frac{13}{22}, \frac{7}{11}, \frac{15}{22}, \frac{8}{11}, \frac{9}{11}, \frac{10}{11}; 1]$)

$$C_6 = [0; \frac{1}{16}, \frac{2}{16}, \frac{5}{32}, \frac{3}{16}, \frac{7}{32}, \frac{4}{16}, \frac{9}{32}, \frac{5}{16}, \frac{11}{32}, \frac{6}{16}, \frac{13}{32}, \frac{7}{16}, \frac{15}{32}, \frac{8}{16}, \frac{9}{16}; \frac{10}{16}] ;$$

(equivalent to $C_6^* = [0; \frac{1}{10}, \frac{2}{10}, \frac{5}{20}, \frac{3}{10}, \frac{7}{20}, \frac{4}{10}, \frac{9}{20}, \frac{5}{10}, \frac{11}{20}, \frac{6}{10}, \frac{13}{20}, \frac{7}{10}, \frac{15}{20}, \frac{8}{10}, \frac{9}{10}; 1]$)

$$C_7 = [0; \frac{1}{16}, \frac{3}{32}, \frac{2}{16}, \frac{5}{32}, \frac{3}{16}, \frac{7}{32}, \frac{4}{16}, \frac{9}{32}, \frac{5}{16}, \frac{11}{32}, \frac{6}{16}, \frac{13}{32}, \frac{7}{16}, \frac{15}{32}, \frac{8}{16}; \frac{9}{16}] ;$$

(equivalent to $C_7^* = [0; \frac{1}{9}, \frac{3}{18}, \frac{2}{9}, \frac{5}{18}, \frac{3}{9}, \frac{7}{18}, \frac{4}{9}, \frac{9}{18}, \frac{5}{9}, \frac{11}{18}, \frac{6}{9}, \frac{13}{18}, \frac{7}{9}, \frac{15}{18}, \frac{8}{9}; 1]$)

Thus, if $n \equiv 6 \pmod 8$, $n \geq 46$, then the initial configuration is C_0 (so, first evaluate f_{14} at $\frac{1}{16}, \frac{2}{16}, \frac{3}{16}, \frac{4}{16}, \frac{5}{16}, \frac{6}{16}, \frac{7}{16}, \frac{8}{16}, \frac{9}{16}, \frac{10}{16}, \frac{11}{16}, \frac{12}{16}, \frac{13}{16}, \frac{14}{16}, \frac{15}{16}$). If $n \equiv \ell \pmod 8$, $n>46$, $-1 \leq \ell < 6$, then the initial configuration is $C^*_{6-\ell}$. For example, if $n \equiv -1 \equiv 7 \pmod 8$, $n \geq 47$, then the initial conficuration is C_7^* (so, first evaluate f_{14} at $\frac{1}{9}, \frac{3}{18}, \frac{2}{9}, \frac{5}{18}, \frac{3}{9}, \frac{7}{18}, \frac{4}{9}, \frac{9}{18}, \frac{5}{9}, \frac{11}{18}, \frac{6}{9}, \frac{13}{18}, \frac{7}{9}, \frac{15}{18}, \frac{8}{9}$). Etc...

REMARK. Recall Example 7a, which details $\bar{S}_1 = \bar{S}_1(n)$. The initial configuration is $A_0 = A_0(n) = [0; \frac{F(n-2)}{F(n)}, \frac{F(n-1)}{F(n)}; 1]$; so $A_0(n_1) \neq A_0(n_2)$ if $n_1 \neq n_2$. Hence, $\bar{S}_1(n_1) = \bar{S}_1(n_2)$ iff $n_1 = n_2$. Also, for $k=3,4,5$, $\bar{S}_k(n_1) = \bar{S}_k(n_2)$ iff $n_1 = n_2$. In contrast, $\bar{S}_0(n_1) = \bar{S}_0(n_2)$ for all n_1, n_2; $\bar{S}_2(n_1) = \bar{S}_2(n_2)$ iff $n_1 = n_2 \pmod 2$; $\bar{S}_6(n_1) = \bar{S}_6(n_2)$ iff

$n_1 \equiv n_2 \pmod 4$; while Remark 13 illuminates that $\bar{S}_{14}(n_1)=\bar{S}_{14}(n_2)$ iff $n_1 \equiv n_2 \pmod 8$. In general, for <u>any</u> k of the form $2(2^m-1)$, $\bar{S}_k(n_1)=\bar{S}_k(n_2)$ iff $n_1 \equiv n_2 \pmod{2^m}$. The author will exhibit elsewhere a proof of this result, and of other findings on $k=2(2^m-1)$.

REFERENCES

1. R. Bellman, Dynamic Programming. Princeton University Press, Princeton, 4-th Edition, 1965.

2. R.S. Booth, Location of zeros of derivatives. SIAM J. Appl. Math. 15 (1967), 1496-1501.

3. R.S. Booth, Random search for zeroes. J. Math. Anal. Appl. 20 (1967), 239-257.

4. R.S. Booth, Location of zeros of derivatives. II. SIAM J. Appl. Math. 17 (1969), 409-415.

5. R.S. Booth, An odd order search problem. SIAM J. Alg. Disc. Meth. 3 (1982), 135-143.

6. E. Isaacson and H.B. Keller, Analysis of numerical methods. John Wiley and Sons, Inc., New York, London, Sydney, 1966.

7. J. Kiefer, Sequential minimax search for a maximum. Proc. Amer. Math. Soc. 4 (1953), 502-506.

8. J. Kiefer, Optimal sequential search and approximation methods under minimum regularity assumptions. SIAM J. Appl. Math. 5 (1957), 105-136.

9. R.J. Wallace, Sequential search for zeroes of derivatives. In: W. Walter (ed.), General Inequalities 4 (pp.151-167), Birkhäuser Verlag, Basel, 1984.

10. R.J. Wallace, The maximal solution of a restricted subadditive inequality in numerical analysis. Aequationes Mathematicae. To appear.

Roger J. Wallace, Department of Quantitative Methods, Victoria College, Prahran, Victoria, 3181, Australia

Inequalities of Functional Analysis

The music room

International Series of
Numerical Mathematics, Vol. 80

POSITIVITY IN ABSOLUTE SUMMABILITY

Wolfgang Beekmann and Karl Zeller

Abstract. For the investigation of absolute summability domains we employ two positivity concepts and two kinds of sectional operators. In particular we obtain a basic positivity result for Cesàro methods and subsequently two known results (of Hardy-Bohr type) concerning summability factors.

1. INTRODUCTION

In a previous paper [5] we have dealt with applications of positivity in ordinary summability. Now we extend these investigations to absolute summability, introducing some new ideas. In the first sections we explain notation and notions: summability methods, positivity (based on the terms or the partial sums of the transformed series), sectional operators P_n (which are positive in our context) and $Q_n = P_n - P_{n-1}$. We are interested in cases where $P_n u \to u$ for all u in a domain. This implies certain convergence results while the Q_n are relevant for absolute convergence. Next we concentrate upon Cesàro methods of non-negative order α and obtain a fundamental positivity theorem in section 6 (the proof distinguishes the cases $0 \le \alpha \le 1$ and $\alpha \ge 1$). From this we deduce two known results about Cesàro summability factors. There are many papers dealing with such factors. We mention Hardy [9], Bohr [6], Andersen [1], Bosanquet [7], and (for the transition to absolute summability domains) Kogbetliantz [11], Peyerimhoff [14], Baron [4], further Bosanquet-Das [8].

This paper is in final form and no version of it will be submitted for publication elsewhere.

2. NOTATION

We consider infinite tuples $x = (x_0, x_1, \ldots)$ with real coordinates x_k and use the following notations adapted to our purposes. A series is a pair of tuples (u,s) (shortly: u) satisfying

$$u_0 + \ldots + u_n = s_n \text{ for } n = 0,1,\ldots ;$$

A summability method M is given by a mapping which assigns to every (or some) series (u,s) another series (v,t) and then transfers properties from the latter to the former. Thus a series (u,s) is M-summable or absolutely M-summable if the series (v,t) is convergent or absolutely convergent, respectively; this determines the respective summability domains M_c and M_a (where the elements are usually considered as tuples u rather than as pairs (u,s)). Further a series (u,s) is t-positive or v-positive (in the M-sense) if

$$t_n \geq 0 \quad \text{or} \quad v_n \geq 0 \quad \text{for } n = 1,\ldots,$$

respectively.

In the following M is given by a triangle matrix, primarily in the form

$$v = Bu : v_n = \sum_{k=0}^{n} b_{nk} u_k \text{ (where } b_{nn} \neq 0;\ n=0,1,\ldots) .$$

But also the two forms

$$t = As \quad \text{and} \quad t = Cu$$

are used, where A,B and C are related by

$$A = SBS^{-1}, \quad C = SB = AS$$

with

$$S = \begin{pmatrix} 1 & & & & \\ 1 & 1 & & & \\ 1 & 1 & 1 & & \\ \cdot & \cdot & \cdot & \cdot & \\ \cdot & \cdot & \cdot & \cdot & \cdot \end{pmatrix}, \quad S^{-1} = \begin{pmatrix} 1 & & & & \\ -1 & 1 & & & \\ & -1 & 1 & & \\ & & \cdot & \cdot & \\ & & & \cdot & \cdot \end{pmatrix}$$

(zeros otherwise). Now M_c and M_a are BK-spaces with the t-norm and the v-norm respectively:

$$\|u\|_t = \|t\|_\infty = \sup|t_n|, \quad \|u\|_v = \|v\|_1 = \sum_{k=0}^{\infty} |v_k| .$$

We always assume that each tuple (series)

$$e^0 = (1,0,0,\ldots) , \ e^1 = (0,1,0,\ldots) , \ e^2 ,\ldots$$

is contained in M_c and M_a .

3. OPERATORS

Given a triangle matrix P we consider the linear operator P_n given by

$$P_n u = p_{no} u_o e^o + \ldots + p_{nn} u_n e^n .$$

Each P_n maps the BK-spaces M_c and M_a into themselves (these spaces contain all e^k by assumption). We are especially interested in the case where

$$P_n u \to u \quad \text{for each } u \in M_c \text{ or } u \in M_a$$

(in the sense of the respective norm). Then the e^k constitute a Toeplitz basis, and the P_n can be called summing or sectional operators. The convergence mentioned above is present if and only if we have convergence for all u of a fundamental set and the norms of the P_n are uniformly bounded. The latter property will be verified via positivity. Each operator P_n will be described by the three matrices (according as u,t or v are used):

$$P_n^u = \text{diag}(p_{no},\ldots,p_{nn},0,0,\ldots) ,$$

$$P_n^t = CP_n^u C^{-1}, \quad P_n^v = BP_n^u B^{-1} .$$

The operator P_n is called t-positive if it maps every t-positive element into a t-positive element; this is equivalent to the fact that all entries in the matrix P_n^t are positive (in the wider sense); analogously we treat v-positivity. Thus relations

$$P_n^t \geq 0 \quad \text{and} \quad P_n^v \geq 0$$

are relevant. In case of such a positivity it is rather easy to describe the norm of P_n (with respect to the t-norm or the v-norm):

$$\|P_n\|_t = \|P_n^t e\|_\infty , \text{ where } e = (1,1,\ldots) ,$$

or

$$\|P_n\|_v = \sup_k \|P_n^v e^k\|_1 ,$$

respectively. These norms of P_n are equal if P_n is both t-positive and v-positive (since $v = e^o$ corresponds to $t = e$, and $v = e^1$ to $t = e-e^o$ and so on).

4. COMPOSED OPERATORS

In the setting above we consider the domain M_a which is a BK-space (containing all e^k by assumption); the domain M_c can be treated in an analogous manner. The continuous (and positive) linear functionals L in M_a are given by

$$L(u) = \sum_{m=0}^{\infty} h_m v_m \quad \text{where } \sup_m |h_m| < \infty \quad (\text{and } h_m \geq 0).$$

In particular,

$$L(e^k) = f_k \quad \text{with} \quad f_k = \sum_{m=k}^{\infty} h_m b_{mk}$$

holds. The P_n (as finite combinations) are continuous linear maps of M_a into itself. Hence also the linear functionals

$$L(P_n u) = \sum_{k=0}^{n} p_{nk} u_k f_k \quad \text{where} \quad f_k = L(e^k)$$

are continuous on M_a. Further we have

$$L(P_n u) \to L(u) \quad \text{if} \quad P_n u \to u,$$

i.e.

$$\lim_n \sum_{k=0}^{n} p_{nk} u_k f_k = L(u) .$$

Thus, for all u such that $P_n u \to u$, L(u) is represented by summability factors (where P is utilized as a series-to-sequence transform).

In another approach we consider operators given by (infinite) linear combinations (assuming suitable convergence):

$$\sum_{n=0}^{\infty} g_n P_n .$$

Such an operator provides a map

$$M_a \to M_a : (u_k) \to (u_k f_k), \text{ where } f_k = \sum_{n=k}^{\infty} g_n p_{nk}$$

Thus the f_k are factors which produce absolute summability. To obtain more factors of the latter type we'll investigate in section 9 the operators

$$Q_n = P_n - P_{n-1}$$

for the Cesàro case: These operators are in general not positive, but have properties like diapositivity and can be estimated rather easily.

5. CESÀRO METHOD

In the following $M = M(\alpha)$ is the Cesàro method of a fixed order $\alpha \geq 0$. We describe the corresponding matrices $A = A(\alpha)$, $B = B(\alpha)$, $C = C(\alpha)$. First we note

$$A = \mathrm{diag}(1/\binom{n+\alpha}{n})S, \quad B = S^{-1}AS \quad , \quad C = AS \ .$$

The elements of the triangles A and C are given by

$$a_{nk} = \binom{n-k+\alpha-1}{n-k}/\binom{n+\alpha}{n} \qquad (0 \leq k \leq n),$$

$$c_{nk} = \binom{n-k+\alpha}{n-k}/\binom{n+\alpha}{n} \qquad (0 \leq k \leq n).$$

For the elements of B we note the useful relation (valid for any Hausdorff method, cf. Knopp-Lorentz [10])

$$b_{nk} = a_{nk}\frac{k}{n} \qquad (0 \leq k \leq n, \text{ with } b_{oo} = 1).$$

There are similar formulae for the inverses:

$$A^{-1} = S^{-1}\mathrm{diag}(\binom{n+\alpha}{n}), \quad B^{-1} = S^{-1}A^{-1}S, \quad C^{-1} = S^{-1}A^{-1}$$

Hence the elements are given by

$$a'_{kr} = \binom{k-r-\alpha-1}{k-r}\binom{r+\alpha}{r} \qquad (0 \leq r \leq k),$$

$$c'_{kr} = \binom{k-r-\alpha-2}{k-r}\binom{r+\alpha}{r} \qquad (0 \leq r \leq k),$$

$$b'_{kr} = a'_{kr}\frac{r}{k} \qquad (0 \leq r \leq k, \text{ with } b'_{oo} = 1).$$

In particular,

$$b_{no} = 0 \quad (n=1,2,\ldots), \qquad b'_{ko} = 0 \quad (k=1,2,\ldots).$$

Important for our purposes is the fact that the following inequality is true for $m,n = 0,1,\ldots$ and $0 \leq \gamma \leq \delta$:

$$\sum_{k=0}^{\min(m,n)} \binom{m-k+\gamma}{m-k}\binom{n-k+\delta}{n-k}\binom{k-\gamma-2}{k} > 0 \ ;$$

see Lorentz-Zeller [13], Askey-Gasper-Ismail [3], compare also Beekmann-Zeller [5]. The case $\gamma = \delta$ is essential. There is a relation to hypergeometric functions and to another inequality by Askey-Gasper [2] which was used in the proof of the Bieberbach conjecture by deBranges (cf. Pommerenke [15]).

6. MAIN RESULT

Again we consider the Cesàro method and the domain M_a .

THEOREM 1. _Suppose_ $M = M(\alpha)$ _is the Cesàro method of a fixed order_ $\alpha \geq 0$ _and_ $P = C(\alpha)$ _the corresponding series-to-sequence matrix. Then each operator_ P_n _(given by a row of_ P) _is v-positive and has norm_ 1 _with respect to the v-norm. Further_

$$P_n u \to u \text{ for each } u \in M_a$$

holds in the sense of the v-norm.

The corresponding statements concerning t-positivity, t-norm and M_c are also true (cf. [5]).

To prove the positivity property we first remark that, for fixed n, P_n^v has trapezoidal shape (elements = 0 in the positions (m,r) with $r > \min(m,n)$). Also, the element in the position (r,r) with $0 \leq r \leq n$ is $p_{nr} > 0$, and the element in the position (m,n) with $m \geq n$ is $b_{mn} > 0$.

We now distinguish two cases. If $0 \leq \alpha \leq 1$, then B^{-1} is dia-positive (elements > 0 in the main diagonal, ≤ 0 below it). The monotonicity property $p_{no} \geq p_{n1} \geq \ldots \geq p_{nn} \geq 0$ and $BB^{-1} = I$ then yield $BP_n^u B^{-1} \geq 0$. To prove positivity in the case $\alpha \geq 1$, we use the inequality in section 5 with $\gamma = \alpha - 1$ and $\delta = \alpha$. (Observe that in the sums for the elements of $BP_n^u B^{-1}$ the factors k/m connecting b_{mk} with a_{mk} and the factors r/k connecting b'_{kr} with a'_{kr} cancel in an appropriate way.)

For the sequence of column-sums of P_n^v we get

$$eP_n^v = e^o + e^1 + \ldots + e^n ,$$

since $eB = e$, $eP_n^u = e^n C$ and $C = SB$ lead to

$$eP_n^v = (eP_n^u)B^{-1} = (e^n C)B^{-1} = e^n S = e^o + \ldots + e^n .$$

Thus $eP_n^v e^k = 1$ and $\| P_n^v e^k \|_1 = 1$ for $0 \leq k \leq n$ (= 0 for $k > n$) which yields the norm statement. (Since the operator P_n is also t-positive, this result can even shorter be derived from the remark at the end of section 3 and $P_n^t e = e$.)

The convergence statement now follows easily via perfectness or weak convergence.

7. SUMMABILITY FACTORS

Here and in the next section we prove two known results about factors (cf. references in the introduction, in particular [14], [4]). The convergence result stated in Theorem 1 leads easily to

PROPOSITION 2. Suppose M_a is the absolute summability domain of the Cesàro method $M = M(\alpha)$ for an $\alpha \geq 0$ and $C = C(\alpha)$ is the corresponding series-to-sequence matrix. Then, for every continuous linear function L on the BK-space M_a,

$$L(u) = \lim_n \sum_{k=0}^{n} c_{nk} u_k f_k \quad (u \in M_a \ , \ f_k = L(e^k)).$$

Thus one can represent L by summability factors (Cesàro summability of the same order). On the other hand, matrix summability factors for an BK-space or FK-space determine a continuous linear functional (by FK-principles). Thus we have a characterization of summability factors mapping M_a into M_c (even into M_a as we shall see later). This characterization is usually described by the following formulae which give necessary and sufficient conditions for factors f_k of the indicated type. The first formula and condition is that the f_k admit a representation

$$f_k = \sum_{m=k}^{\infty} h_m b_{mk} \qquad (\text{where } \sup_m |h_m| < \infty) \ ;$$

here $B = B(\alpha)$ is the corresponding series-to-series matrix. By an inversion process one finds that this is equivalent to

$$\Delta^{\alpha}(f_k/k) = O(k^{-1}) \text{ and } \sup_k |f_k| < \infty \ .$$

And the latter can be modified by some computations to

$$\Delta^{\alpha} f_k = O(k^{-\alpha}) \quad \text{and } \sup_k |f_k| < \infty \ .$$

The reader may consult Kogbetliantz [11], Peyerimhoff [14], Baron [4], Zeller-Beekmann [16] p.7. and 105 (and the literature quoted there), Bosanquet-Das [8].

8. ABSOLUTE SUMMABILITY

The aim of this section is to prove

PROPOSITION 3. <u>The summability appearing in theorem</u> 2 <u>is even absolute</u>.

This statement is known (see e.g. Peyerimhoff [14], Baron [4]). We discuss how it can be proved in our context.

For fixed n, consider the operator $Q_n = P_n - P_{n-1}$ $(P_{-1} = 0)$ mapping M_a into M_a. It is connected with the matrix

$$Q_n^v = B(P_n^u - P_{n-1}^u)B^{-1} = BB_n^uB^{-1} \text{ with } B_n^u = \operatorname{diag}(b_{no},\ldots,b_{nn},0,\ldots),$$

since $c_{nk} - c_{n-1,k} = b_{nk}$. As in section 6 we see that Q_n^v has trapezoidal shape with elements $= b_{nr} > 0$ in the position (r,r) for $0 \le r \le n$, and $= b_{mn} > 0$ in the position (m,n) for $m \ge n$. For the sequence of column-sums we now get

$$eQ_n^v = eP_n^v - eP_{n-1}^v = e^n \text{ for } n \ge 1.$$

As before we distinguish the two cases $0 \le \alpha \le 1$ and $\alpha \ge 1$.

In the first case the diapositivity of B^{-1} and the monotonicity property $0 \le b_{no} < b_{n1} < \ldots < b_{nn}$ yield that the elements of Q_n^v inside the trapezoid (positions (m,r) with $r < \min(m,n)$) are non-positive (see also Kuttner-Maddox [12], Lemma 5). Thus the norm of the r-th column of Q_n^v is

$$\| Q_n^v e^r \|_1 = 2b_{nr} \text{ for } r \neq n, \text{ and } =1 \text{ for } r = n$$

(use $eQ_n^v = e^n$ for $n \ge 1$), whence

$$\sum_{n=0}^{\infty} \| Q_n^v e^r \|_1 = 1 + \sum_{n=r+1}^{\infty} 2b_{nr} = 3 - 2b_{rr} < 3 .$$

In the case $\alpha \ge 1$ we use the identity

$$b_{nk} = a_{nk} - (1-\gamma_n)c_{n-1,k}$$

with $c_{-1,k} = 0$, $\gamma_n = n/(n+\alpha) = 1 - \alpha/(n+\alpha))$. This leads to

$$(*) \qquad Q_n^v = A_n^v - (1-\gamma_n)P_{n-1}^v ,$$

where $A_n^v = BA_n^uB^{-1}$, $A_n^u = \operatorname{diag}(a_{no},\ldots,a_{nn},0,\ldots)$. In (*), $A_n^v \ge 0$ by the inequality of section 5, and $P_{n-1}^v \ge 0$ by theorem 1. To estimate the column-norms we exhibit the diagonal element to get (for $r < n$)

$$\|Q_n^v e^r\|_1 = b_{nr} + \|Q_n^v e^r - b_{nr} e^r\|_1$$

$$\le b_{nr} + \|A_n^r e^r - a_{nr} e^r\|_1 + (1-\gamma_n)\|P_{n-1}^v e^r - c_{n-1,r} e^r\|_1 ,$$

where we used (*) in the last estimate. Since, for $n > r$, the column-sums $Q_n^v e^r$ and $P_{n-1}^v e^r$ are 0 and 1, respectively, we conclude (employing positivity and (*)) that

$$\|A_n^v e^r - a_{nr} e^r\|_1 = \|A_n^v e^r\|_1 - a_{nr}$$

$$= (1-\gamma_n)\|P_{n-1}^v e^r\|_1 - (b_{nr} + (1-\gamma_n)c_{n-1,r})$$

$$= (1-\gamma_n)(1-c_{n-1,r}) - b_{nr}$$

and that

$$\|P_{n-1}^v e^r - c_{n-1,r} e^r\|_1 = \|P_{n-1}^v e^r\|_1 - c_{n-1,r} = 1 - c_{n-1,r} .$$

Hence

$$\|Q_n^v e^r\|_1 \le 2(1-\gamma_n)(1-c_{n-1,r})$$

$$\le 2(\alpha+1)r/n(n+1) \text{ for } 0 \le r \le n$$

resulting from $1 - \gamma_n < (\alpha+1)/(n+1)$ and $1 - c_{n-1,r} \le r/n$. Thus

$$\sum_{n=0}^{\infty} \|Q_n^v e^r\|_1 = \sum_{n=r}^{\infty} \|Q_n^v e^r\|_1 = 1 + \sum_{n=r+1}^{\infty} \|Q_n^v e^r\|_1 < 2\alpha + 3 .$$

This proves the theorem, since for any bounded linear functional L on M_a and any $u \in M_a$ we have

$$\sum_{n=0}^{\infty} |L(P_n u) - L(P_{n-1} u)| = \sum_{n=0}^{\infty} |L(Q_n u)| \le \sum_{n=0}^{\infty} \|L\| \|Q_n u\|_v$$

$$\le \|L\| \sup_r \sum_{n=0}^{\infty} \|Q_n^v e^r\|_1 \|u\|_v < \infty .$$

The proof shows a pattern which is familiar in ordinary Cesàro summability: diapositivity arguments in the case $0 < \alpha \le 1$, a decomposition procedure in the case $\alpha \ge 1$, which here, in view of bounded variation properties, means a decomposition into a suitable difference of positive parts. Moreover, the proof works, since the diagonal elements of P_n^v are of the same size as the (absolute) sum of all other elements of their respective columns, and P_n^v tends to the unit matrix as $n \to \infty$.

9. REMARKS

Some results in summability theory have been obtained by rather complicated estimates. To systematize and simplify proofs, structural properties have been emphasized and successfully used in the sequel, and many of them could be interpreted in a functional analytic setting; we mention mean value properties as a typical example.

Positivity considerations also belong to this realm. We have indicated that they are useful not only in ordinary, but also in absolute summability, in particular in the Cesàro case. Our investigations give some insight why in this case summability factors of different kind (ordinary and absolute) are rather easy to handle and related.

Extensions are possible in several directions, for instance to factors of type $M_c \to M_a$ or for strong summability or of the sequence-sequence type. Also other methods could be included such as Cesàro methods of negative order, or generalized Cesàro methods, or, more generally, methods of the Cesàro-Abel type (cf. [16], Chapter VI).

REFERENCES

[1] A. F. Andersen, Studier over Cesàro's summabilitetsmetode. Dissertation, Kopenhagen, 1921.

[2] R. Askey and G. Gasper, Positive Jacobi polynomial sums, II. Amer. J. Math. 98 (1976), 709-737.

[3] R. Askey, G. Gasper and M. E.-H. Ismail, A positive sum from summability theory. J. Approx. Th. 13 (1975), 413-420.

[4] S. Baron, Neue Beweise der Hauptsätze für Summierbarkeitsfaktoren. Izvestija Akad. Nauk Estonsk. SSR, ser. rechn. fiz.-mat. Nauk 9, 47-67, deutsche Zusammenfassung 68 (1960) [Russisch].

[5] W. Beekmann and K. Zeller, Positivity in summability. ISNM, Vol. 71 (1984), 111-117.

[6] H. Bohr, Sur la série de Dirichlet. C.R. 148 (1909), 75-80.

[7] L. S. Bosanquet, Note on convergence and summability factors (III). Proc. London Math. Soc. (2), 50 (1949), 482-496.

[8] L. S. Bosanquet and G. Das, Absolute summability factors for Nörlund means. Proc. London Math. Soc. (3) 38 (1979), 1-52.

[9] G. H. Hardy, Generalisation of a theorem in the theory of divergent series. Proc. London Math. Soc. (2) 6 (1908), 255-264.

[10] K. Knopp and G. G. Lorentz, Beiträge zur absoluten Limitierung. Archiv. Math. 2 (1949), 10-16.

[11] E. Kogbetliantz, Sur les séries absolument sommables par la méthode des moyennes arithmétiques. Bull. Sci. Math. (2) 49 (1925), 234-251.

[12] B. Kuttner and I. J. Maddox. Strong Cesàro summability factors. Quart. J. Math. Oxford (2), 21 (1970), 37-59.

[13] G. G. Lorentz und K. Zeller, Abschnittslimitierbarkeit und der Satz von Hardy-Bohr. Archiv Math. 15 (1964), 208-213.

[14] A. Peyerimhoff, Summierbarkeitsfaktoren für absolut Cesàro-summierbare Reihen. Math. Z. 59 (1954), 417-424.

[15] Ch. Pommerenke, The Bieberbach conjecture. Math. Intelligencer 7, No. 2 (1985), 23-25 and 32.

[16] K. Zeller und W. Beekmann, Theorie der Limitierungsverfahren. Springer-Verlag, Berlin-Heidelberg-New York, 2nd edition, 1970.

Wolfgang Beekmann, Fachbereich Mathematik und Informatik, Fernuniversität, Gesamthochschule, Postfach 940, D-5800 Hagen

Karl Zeller, Mathematische Fakultät, Universität Tübingen, Auf der Morgenstelle 10, D-7400 Tübingen

International Series of
Numerical Mathematics, Vol. 80

FOURIER INEQUALITIES WITH A_p-WEIGHTS

J.J. Benedetto, H.P. Heinig and R. Johnson

Abstract. In this paper weighted Fourier inequalities are established with weights in the A_p-class of Muckenhoupt. Specifically, for even, non-decreasing weights on $(0,\infty)$ the weight conditions are necessary and sufficient for the Paley-Titchmarsh inequality and its extensions. For decreasing weights the inequalities fail.

1. INTRODUCTION

In this paper we consider the weighted Fourier transform inequality

$$(1.1)\qquad \int_{-\infty}^{\infty} |\hat{f}(x)|^p |x|^{p-2}\omega(1/x)dx \le C\int_{-\infty}^{\infty} |f(x)|^p\omega(x)dx\ , \quad 1<p\le 2\ ,$$

and its extensions. Here ω is a non-negative weight function, C a constant independent of f and $\hat{f}$ the Fourier transform defined by

$$\hat{f}(x) = \int_{-\infty}^{\infty} e^{ixy}f(y)dy\ , \quad x\in\mathbb{R}\ .$$

If ω is an even function and non-decreasing on $(0,\infty)$, then Theorem 3.1 asserts that (1.1) holds if and only if $\omega\in A_p$. A_p stands for the Muckenhoupt weight class and consists of all non-negative locally integrable functions ω, such that for all intervals $(a,b)\subset\mathbb{R}$

$$\left(\int_a^b \omega(x)dx\right)^{1/p}\left(\int_a^b \omega(x)^{1-p'}dx\right)^{1/p'} \le (b-a)C$$

holds, where $1<p<\infty$ and p' is the conjugate index of p defined

---

This paper is in final form and no version of it will be submitted for publication elsewhere.

by $1/p' + 1/p = 1$. If $a = 0$ we denote the corresponding weight class by $A_p(0)$.

Since $|x|^\alpha \in A_p$ for $-1 < \alpha < p-1$ and if one restricts $\alpha \in [0,p-1)$, the result reduces to Pitt's theorem [9], while the case $\alpha = 0$ yields the Paley-Titchmarsh inequality (first proved in [4]). The generalizations of this result given in section 3 yield similar characterizations for mixed normed estimates. A special case is a weighted Hausdorff-Young inequality with $\omega^{p'-1} \in A_2$. For even weights, monotonic on $(0,\infty)$, one also obtains that $\omega \in A_2$ if and only if $\omega(1/x)$ is in A_2.

In section 4 it is shown that these results are sharp in the sense that if ω is even and decreasing on $(0,\infty)$, then (1.1) fails. On the other hand an example of a weight is given which is not <u>essentially increasing</u> but for which (1.1) holds with $p = 2$. A weight ω is an essentially increasing function if there exists an increasing function u and positive constants C_1, C_2, such that $C_1 \le \omega(x)/u(x) \le C_2$.

In order to illustrate that our methods are not confined to integral inequalities, we conclude by giving a sharp integrability theorem for the Fourier cosine series.

Inequalities similar to (1.1) for functions with vanishing moments were recently given in [8]. These results yield for example a Pitt inequality with α outside the usual range. In section 2 we prove two weight mixed norm inequalities also of functions with vanishing moments. Here the weights belong to the $F^*_{p,q}$ weight class considered in [2], and on specializing the results yield again Pitt type results.

Throughout C denotes a constant which may be different at different occurrences, but independent of the function f. p', as already observed, denotes always the conjugate index of p, and similarly for other letters.

2. INEQUALITIES OF FUNCTIONS WITH VANISHING MOMENTS

We begin with some lemmas.

LEMMA 2.1. *Define the operators* P *and* T *by*

$$(Pf)(x) = \operatorname{sgn} x \int_0^x f(t)dt , \quad (Tf)(x) = xf(x) , \quad x \in \mathbb{R} ,$$

where $f \in S$ (Schwartz function) and sgn x is 1 if $x \geq 0$ and -1 if $x < 0$. If $\int f(x)dx = 0$ then $[P(\widehat{Tf})](x) = -i \operatorname{sgn} x \hat{f}(x)$; and if $\int x^j f(x)dx = 0$, $j = 0,1,2,\dots,k-1$, then $[P^k(\widehat{T^k f})](x) = (-i \operatorname{sgn} x)^k \hat{f}(x)$, where $P^k = P(P^{k-1})$ and $T^k = T(T^{k-1})$, $k = 1,2,\dots$.

Proof. Since $\hat{f}(0) = 0$

$$[P(\widehat{Tf})](x) = \operatorname{sgn} x \int_0^x \left(\int_{-\infty}^{\infty} e^{iyt} tf(t)dt\right) dy$$

$$= \operatorname{sgn} x \int_{-\infty}^{\infty} tf(t) \left(\int_0^x e^{iyt}dy\right) dt$$

$$= \operatorname{sgn} x \int_{-\infty}^{\infty} (-i)f(t)(e^{ixt}-1)dt$$

$$= -i \operatorname{sgn} x \hat{f}(x) .$$

If $\int xf(x)dx = 0$ then $(\widehat{Tf})(0) = 0$ and by what we have shown $[P(\widehat{T^2 f})](x) = [P(\widehat{T(Tf)})](x) = -i \operatorname{sgn} x (\widehat{Tf})(x)$. Hence

$$[P^2(\widehat{T^2 f})](x) = \operatorname{sgn} x \int_0^x -i \operatorname{sgn} t(\widehat{Tf})(t)dt = -i \int_0^x (\widehat{Tf})(t)dt$$

$$= -i \operatorname{sgn} x[P(\widehat{Tf})](x) = (-i \operatorname{sgn} x)^2 \hat{f}(x) .$$

The general case follows now by induction.

LEMMA 2.2. Let P be as before and $1 < p \leq q < \infty$, then

$$\left(\int_{-\infty}^{\infty} u(x)|(Pf)(x)|^q dx\right)^{1/q} \leq C\left(\int_{-\infty}^{\infty} v(x)|f(x)|^p dx\right)^{1/p} \tag{2.1}$$

(u, v non-negative locally integrable functions) if and only if

$$\sup_{s>0} \left(\int_{|x|>s} u(x)dx\right)^{1/q} \left(\int_{|x|\leq s} v(x)^{1-p'} dx\right)^{1/p'} < \infty . \tag{2.2}$$

Proof. Since

$$\int_{-\infty}^{\infty} u(x)|(Pf)(x)|^q dx = \int_0^{\infty} u(x) \left|\int_0^x f(t)dt\right|^q dx + \int_0^{\infty} u(-x) \left|\int_0^x f(-t)dt\right|^q dx$$

the sufficiency of the result follows from [3].

To show that (2.1) implies (2.2) let $f(x) = v(x)^{1-p'}\chi_{(-s,s)}(x)$, $s > 0$ in (2.1), where χ_E is the characteristic function of $E \subset \mathbb{R}$. Then reducing the left side of (2.1) and using

$$|\operatorname{sgn} x \int_0^x \chi_{(-s,s)}(t)v(t)^{1-p'}dt| \geq \frac{1}{2} \int_{|t|\leq s} v(t)^{1-p'}dt$$

one obtains (2.2).

Let g be a Lebesgue measurable function defined on $\mathbb{R}$. The <u>distribution function</u> D_g of g is defined by $D_g(\lambda) = |\{x \in \mathbb{R}: |g(x)| > \lambda\}|$, $\lambda > 0$, where $|E|$ denotes the Lebesgue measure of the set $E \subset \mathbb{R}$. The <u>equimeasurable decreasing rearrangement</u> of g is defined by $g^*(x) = \inf\{\lambda > 0: D_g(\lambda) \leq x\}$.

We now define the weight class $F^*_{p,q}$, $1 < p \leq q < \infty$, to be the collection of all pairs of non-negative locally integrable functions (u,v) on $\mathbb{R}$ such that

$$(2.3) \qquad \sup_{s>0} \left[\int_0^{1/s} u^*(t)dt\right]^{1/q} \left[\int_0^s (1/v)^*(t)^{p'-1}dt\right]^{1/p'} < \infty$$

and write $(u,v) \in F^*_{p,q}$. We also write $(u,v) \in F_{p,q}$ if condition (2.3) holds without rearrangements on u and 1/v.

It was shown in [2] that if $(u,v) \in F^*_{p,q}$, $1 < p \leq q < \infty$, then $\|\hat{f}\|_{q,u} \leq C\|f\|_{p,v}$, where $\|g\|_{r,\omega} = \|g\omega^{1/r}\|_r$ and $\|\cdot\|_r$ is the usual Lebesgue norm. Moreover, if $\|\hat{f}\|_{q,u} \leq C\|f\|_{p,v}$, where u and v are even functions, then $(u,v) \in F_{p,q}$. We shall make use of these facts in the sequel.

The first inequality for functions with vanishing moment is:

THEOREM 2.3. <u>Suppose</u> $(u,v) \in F^*_{p,q}$, $1 < p \leq q < \infty$, <u>with</u> u <u>even and non-increasing on</u> $(0,\infty)$. <u>If</u> $f \in S$ <u>and</u> $\hat{f}(0) = 0$, <u>then</u>

$$(2.4) \qquad \left(\int_{-\infty}^{\infty} |\hat{f}(x)/x|^q u(x)dx\right)^{1/q} \leq C\left(\int_{-\infty}^{\infty} |xf(x)|^p v(x)dx\right)^{1/p}.$$

<u>Conversely, if</u> u <u>and</u> v <u>are even functions and</u> (2.4) <u>holds for all</u> f <u>whose first moment vanishes, then</u> $(u,v) \in F_{p,q}$.

<u>Proof.</u> By Lemma 2.1, $\hat{f}(x) = i \operatorname{sgn} x[P(\widehat{Tf})](x)$, so that

$$\left(\int_{-\infty}^{\infty} |\hat{f}(x)/x|^q u(x)dx\right)^{1/q} = \left(\int_{-\infty}^{\infty} |[P(\widehat{Tf})](x)/x|^q u(x)dx\right)^{1/q},$$

and by Lemma 2.2, this is dominated by

$$\left(\int_{-\infty}^{\infty} u(x)|(\widehat{Tf})(x)|^q dx\right)^{1/q} \tag{2.5}$$

if and only if for all $s > 0$

$$\left(\int_{|x|>s} |x|^{-q}u(x)dx\right)^{1/q}\left(\int_{|x|<s} u(x)^{1-q'}dx\right)^{1/q'} \le C\,. \tag{2.6}$$

But u is even and decreasing on $(0,\infty)$, so that the left side is bounded by $Cu(s)^{1/q}s^{-1+(1/q)}u(s)^{(1/q')-1}s^{1/q'} = C$. Also by [2, Theorem 1.1], (2.5) is dominated by $\|Tf\|_{p,v}$, whenever $(u,v)\in F^{*}_{p,q}$. Since $(Tf)(x) = xf(x)$, the first part is proved.

Conversely, let $f(x) = v(x)^{1-p'}/x$ if $\varepsilon \le x \le s$, $s > 0$, and $f(x) = 0$ for $0 < x < \varepsilon$. Extend f as an odd function to $\mathbb{R}$, then the integral of f is zero and

$$\hat{f}(x) = 2i\int_{\varepsilon}^{s} ((\sin xy)/y)v(y)^{1-p'}dy\,.$$

Substituting this into (2.4) and reducing the left side yields

$$\left(\int_{\varepsilon}^{1/s} u(x)\left|\int_{\varepsilon}^{s}((\sin xy)/(xy))v(y)^{1-p'}dy\right|^q dx\right)^{1/q} \le C\left(\int_{\varepsilon}^{s} v(x)^{1-p'}dx\right)^{1/p}$$

since u and v are even. Now $\varepsilon \le x \le s$ and $\varepsilon \le y \le 1/s$ imply $\varepsilon^2 \le xy \le 1$, and since $(\sin xy)/(xy) \ge \cos xy \ge \cos 1$ this shows that

$$\left(\int_{\varepsilon}^{1/s} u(x)dx\right)^{1/q}\left(\int_{\varepsilon}^{s} v(y)^{1-p'}dy\right)^{1-(1/p)} \le C\,,$$

where C is independent of ε. Now let $\varepsilon \to 0+$, then this implies that $(u,v)\in F_{p,q}$.

Observe that the condition u even and non-increasing in the first part of Theorem 2.3 can be replaced by the condition $u \in A_q$, because then by [5, Lemma 1] (2.6) holds since the first term is bounded by

$$Cs^{-1}\left(\int_{|x|<s} u(x)dx\right)^{1/q}\,.$$

Note also that $u(x) = |x|^{\alpha}$ and $v(x) = |x|^{\beta}$, $-1 < \alpha < 0$, $0 < \beta < p-1$ with $(1-\beta)/p = (1+\alpha)/q$ satisfy the condition of Theorem 2.3. Therefore (2.4) yields the Pitt type inequality

$$\left(\int_{-\infty}^{\infty} |\hat{f}(x)|^q |x|^{\alpha-q}dx\right)^{1/q} \le C\left(\int_{-\infty}^{\infty} |f(x)|^p |x|^{\beta+p}dx\right)^{1/p}\,.$$

If one requires that higher moments of f vanish, then a more general result can be obtained.

THEOREM 2.4. Suppose $(u,v) \in F^*_{p,q}$, $1 < p \le q < \infty$, with u even and non-increasing on $(0,\infty)$. If $f \in S$ and $\int x^j f(x)dx = 0$, $j = 0,1,\ldots,k-1$, then

$$\left(\int_{-\infty}^{\infty} |\hat{f}(x)/x^k|^q u(x)dx\right)^{1/q} \le C\left(\int_{-\infty}^{\infty} |f(x)|^p |x|^{kp} v(x)dx\right)^{1/p} .$$

Proof. By Lemma 2.1 and Lemma 2.2

$$\left(\int_{-\infty}^{\infty} |\hat{f}(x)/x^k|^q u(x)dx\right)^{1/q} = \left(\int_{-\infty}^{\infty} |[P^k(\widehat{T^k f})](x)|^q |x|^{-kq} u(x)dx\right)^{1/q}$$

(2.7)

$$\le C\left(\int_{-\infty}^{\infty} |[P^{k-1}(\widehat{T^k f})](x)|^q |x|^{(1-k)q} u(x)dx\right)^{1/q}$$

if and only if

$$\left(\int_s^{\infty} x^{-kq} u(x)dx\right)^{1/q} \left(\int_0^s [x^{(1-k)q} u(x)]^{1-q'} dx\right)^{1/q'} \le C .$$

But u is non-increasing on $(0,\infty)$, so that the left side of this expression is bounded by

$$Cu(s)^{1/q} s^{(1/q)-k} u(s)^{(1/q')-1} s^{((q'-1)(k-1)q/q')+(1/q')} = C .$$

Therefore (2.7) is satisfied. Applying Lemma 2.2 (k-1) times to (2.7) one obtains

$$\left(\int_{-\infty}^{\infty} |\hat{f}(x)/x^k|^q u(x)dx\right)^{1/q} \le C\left(\int_{-\infty}^{\infty} |(\widehat{T^k f})(x)|^q u(x)dx\right)^{1/q}$$

$$\le C\left(\int_{-\infty}^{\infty} |(T^k f)(x)|^p v(x)dx\right)^{1/p} .$$

Here the last inequality follows from [2, Theorem 1.1]. Since $(T^k f)(x) = x^k f(x)$ the result follows.

3. MAIN RESULTS

We now give a number of Fourier inequalities with A_p weights. The first result is a generalization of the Paley-Titchmarsh inequality.

THEOREM 3.1. _Let_ ω _be an even weight function, non-decreasing on_ $(0,\infty)$. _For_ $1<p\le 2$

$$\int_{-\infty}^{\infty}|\hat{f}(x)|^p|x|^{p-2}\omega(1/x)dx \le C\int_{-\infty}^{\infty}|f(x)|^p\omega(x)dx \tag{3.1}$$

if and only if $\omega\in A_p$.

Proof. Let $\omega\in A_p$, then by [5, Lemma 1]

$$\int_s^{\infty}x^{-p}\omega(x)dx \le Cs^{-p}\int_0^s\omega(x)dx$$

for each $s>0$. Therefore

$$\left(\int_s^{\infty}x^{-p}\omega(x)dx\right)^{1/p}\left(\int_0^s\omega(x)^{1-p'}dx\right)^{1/p'} \le C \tag{3.2}$$

and replacing x by 1/x in the first integral of (3.2) it follows that $(\tilde{\omega},\omega)\in F_{p,p}$, where $\tilde{\omega}(x)=|x|^{p-2}\omega(1/x)$. Since $\tilde{\omega}$ is non-decreasing, [1, Theorem 1] implies (3.1).

Conversely, if (3.1) is satisfied with ω even, then by [1, Theorem 2] $(\tilde{\omega},\omega)\in F_{p,p}$ or equivalently, (3.2) holds. But ω is non-decreasing on $(0,\infty)$, so that for each $s>0$

$$\begin{aligned}\left(\int_s^{\infty}x^{-p}\omega(x)dx\right)^{1/p} &\ge \omega(s)^{1/p}s^{-1/p'}(p-1)^{-1/p}\\ &\ge (p-1)^{-1/p}s^{-1}\left(\int_0^s\omega(x)dx\right)^{1/p}\end{aligned}$$

and therefore $\omega\in A_p(0)$. Now let $0<\theta<1$. Then as above

$$\begin{aligned}(p-1)^{1/p}\left(\int_s^{\infty}x^{-p}\omega(x)dx\right)^{1/p} &\ge \omega(s)^{1/p}s^{-1/p'}\\ &\ge s^{-1}(1-\theta)^{-1/p}\left(\int_{\theta s}^{s}\omega(x)dx\right)^{1/p}\end{aligned}$$

and this together with (3.2) implies that for every $s>0$

$$s^{-1}(1-\theta)^{-1/p}\left(\int_{\theta s}^{s}\omega(x)dx\right)^{1/p}\left(\int_0^s\omega(x)^{1-p'}dx\right)^{1/p'} \le C\,.$$

If we show that

$$\left(\int_0^s\omega(x)^{1-p'}dx\right)^{1/p'} \ge (1-\theta)^{-1/p'}\left(\int_{\theta s}^{s}\omega(x)^{1-p'}dx\right)^{1/p'}, \tag{3.3}$$

then this together with the previous inequality shows that $\omega\in A_p$

with $b = s$ and $a = \theta s$, $b-a = s(1-\theta)$.

To prove (3.3) observe that

$$\int_{\theta s}^{s} \omega(x)^{1-p'}dx \le \omega(\theta s)^{1-p'}s(1-\theta) \le ((1-\theta)/\theta)\int_{0}^{\theta s} \omega(x)^{1-p'}dx$$

and adding $((1-\theta)/\theta)\int_{\theta s}^{s} \omega(x)^{1-p'}dx$ to both sides of this inequality yields

$$\theta^{-1}\int_{\theta s}^{s} \omega(x)^{1-p'}dx \le ((1-\theta)/\theta)\int_{0}^{s} \omega(x)^{1-p'}dx .$$

This implies (3.3). A similar calculation works for intervals (a,b), where $a < 0 < b$, by splitting the integral as $\int_a^0 + \int_0^b$; and hence the theorem is proved.

The proof of Theorem 3.1 actually gives at once the following:

COROLLARY 3.2. Let ω be even and non-decreasing on $(0,\infty)$. Then $\omega \in A_p$, $1 < p < \infty$ if and only if (3.2) holds.

By duality, one also finds a necessary and sufficient condition for a decreasing function to belong to A_p.

PROPOSITION 3.3. Let ω be even and monotonic on $(0,\infty)$. Then $\omega \in A_2$ if and only if $\omega(1/x) \in A_2$.

Proof. We suppose ω is non-decreasing. By Theorem 3.1, $\omega \in A_2$ is equivalent to (3.1) with $p = 2$. But by [1, Theorems 1 and 2] (see also [2, Theorem 1.1]) this is equivalent to $(1/u,\omega) \in F_{2,2}$, where $u(x) = 1/\omega(1/x)$, or

$$\left(\int_0^{1/s} u(x)^{-1}dx\right)^{1/2}\left(\int_0^{s} \omega(x)^{-1}dx\right)^{1/2} \le C$$

for each $s > 0$. Replacing x by $1/x$ in the second integral this becomes

$$\left(\int_0^{1/s} u(x)^{-1}dx\right)^{1/2}\left(\int_{1/s}^{\infty} x^{-2}u(x)dx\right)^{1/2} \le C$$

and this by Corollary 3.2 with $p = 2$ is equivalent to $u \in A_2$. It is well known that $u \in A_2$ if and only if $1/u \in A_2$.

The proof for ω non-increasing is similar.

Theorem 3.1 has the following extension:

THEOREM 3.4. *Suppose* ω *is even, non-decreasing on* $(0,\infty)$ *and* $1 < p \leq q \leq p' < \infty$, *then*

$$(3.4) \qquad \left(\int_{-\infty}^{\infty} |f(x)|^q |x|^{(q/p')-1}\omega(1/x)^{q/p}dx\right)^{1/q} \leq C\left(\int_{-\infty}^{\infty} |f(x)|^p \omega(x)dx\right)^{1/p}$$

if and only if $\omega^{q/p} \in A_{1+(q/p')}$.

Proof. We only sketch the proof since it is quite similar to that of Theorem 3.1.

Let $u = \omega^{q/p}$ and $r = 1+(q/p')$ then $u \in A_r$ and [5, Lemma 1] implies that for each $s > 0$

$$\left(\int_s^{\infty} x^{-r}u(x)dx\right)^{1/r}\left(\int_0^s u(x)^{1-r'}dx\right)^{1/r'} \leq C .$$

Replacing x by 1/x in the first integral and raising the inequality to the (r/q)-th power one obtains $(\tilde{\omega},\omega) \in F_{p,q}$ with $\tilde{\omega}(x) = |x|^{(q/p')-1}\omega^{q/p}(1/x)$ since $r-2 = (q/p')-1$, $r/(qr') = 1/p'$ and $q(q-r')/p = 1-p'$. But $\tilde{\omega}$ is non-increasing on $(0,\infty)$ so that (3.4) follows from [1, Theorem 1].

Conversely, if (3.4) holds then [1, Theorem 2] implies

$$(3.5) \qquad \left(\int_0^{1/s} \omega(1/x)^{q/p}x^{(q/p')-1}dx\right)^{1/q}\left(\int_0^s \omega(x)^{1-p'}dx\right)^{1/p'} \leq C .$$

But ω non-decreasing on $(0,\infty)$ shows that the first product dominates

$$(3.6) \qquad \omega(s)^{1/p}s^{-1/p'} \geq s^{-(1/p'+1/q)}\left(\int_0^s \omega(x)^{q/p}dx\right)^{1/q} .$$

Substituting this estimate into (3.5) and raising the resulting expression to the (q/r)-th power shows that $\omega^{q/p} \in A_r(0)$. If $0 < \theta < 1$ one obtains an inequality (3.5) with a factor $(1-\theta)^{-1/q}$ on the right and the lower integral limit is θs. Also, as in the proof of Theorem 3.1

$$\int_{\theta s}^{s} \omega(x)^{1-p'}dx \leq (1-\theta)\int_0^s \omega(x)^{1-p'}dx$$

so that substituting these estimates in (3.5) with $s = b$, $\theta s = a$ yields

$$(b-a)^{-r/q}\left[\int_a^b \omega(x)^{q/p}dx\right]^{1/q}\left[\int_a^b \omega(x)^{(q/p)(1-r')}dx\right]^{1/p'} \le C$$

and this implies that $\omega^{q/p} \in A_r$.

With $q = p'$ we obtain the following weighted Hausdorff-Young inequality:

COROLLARY 3.5. *Suppose* ω *is even, non-decreasing on* $(0,\infty)$ *and* $1 < p \le 2$, *then*

$$\left[\int_{-\infty}^{\infty} |\hat{f}(x)|^{p'} \omega(1/x)^{p'-1} dx\right]^{1/p'} \le C\left[\int_{-\infty}^{\infty} |f(x)|^p \omega(x)dx\right]^{1/p}$$

if and only if $\omega^{p'-1} \in A_2$.

The dual of Theorem 3.4 is the last result of this section.

THEOREM 3.6. *Suppose* ω *is even, non-decreasing on* $(0,\infty)$ *and* $1 < p \le q \le p' < \infty$, *then*

$$\left[\int_{-\infty}^{\infty} |\hat{f}(x)|^{p'} \omega(x)^{1-p'} dx\right]^{1/p'}$$

$$\le C\left[\int_{-\infty}^{\infty} |f(x)|^{q'} |x|^{(q'/p)-1} \omega(1/x)^{-q'/p} dx\right]^{1/q'}$$

if and only if $\omega^{1-p'} \in A_{1+(p'/q)}$.

Proof. We assume that f is simple, for then the general case follows by approximating.

Now using the fact that $u \in A_r$, $1 < r < \infty$, if and only if $u^{1-r'} \in A_{r'}$, we see that $\omega^{1-p'} \in A_{1+(p'/q)}$ if and only if $\omega^{q/p} \in A_{1+(q/p')}$. Also, by duality

$$\left[\int_{-\infty}^{\infty} |\hat{f}(x)|^{p'} \omega(x)^{1-p'} dx\right]^{1/p'} = \sup \left|\int_{-\infty}^{\infty} \hat{f}(x)h(x)dx\right| ,$$

where the supremum is taken over all simple h with $\|h\|_{p,\omega} \le 1$. Now Parseval's equation, Hölder's inequality and Theorem 3.4 show that

$$\left|\int_{-\infty}^{\infty} \hat{f}(x)h(x)dx\right| = \left|\int_{-\infty}^{\infty} f(x)\hat{h}(x)dx\right|$$

$$\le \left(\int_{-\infty}^{\infty} |\hat{h}(x)|^q |x|^{(q/p')-1} \omega(1/x)^{q/p} dx\right)^{1/q} \times$$

$$\left(\int_{-\infty}^{\infty} |f(x)|^{q'} |x|^{q'(1/q-1/p')} \omega(1/x)^{-q'/p} dx\right)^{1/q'}$$

$$\le \left(\int_{-\infty}^{\infty} |h(x)|^p \omega(x) dx\right)^{1/p} \left(\int_{-\infty}^{\infty} |f(x)|^{q'} |x|^{(q'/p)-1} \omega(1/x)^{-q'/p} dx\right)^{1/q'},$$

and this implies the inequality.

Conversely, if the inquality holds with ω even, then [1, Theorem 2] shows that for each $s > 0$

$$\left(\int_0^s \omega(x)^{1-p'} dx\right)^{1/p'} \left(\int_0^{1/s} |x|^{(q/p')-1} \omega(1/x)^{q/p} dx\right)^{1/q} \le C .$$

The same argument as the proof of Theorem 3.4 shows that $\omega^{q/p} \in A_{1+(q/p')}$ which is equivalent to $\omega^{1-p'} \in A_{1+(p'/q)}$.

In [7] Rooney proved the Fourier inequality

$$\left(\int_{-\infty}^{\infty} |\hat{f}(x)|^q |x|^{(q/r')-1} dx\right)^{1/q} \le C\left(\int_{-\infty}^{\infty} |f(x)|^p |x|^{(p/r)-1} dx\right)^{1/p} ,$$

where $1 < p < \infty$, $1 < r \le \min(p,p')$, and $p \le q \le r'$. We note that in the case $1 < p \le 2$ his inequality follows from Theorem 3.4 with $\omega(x) = |x|^{(p/r)-1}$ while the case $p > 2$ follows from Theorem 3.6, only there one replaces p' by q and q' by p, $p > 2$ and takes $\omega(x) = |x|^{(q'/r)-1}$.

The restriction $1 < p \le 2$ arises in Theorem 3.1 only to guarantee that $|x|^{p-2}\omega(1/x)$ is non-increasing. The result holds for $p > 2$ if we assume that $|x|^{p-2}\omega(1/x)$ is non-increasing since this implies that ω is non-decreasing. Similar remarks apply to the other results.

4. ADDITIONAL RESULTS

The first result in this section shows that the weight ω in Theorem 3.1 cannot be replaced by a decreasing function.

PROPOSITION 4.1. *If ω is a non-increasing even function for which inequality* (3.1) *holds, then ω is equivalent to a constant a.e.*

Proof. Fix $f \in L^p \cap L^p_\omega$, $1 < p \le 2$, and define f_δ, $\delta > 0$, by $f_\delta(x) = \delta^{-1} f([x-x_0]/\delta)$ where $x_0 > 0$. Then $\hat{f}_\delta(x) = e^{ixx_0}\hat{f}(\delta x)$ so that (3.1) takes the form

$$\int_{-\infty}^{\infty} |\hat{f}(\delta x)|^p |x|^{p-2}\omega(1/x)dx \le C \int_{-\infty}^{\infty} |\delta^{-1} f([x-x_0]/\delta)|^p \omega(x)dx .$$

If $y = \delta x$ in the first and $t = (x-x_0)/\delta$ in the second integral one obtains

$$\int_{-\infty}^{\infty} |\hat{f}(y)|^p |y|^{p-2}\omega(\delta/y)dy \le C \int_{-\infty}^{\infty} |f(t)|^p \omega(t\delta + x_0)dt ,$$

so that by Fatou's lemma as $\delta \to 0+$, $\omega(0+) \le C\omega(x_0)$. But since $\omega(x_0) \le \omega(0+)$ the result follows.

A similar result can be proved for Theorem 3.4.

It is easy to see that the theorems of section 3 are still valid if the weights ω are replaced by even functions which are essentially increasing on $(0,\infty)$. One might therefore conjecture that these are precisely the functions required in these theorems. The following example shows however that this is not the case.

EXAMPLE. Define ω by

$$\omega(x) = \begin{cases} (3x/2)^{\alpha+\gamma} & \text{if } 0 < x \le 2/3 \\ 2^\gamma |x^{-1}-1|^\gamma & \text{if } 2/3 < x \le 2 \\ (x/2)^\alpha & \text{if } 2 < x , \end{cases}$$

where $\alpha > 0$, $\gamma > 0$ and $0 < \alpha+\gamma < 1$. Extend ω as an even function to $(-\infty,\infty)$. Then ω is not essentially increasing on $(0,\infty)$. We show, however, that ω satisfies (3.1) with $p = 2$. For this we show that $(\bar{\omega},\omega) \in F^*_{2,2}$, where $\bar{\omega}(x) = \omega(1/x)$, for then [2, Theorem 1.1] shows that (3.1) with $p = 2$ is satisfied.

We now compute $\bar{\omega}^*$ and then $(1/\omega)^*$. Since

$$\bar{\omega}(x) = \begin{cases} (2x)^{-\alpha} & \text{if } 0 < x < 1/2 \\ 2^\gamma |x-1|^\gamma & \text{if } 1/2 \le x < 3/2 \\ [3/(2x)]^{\alpha+\gamma} & \text{if } 3/2 \le x < \infty \end{cases}$$

the distribution function of $\bar{\omega}$ is $D_{\bar{\omega}}(\lambda) = |\{x \in (0,1/2): (2x)^{-\alpha} > \lambda\}| +$

$|\{x \in [1/2,3/2]: 2^\gamma|x-1|^\gamma > \lambda\}| + |\{x \geq 3/2: [3/(2/x)]^{\alpha+\gamma} > \lambda\}| = D^1 + D^2 + D^3$, respectively. If $\lambda \geq 1$, then $D^2 + D^3 = 0$ and $D^1 = \lambda^{-1/\alpha}/2$. If $0 < \lambda < 1$, then $D^1 = 1/2$, $D^2 = 1-\lambda^{1/\gamma}$ and $D^3 = 3(\lambda^{-1/(\alpha+\gamma)}-1)/2$ so that $D_{\bar{\omega}}(\lambda) \leq 3\lambda^{-1/(\alpha+\gamma)}/2$ if $0 < \lambda < 1$. Therefore

$$\bar{\omega}^*(t) \leq \begin{cases} (2t)^{-\alpha} & \text{if } 0 < t \leq 1/2 \\ (2t/3)^{-(\alpha+\gamma)} & \text{if } t > 1/2 . \end{cases} \tag{4.1}$$

Also, since

$$(1/\omega)(x) = \begin{cases} (3x/2)^{-(\alpha+\gamma)} & \text{if } 0 < x \leq 2/3 \\ 2^{-\gamma}|x^{-1}-1|^{-\gamma} & \text{if } 2/3 < x \leq 2 \\ (x/2)^{-\alpha} & \text{if } 2 < x \end{cases}$$

the distribution function of $1/\omega$ is $D_{1/\omega}(\lambda) = |\{x \in (0,2/3]: (3x/2)^{-(\alpha+\gamma)} > \lambda\}| + |\{x \in (2/3,1]: 2^{-\gamma}|x^{-1}-1|^{-\gamma} > \lambda\}| + |\{x \in (1,2]: 2^{-\gamma}|x^{-1}-1|^{-\gamma} > \lambda\}| + |\{x > 2: (x/2)^{-\alpha} > \lambda\}| = D_1 + D_2 + D_3 + D_4$, respectively. If $\lambda \geq 1$, straightforward calculations show that $D_1 = 2\lambda^{-1/(\alpha+\gamma)}/3$, $D_2 = 1-1/(1+\lambda^{-1/\gamma}/2)$, $D_3 = -1+1/(1-\lambda^{-1/\gamma}/2)$ and $D_4 = 0$. But since $D_2 + D_3 \leq 4\lambda^{-1/\gamma}/3 \leq 4\lambda^{-1/(\alpha+\gamma)}/3$ it follows that for $\lambda \geq 1$, $D_{1/\omega}(\lambda) \leq 2\lambda^{-1/(\alpha+\gamma)}$. If $0 < \lambda < 1$, $D_1 = 2/3$, $D_2 = 1/3$, $D_3 = 1$ and $D_4 = 2(\lambda^{-1/\alpha}-1)$, so that $D_{1/\omega}(\lambda) \leq 2\lambda^{-1/\alpha}$. Hence

$$(1/\omega)^*(t) \leq \begin{cases} (t/2)^{-(\alpha+\gamma)} & \text{if } 0 < t \leq 2 \\ (t/2)^{-\alpha} & \text{if } t > 2 . \end{cases} \tag{4.2}$$

Now substituting (4.1) and (4.2) into

$$\left(\int_0^{1/s} \bar{\omega}^*(t)dt\right)\left(\int_0^s (1/\omega)^*(t)dt\right)$$

and considering the cases $0 < s \leq 1/2$ and $1/2 < s < \infty$ separately it follows that the integral product is bounded for all $s > 0$ and therefore $(\bar{\omega},\omega) \in F^*_{2,2}$.

We conclude with a weighted integrability theorem for the Fourier cosine series which generalizes a corresponding result of Igari [6].

THEOREM 4.2. _Suppose_ $\{\lambda_n\}_{n=1}^\infty$ _is a decreaing sequence such that_ $\lim \lambda_n = 0$ _and let_ $f(x) = \sum \lambda_n \cos nx$. _For_ ω _a non-decreasing_ *function in* A_p, $1 < p \le 2$, *the series* $\sum \lambda_n^p n^{p-2} \omega(\pi/n)$ *converges if and only if* $f \in L_\omega^p(0,2\pi)$.

Proof. Let

$$F(x) = \int_0^x f(t)dt = \sum_{n=1}^{\infty} n^{-1} \lambda_n \sin nx ,$$

then following the proof of [6, Theorem 2], $F(\pi/n) \ge C\lambda_n$, $n = 2, \dots$; and hence

$$\sum_{n=2}^{\infty} \lambda_n^p n^{p-2} \omega(\pi/n) \le C \sum_{n=2}^{\infty} \int_{\pi/n}^{\pi/(n-1)} x^{-p} \omega(x) |F_1(x)|^p dx$$

$$\le C \int_0^{\pi} \omega(x) |f(x)|^p dx ,$$

where $F_1(x) = \int_0^x |f(t)|dt$. Here the last inequality follows from Lemma 2.2 with $q = p$ if and only if

$$\sup_{s \in (0,\pi)} \left(\int_s^{\pi} x^{-p} \omega(x) dx \right)^{1/p} \left(\int_0^s \omega(x)^{1-p'} dx \right)^{1/p'} \equiv C < \infty .$$

But by Corollary 3.2 this holds if $\omega \in A_p$.

Conversely, from [6, Theorem 2],

$$|f(x)| \le \left| \sum_{m=1}^{n} \lambda_m \right| + \left| \sum_{m=n+1}^{\infty} \lambda_m \cos mx \right| \le P_n + \pi\lambda_n/x ,$$

where $P_n = \sum_{m=1}^{n} \lambda_m$, and therefore $|f(x)| \le CP_n$ for $\pi/(n+1) \le x < \pi/n$. Let $g(x) = \lambda_n$ if $n-1 \le x < n$, $n = 1, 2, \dots$, and $P(x) = \int_0^x g(t)dt$, then

$$\int_0^{\pi} \omega(x) |f(x)|^p dx = \sum_{n=1}^{\infty} \int_{\pi/(n+1)}^{\pi/n} \omega(x) |f(x)|^p dx$$

$$\le C \sum_{n=1}^{\infty} n^{-2} \omega(\pi/n) P_n^p$$

$$\le C\omega(\pi)\lambda_1^p + C \sum_{n=2}^{\infty} \int_{n-1}^{n} x^{-2} \omega(\pi/x) P(x)^p dx$$

$$\le C\omega(\pi)\lambda_1^p + C \int_0^{\infty} \omega(\pi/x) x^{p-2} |g(x)|^p dx$$

$$= C\omega(\pi)\lambda_1^p + C \sum_{n=1}^{\infty} \omega(\pi/n) n^{p-2} \lambda_n^p .$$

Here the last inequality follows from Lemma 2.2 with $p = q$ if and only if

$$\sup_{s>0} \left[\int_s^\infty x^{-2}\omega(\pi/x)dx\right]^{1/p} \left[\int_0^s [x^{p-2}\omega(\pi/x)]^{1-p'}dx\right]^{1/p'} < \infty$$

which with $y = \pi/x$ is equivalent to

$$(4.3) \quad \sup_{s>0} \left[\int_0^{\pi/s} \omega(y)dy\right]^{1/p} \left[\int_{\pi/s}^\infty y^{-p'}\omega(y)^{1-p'}dy\right]^{1/p'} < \infty .$$

But $\omega \in A_p$ if and only if $\omega^{1-p'} \in A_{p'}$, and since by Corollary 3.2 (4.3) is equivalent to $\omega^{1-p'} \in A_{p'}$, the result follows.

REFERENCES

1. J.J. Benedetto and H.P. Heinig, Weighted Hardy spaces and the Laplace transform. Harmonic Anal. Conf., Cortona, Italy, 1982. Lect. Notes in Math. 992, Springer Verlag, 240-277.

2. J.J. Benedetto, H.P. Heinig and R. Johnson, Weighted Hardy spaces and the Laplace transform II. Math. Nachr. (to appear).

3. J.S. Bradley, Hardy inequalities with mixed norms. Canad. Math. Bull. 25 (1978), 405-408.

4. G.H. Hardy and J.E. Littlewood, Some new properties of Fourier constants. Math. Ann. 97 (1926-27), 159-209.

5. R. Hunt, B. Muckenhoupt and R. Wheeden, Weighted norm inequalities for the conjugate function and Hilbert transform. Trans. Amer. Math. Soc. 176 (1973), 227-251.

6. S. Igari, Some integrability theorems of trigonometric series and monotone decreasing functions. Tôhoku Math. J. 12 (1) (1960), 139-146.

7. P.G. Rooney, Generalized H_p-spaces and Laplace transforms. Abstract spaces and approximation. In: P.L. Butzer and B.Sz. Nagy (ed.), Proc. Conf. Oberwolfach, 1966, 258-269.

8. C. Sadosky and R.L. Wheeden, Some weighted norm inequalities for the Fourier transform of functions with vanishing moments. Manuscript (18 pages).

9. E.M. Stein, Interpolation of linear operators. Trans. Amer. Math. Soc. 83 (1956), 482-492.

J.J. Benedetto, Department of Mathematics, University of Maryland, College Park, Maryland 20742, U.S.A.

H.P. Heinig, Department of Mathematics, McMaster University, Hamilton, Ontario, L8S 4K1, Canada

R. Johnson, Department of Mathematics, University of Maryland, College Park, Maryland 20742, U.S.A.

The authors were supported by NSF and NSERC grants.

International Series of
Numerical Mathematics, Vol. 80

EXPERIMENTING WITH OPERATOR INEQUALITIES USING APL

Achim Clausing

Abstract. In this paper, some results concerning differential operator inequalities are presented which have first been found "experimentally" using a set of functions written in APL. By interactively using a computer, inequalities for the coefficients of certain polynomials related to Pólya operators as well as eigenvalue inequalities have been obtained.

"It will seem not a little paradoxical to ascribe a great importance to observations even in that part of the mathematical sciences which is usually called Pure Mathematics [...]. As we must refer the numbers to the pure intellect, we can hardly understand how observations and quasi-experiments can be of use in investigating the nature of the numbers. Yet, in fact, as I shall show here with very good reasons, the properties of the numbers known today have been mostly discovered by observation [...].

Indeed, we should use such a discovery as an opportunity to investigate more exactly the properties discovered and to prove or disprove them; in both cases we may learn something useful."

(L. Euler, Specimen de usu observationum in mathesi pura. [6], cf. G. Pólya [10], p. 3)

1. INTRODUCTION

G. Pólya, one of the founders of inequality theory, was also an engaged promoter of what is now offen called experimental mathematics. On many occasions he argued that attention should be given to the heuristic process (cf. e.g. [10]). Today,

---

This paper is in final form and no version of it will be submitted for publication elsewhere.

computers are a main tool for heuristics, even in pure mathematics, since they allow for making observations, in the sense of the preceding quotation from Euler, on a larger scale and in a more systematic way than before.

In this paper, I will report on an experimental exploration of Pólya operators, a class of ordinary linear differential operators defined by an operator inequality. The "experiments" have led to new inequalities concerning certain polynomials, Green's functions, and eigenvalues related to these operators.

The computations have been carried out using an APL workspace. The choice of APL instead of one of the more customary programming languages was made for two reasons. The main argument was that investigating linear operators involves the manipulation of rectangular data which is the basic data type supported by APL. In fact, APL is the only major language to have this feature. Furthermore, APL is an interpreted language that is used interactively. The ensuing "dialogue with the computer", in this case, turned out to be a very valuable help for guiding the exploration process. The decision to use APL was influenced by Grenanders book [7], where information about APL as a tool for heuristics can be found.

2. POLYA OPERATORS

The objects of this paper will be certain ordinary linear differential operators of the form

$$\mathcal{L}: C^p[a,b] \to C[a,b] \times \mathbb{R}^p, \qquad \mathcal{L}(f) = (Lf, \delta_1(f), \ldots, \delta_p(f)), \tag{1}$$

where $p \in \mathbb{N}$, all functions are real-valued, and L, resp. $\delta_1, \ldots, \delta_p$, denotes a differential expression

$$L = \sum_{i=0}^{p} a_i D^i \qquad (D = \frac{d}{dx}, \; a_1, \ldots, a_p \in C[a,b]),$$

resp. a set of boundary conditions for L (see below for details). If, for all $f \in C^p[a,b]$,

$$Lf = 0, \; \delta_i(f) = 0 \qquad (i = 1, \ldots, p) \Rightarrow f = 0 \tag{2}$$

holds, then the operator $\mathcal{L}$ is invertible and its inverse operator can be written as follows:

$$\mathcal{L}^{-1}(g,c_1,\ldots,c_p) = \sum_{i=1}^{p} c_i G_i + \int_a^b G(\cdot,t)g(t)dt. \tag{3}$$

Here, $G_1,\ldots,G_p \in N_L = \{ y \in C^p[a,b] \mid Ly = 0\}$ is the base of N_L which is uniquely determined by

$$\delta_i(G_j) = \delta_{ij} \qquad (i,j = 1,\ldots,p), \tag{4}$$

and $G(x,t)$ is the Green's kernel associated with $\mathcal{L}$.

The "inequality version" of condition (2) is

$$Lf \geqq 0,\ \delta_i(f) \geqq 0 \quad (i = 1,\ldots,p) \Rightarrow f \geqq 0. \tag{5}$$

An operator $\mathcal{L}$ satisfying (5) is called inverse-positive (cf. [11]). Of course, any inverse-positive operator is invertible, and it is trivial to see that an invertible operator $\mathcal{L}$ is inverse-positive if and only if

$$\begin{aligned} &G(x,t) \geqq 0 \qquad (x,t \in [a,b]), \\ &G_i(x) \geqq 0 \qquad (x \in [a,b],\ i = 1,\ldots,p). \end{aligned} \tag{6}$$

The theory of inverse-positive operators is not as well-developed as it deserves to be. It has close ties with potential theory, inequality theory, function theory, and the theory of ordinary, linear and nonlinear, differential equations (cf. [11]). It is this broad range of applications that motivates the study of Pólya operators ([2], [3], [4]), a large subclass of inverse-positive differential operators.

For brevity, and since they are the most natural example, we consider in this paper only those Pólya operators for which

$$L = \sigma_L D^p \qquad (\sigma_L \in \{0,1\}) \tag{7}$$

holds. The base interval will be $[0,1]$. The boundary conditions $\delta_1,\ldots,\delta_p$ are said to be of *Rolle type* if they look as follows:

$$\delta_i = u_i\, y_{x_i}^{(j_i)} + v_i y_{x_i}^{(j_i+1)} \qquad (i = 1,\ldots,p), \tag{8}$$

where

$j_i \in \{0,\ldots,p-1\}$ (the *order* of δ_i), $j_i \neq j_k$ if $i \neq k$,

$x_i \in \{0,1\}$, $u_i \neq 0$,

$$(9) \qquad u_i v_i \begin{cases} \leqq 0 & \text{if} \quad x_i = 0, \\ \geqq 0 & \text{if} \quad x_i = 1. \end{cases}$$

For a function $f \in C^{(j_i+1)}[a,b]$ satisfying $\delta_i(f) = 0$, the derivative $f^{(j_i)}(x)$ either has a zero at x_i or moves away from the x-axis as x moves away from x_i into the interior of $[0,1]$. (More general conditions having this behaviour exist but shall not be considered here.)

DEFINITION. An operator $\mathcal{L}$ satisfying (7) - (9) is called a *Pólya operator* if it is inverse positive. If $v_i = 0$ $(i = 1,\ldots,p)$, then $\mathcal{L}$ is called a *standard* Pólya operator.

Part i of the following explicit description of Pólya operators is essentially due to G. Pólya [9]. As for notation, we let M_j denote the number of boundary conditions such that $j_i \leqq j$. For part ii, we assume the boundary conditions to be numbered such that

$$(10) \qquad j_1 < \ldots < j_q, \; j_{q+1} > \ldots > j_p,$$

where q is the number of boundary conditions for which $x_i = 0$ holds.

THEOREM. ([2], Proposition 4.4) Let $\mathcal{L}$ be an operator as described in (7) - (9). Then

i) $\mathcal{L}$ is nonsingular if and only if Pólya's condition holds:

$$(11) \qquad M_{j-1} \geqq j \qquad (j = 1,\ldots,p)$$

ii) If $\mathcal{L}$ is nonsingular then it is inverse-positive if and only if the signs σ_L and $\sigma_i = \operatorname{sgn}(u_i)$ $(i = 1,\ldots,p)$ are given by

$$\sigma_L = (-1)^{p-q}$$

$$(12) \qquad \sigma_i = \begin{cases} (-1)^{j_i-i+1} & (i = 1,\ldots,q), \\ (-1)^{p-i} & (i = q+1,\ldots,p). \end{cases}$$

In other words, $\mathcal{L}$ is a Pólya operator if and only if (11) and (12) hold.

3. REPRESENTATION BY POLYA SCHEMES

A compact form to represent Pólya operators is by using

incidence schemes: With $\mathcal{L}$ satisfying (7) - (9) we associate a $(2 \times p)$-matrix

$$P = (e_{k,j})_{k=0,1;\, j=0,\dots,p-1}$$

of 0's and 1's in which δ_i is represented by a 1 in row x_i, column j_i. The remaining entries of P are 0. Then we have

$$(13) \qquad q = \sum_{j=0}^{p-1} e_{0,j}, \quad M_j = \sum_{k=0,1} \sum_{i=0}^{j} e_{k,i},$$

where q and M_j are the numbers appearing in Theorem 1. A $(2 \times p)$-matrix of p. 0's and 1's satisfying (7) is called a *Pólya scheme* of order p. Without loss of generality we can assume $|u_i| = 1$ $(1 = 1,\dots,p)$ since only the signs σ_i of the u_i matter. Also, since P determines the signs of the v_i by (12), a Pólya operator can be represented as a pair (P,v) where

$$(14) \qquad \begin{array}{l} P \text{ is a Pólya scheme of order } p, \text{ and} \\ v = (|v_1|, |v_2|,\dots,|v_p|) \in [0,\infty)^p. \end{array}$$

EXAMPLE.

$$P = \begin{pmatrix} 1010 \\ 1100 \end{pmatrix}, \quad v = (v_1,v_2,v_3,v_4) \in [0,\infty)^4$$

represents the operator

$$L = D^4,$$

$$\delta_1 = y_0 - v_1 y_0', \quad \delta_2 = -y_0'' + v_2 y_0''', \quad \delta_3 = -y_1' - v_3 y_1'', \quad \delta_4 = y_1 + v_4 y_1'.$$

The standard Pólya operators in this way are in 1-1-correspondence with the Pólya schemes.

We are now going to describe some properties of Pólya operators that have been found "experimentally", that is, in a way where the heuristics of the results were done using the APL workspace for Pólya operators. Proofs will be given, in the Appendix, for the results in the next section. For further proofs we refer to [4].

4. BASIC POLYNOMIALS

From now on, we assume $\mathcal{L}$ to be a Pólya operator as described above. In particular, the boundary conditions are numbered according to (10). Since $L = \sigma_L D^p$, $G_1,\dots,G_p$ is a base for the space of polynomials of order $\le p-1$.

These bases have some very typical features. Let us look at some examples. We assume $v \equiv 0$ in all three cases below, so that the Pólya scheme determines $\mathcal{L}$.

Ex. 1. $P = \binom{11010}{11000}$, $v = 0$

$$\begin{aligned} G_1(x) &= 1 - 2x^2 + x^4 \\ G_2(x) &= (2x - 3x^2 + x^4)/2 \\ G_3(x) &= (x^2 - 2x^3 + x^4)/12 \\ G_4(x) &= (x^2 - x^4)/2 \\ G_5(x) &= 2x^2 - x^4 \end{aligned}$$

Ex. 2. $P = \binom{11101100}{10100100}$, $v = 0$

$$\begin{aligned} G_1(x) &= (16 - 21x^3 + 7x^6 - 2x^7)/16 \\ G_2(x) &= (16x - 21x^3 + 7x^6 - 2x^7)/16 \\ G_3(x) &= (24x^2 - 29x^3 + 7x^6 - 2x^7)/16 \\ G_4(x) &= (11x^3 - 16x^4 + 7x^6 - 2x^7)/48 \\ G_5(x) &= (15x^3 - 48x^5 + 43x^6 - 10x^7)/5760 \\ G_6(x) &= (x^3 - 3x^6 + 2x^7)/2880 \\ G_7(x) &= (5x^3 - 7x^6 + 2x^7)/96 \\ G_8(x) &= (21x^3 - 7x^6 + 2x^7)/16 \end{aligned}$$

Ex. 3. $P = \binom{0010111}{1110000}$, $v = 0$

$$\begin{aligned} G_1(x) &= (1 - 3x + 3x^2 - x^3)/6 \\ G_2(x) &= (1 - 2x + 2x^3 - x^4)/24 \\ G_3(x) &= (8 - 15x + 10x^3 - 3x^5)/360 \\ G_4(x) &= (5 - 9x + 5x^3 - x^6)/720 \\ G_5(x) &= 1 - x \\ G_6(x) &= 1 \end{aligned}$$

It follows from $G_i \geqq 0$ that the lowest order nonzero coefficient of G_i is always positive. Less obvious is the observation that the signs of the subsequent nonzero coefficients alternate. This pattern can be destroyed by choosing $v \neq 0$, as the following variation of Example 2 shows:

Ex. 2'. $P = \binom{11101100}{10100100}$, $v = (1,0,0,0,0,0,0,0)$

$$\begin{aligned} G_1(x) &= [\text{same as in Ex. 2}] \quad (i \neq 2) \\ G_2(x) &= (8 + 8x - 21x^3 + 7x^6 - 2x^7)/8 \end{aligned}$$

In the context of the theory of total positivity, sign changing patterns are not unexpected. However, it is likely that this one would have gone unnoticed without the help of a computer since it depends on the very particular form of the boundary conditions involved. Our first observation thus is

PROPOSITION 1. If $\mathcal{L}$ is a *standard* Pólya operator, then the signs of the nonzero coefficients of its basic polynomials alternate.

Most features of the basic polynomials reflect, in contrast to the above one, the structure of the Pólya scheme associated with $\mathcal{L}$. Let

$$d_i = \text{degree } (G_i) \qquad (i = 1, \ldots, p).$$

In the first two examples we find $d_i = p - 1$ for $i = 1, \ldots, p$. After experimenting with more cases, it became apparent that this happens if $\mathcal{L}$ is a "prime" Pólya operator in the following sense:

If P_1 and P_2 are Pólya schemes of orders p_1, resp. p_2, then their product $P = P_1 \circ P_2$ is the scheme of order $p = p_1 + p_2$ obtained by writing P_1 to the left of P_2. Since $M_{p-1} = p$ holds for every Pólya scheme of order p, a necessary and sufficient condition for P not being a product of lower order schemes is easily seen to be

$$M_{j-1} > j \qquad (j = 1, \ldots, p-1). \tag{15}$$

PROPOSITION 2. The basic polynomials all have degree $p - 1$ if and only if (15) holds.

Examples 1 and 2 satisfy this condition, while in Example 3

$$P = \binom{0010111}{1110000} \tag{16}$$

can be written as the product

$$P = \binom{0}{1}^2 \circ \binom{10}{10} \circ \binom{1}{0}^3$$

of Pólya schemes satisfying (15).

The general case now is not difficult to guess:

COROLLARY. The degrees d_i can be read from the Pólya scheme P of $\mathcal{L}$ as follows. Let

$$P = P_1 \circ P_2 \circ \ldots \circ P_n$$

be the unique factorization of P into Pólya schemes P_i of orders p_i satisfying (12), and let

$$s_k = p_1 + \ldots + p_k - 1 \qquad (k = 1,\ldots,n).$$

Then

(17) $$d_i = \min \{s_k \mid j_i \leq s_k\}.$$

For example, if P is given by (16) we have $(s_1,\ldots,s_6) = (0,1,3,4,5,6)$ and $(j_1,\ldots,j_7) = (2,4,5,6,2,1,0)$. Hence the degrees of the basic polynomials are $(d_1,\ldots,d_7) = (3,4,5,6,3,1,0)$.

One can continue in this way to gather information about the basic polynomials from the Pólya scheme. Another result that was first found experimentally is

PROPOSITION 3. _For a standard Pólya operator, the number_ r_i _of nonzero terms of the basic polynomial_ G_i _is given by_

(18) $$r_i = d_i - M_{d_i,0} + \begin{cases} 2 & \text{if } i \leq q, \\ 1 & \text{if } i > q, \end{cases}$$

where

$$M_{j,0} = \sum_{i=0}^{j} e_{0,i} \qquad (j = 0,\ldots,p-1).$$

COROLLARY. _If the Pólya scheme_ P _satisfies_ (12), _then_

(19) $$r_i = p - q + \begin{cases} 1 & \text{if } i \leq q, \\ 0 & \text{if } i > q. \end{cases}$$

Proof. In this case, $d_i = p - 1$ $(i = 1,\ldots,p)$ by Proposition 2, hence $M_{d_i,0} = q$ $(i = 1,\ldots,p)$. ◊

Examples 1 and 2 illustrate this phenomenon: The basic polynomials associated with boundary conditions at the left endpoint have one term less than those according to right hand side conditions.

The list of experimentally found features of the basic polynomials G_i is by far not exhausted here. Further results will be given in [4].

5. GREEN'S KERNELS

Using the APL workspace for Pólya operators it is easy to calculate the Green's kernel explicitly. Recall that the Green's kernel together with the basic functions constitute the inverse of a Pólya operator.

The Green's kernel can be calculated from the basic polynomials by using

$$(20)\qquad G(x,t) = \begin{cases} -\sum_{j=q+1}^{p} G_j^*(x)G_j(t) & (x \in [0,t]), \\ \sum_{j=1}^{q} G_j^*(x)G_j(t) & (x \in [t,1]). \end{cases}$$

Here,

$$(21)\qquad \begin{aligned} G_j^*(t) &= \delta_j(\varphi(\cdot,t)), \text{ with} \\ \varphi(x,t) &= \sigma_L \frac{(x-t)_+^{p-1}}{(p-1)!}, \end{aligned}$$

are the *adjunct* polynomials for $\mathcal{L}$ (cf. [12]).

The functions

$$G_t(x) = G(x,t) \qquad (t \in (0,1))$$

are situated "in between" the two basic polynomials G_q and G_{q+1} (provided that $q \neq 0$, $q \neq p$). This "in between"-relation was discussed in [2], we illustrate it in the case of Example 1 of the preceding section:

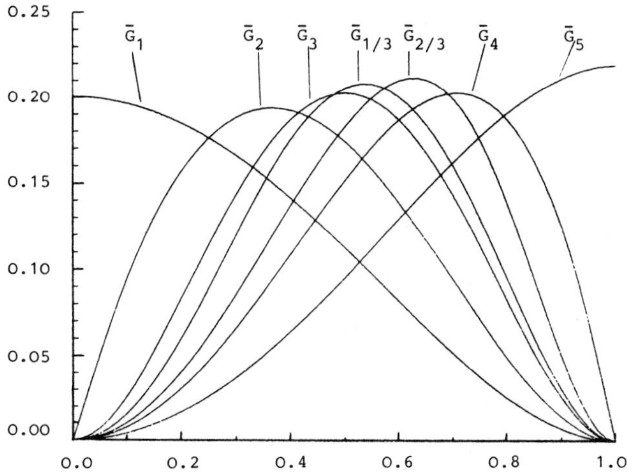

In [2], it was proved that the Green's kernel of every Pólya operator is totally positive. The explicit calculation of a number of Green's kernels did reveal that they share another useful property, illustrated again by the operator of Example 1:

Pólya scheme: $\begin{matrix}1&1&0&1&0\\1&1&0&0&0\end{matrix}$

Unit of z-axis: 0.001

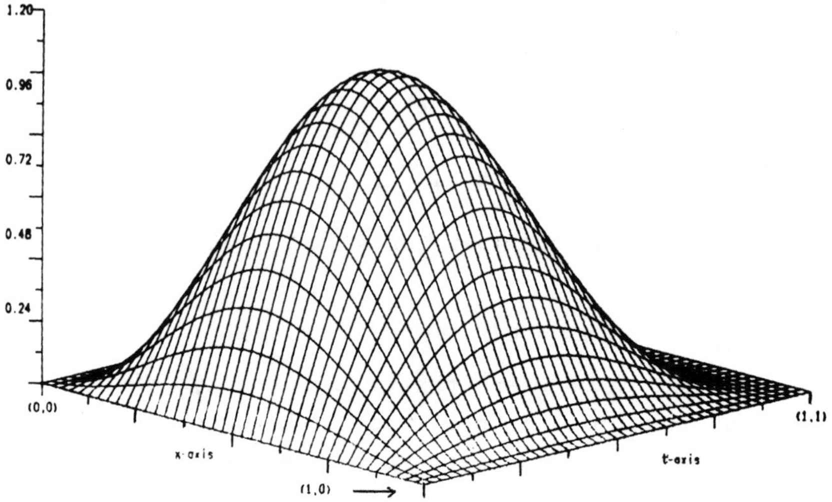

It is obvious in this case that the *level sets*

(22) $\{(x,t) \in [0,1]^2 \mid G(x,t) \geqq c\} \qquad (c \geqq 0)$

are all convex. A function having this property is called *quasiconcave*.

It was known ([1], Section 4) that the Green's kernel of the standard Pólya operator associated with

$$P = \begin{pmatrix}1010\ldots10\\1010\ldots10\end{pmatrix}$$

is quasiconcave. Thus, in this case the result was not discovered by studying examples but the experiment was, in the classical sense, a "question to nature". My original conjecture had been that quasiconcavity of the Green's kernel would not hold without additional assumptions. However, we have

THEOREM. <u>For every Pólya operator, the Green's kernel is</u>

quasiconcave.

This property, as demonstrated in [1], has interesting applications to eigenvalue inequalities. Thus the next step is to calculate eigenvalues of Pólya operators.

6. EIGENVALUE INEQUALITIES

In this section, $\mathcal{L}$ denotes a Pólya operator of order $p \geq 2$ such that $q \neq 0$, $q \neq p$. (We continue to use the notation introduced in Section 2.)

It is known ([3], [4], and [5]) that in this case the eigenvalue problem

$$\text{(23)} \qquad Ly = \lambda y, \quad \delta_j(y) = 0 \qquad (j = 1,\ldots,p)$$

has solutions $\lambda_i(\mathcal{L})$, $i = 0,1,\ldots$, which are strictly positive and simple:

$$\text{(24)} \qquad 0 < \lambda_0(\mathcal{L}) < \lambda_1(\mathcal{L}) < \ldots$$

The eigenvalue $\lambda_0(\mathcal{L})$ is of particular interest in connection with generalized completely convex functions, cf. [4].

In our APL workspace, relation (24) is used for the calculation of the eigenvalues. We proceed as follows: For $\sigma \in \{-1,1\}$, define the *Mikusinski sines*:

$$\text{(25)} \qquad \phi_{\sigma,p,k}(x) = \sum_{n=0}^{\infty} \sigma^n \frac{x^{pn+k}}{(pn+k)!} \qquad (k = 0,\ldots,p-1).$$

They form a base for the solutions y of $\sigma y^{(p)} = y$.

The solutions of (23) thus can be written as

$$\text{(26)} \qquad y(x) = \sum_{k=0}^{p-1} a_k \, \phi_{\sigma,p,k}(\lambda x) =: \sum_{k=0}^{p-1} a_k \, \psi^{(\lambda)}_{\sigma,p,k}(x),$$

so that the eigenvalues λ_i are the zeros of the determinant of the system of linear equations $\delta_i(y) = 0$ $(i = 1,\ldots,p)$, that is, the zeros of the entire function

$$\text{(27)} \qquad F_{\mathcal{L}}(\lambda) = \det(\delta_i(\psi^{(\lambda)}_{\sigma,p,k}))_{i=1,\ldots,p;\ k=0,\ldots,p-1}.$$

The calculation of the zeros of $F_{\mathcal{L}}$ essentially uses three ingredients:

- the in-built array-processing facilities of APL,
- the knowledge that all zeros of $F_{\mathfrak{L}}$ are situated on the positive halfline,
- and the information given by the following result:

PROPOSITION 4. For every Pólya operator $\mathfrak{L}$ of order p with q boundary conditions at $x = 0$ ($q \neq 0$, $q \neq p$), let the *weight* of $\mathfrak{L}$ be

$$(28) \qquad g_{\mathfrak{L}} = \sum_{i=1}^{p} j_i .$$

Then

$$(29) \qquad \lim_{k\to\infty} \frac{\lambda_k^{1/p}(\mathfrak{L}) \sin(1-q/p)\pi}{(k+p/2-g_{\mathfrak{L}}/p)\pi} = 1.$$

(The proof of this result is obtained by writing the Mikusinski sines ϕ_k as exponential polynomials and then using a Laplace expansion of the determinant (27), but it can also be derived from a more general result of Keldysh, cf. [8], p. 80.)

Proposition 4 has proved to be of great practical value. The number

$$(30) \qquad \rho_k = \frac{(k + p/2 - g_{\mathfrak{L}}/p)\pi}{\sin(1-q/p)\pi}$$

is, for standard Pólya operators, usually a good initial guess for the exact value of $\lambda_k^{1/p}(\mathfrak{L})$: For all 185 standard Pólya operators of order $p \leq 5$ satisfying $q \neq 0$, $q \neq 5$, the error in (30) is less than 10^{-6} for $k = 2$ (and hence for $k > 2$, too).

Recall that a Pólya operator is determined by a pair (P,v) where P is a Pólya scheme and $v \in [0,\infty)^p$. A main point of interest is to find the type of dependence of $\lambda_k(\mathfrak{L}) = \lambda_k(P,v)$ on the arguments P and v. (Note that the weight $g_{\mathfrak{L}}$ depends solely on P.)

A first conjecture, suggested by the approximation (31) to $\lambda_k^{1/p}(\mathfrak{L})$ could be that the following eigenvalue inequality holds: If p,q and v are kept fixed, then $\lambda_k(\mathfrak{L})$ decreases with increasing weight $g_{\mathfrak{L}}$.

Let us test this conjecture for $p = 5$, $q = 3$, and $v = 0$. The smallest possible weight then is $g_{\mathfrak{L}} = 4$, attained for $P = \binom{11100}{11000}$ only. The corresponding first eigenvalue is $\lambda_0^{1/5}(P,0) = 5.641174..$

(the approximation (30) yields $\rho_o = 5.6155..$). Weight $g_{\mathfrak{L}} = 5$ is attained by the two schemes $P_1 = \binom{11100}{10100}$ and $P_2 = \binom{11010}{11000}$. The corresponding operators are skew-adjoint and thus will have the same eigenvalues, as must be if the conjecture is true. The common value for $\lambda_o^{1/5}(\mathfrak{L})$ for both is 4.967851.. ((30) yields $\rho_o = 4.9549..$), supporting the conjecture. There are 6 Pólya schemes of weight 6 in this class: $P_3 = \binom{11100}{01100}$, $P_4 = \binom{11010}{10100}$, $P_5 = \binom{11100}{10010}$ and their skew-adjoints (the Pólya scheme $(e^{\#}_{k,j})$ of the skew-adjoint operator is obtained from the scheme $(e_{k,j})$ of the operator by $e^{\#}_{k,j} = e_{1-k,p-1-j}$, $k = 0,1$, $j = 0,\dots,p-1$). However, this time we do not get the same eigenvalues: $\lambda_o^{1/5}(P_3,0) = 4.345501..$, $\lambda_o^{1/5}(P_4,0) = 4.302071..$, $\lambda_o^{1/5}(P_5,0) = 4.274523..$ ((30) yields $\rho_o = 4.2942$). The conjecture is thus disproved. Instead, a modified version is true:

PROPOSITION 5. _Let_ P _and_ P' _be two Pólya schemes of order_ p _having the same number_ q _of ones in the upper row. If we write_ $P < P'$ _to indicate that for the respective degrees_ $j_i \leq j'_i$ _holds for all_ $i = 1,\dots,p$, _then_ $P < P'$ _implies_

$$\lambda_k(P,v) \geq \lambda_k(P',v) \quad (k = 0,1,..\ ; v \in [0,\infty)^p). \tag{31}$$

A complement to this inequality is illustrated by the following figure. It shows the eigenvalue $\lambda_o(P,v)$ for $P = \binom{1010}{0110}$ and $v = (0,x,0,0)$ as a function of $x \in [0,6]$.

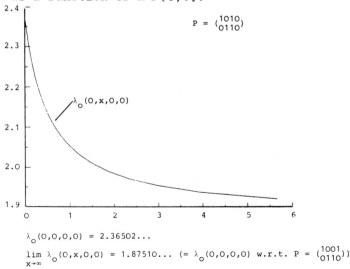

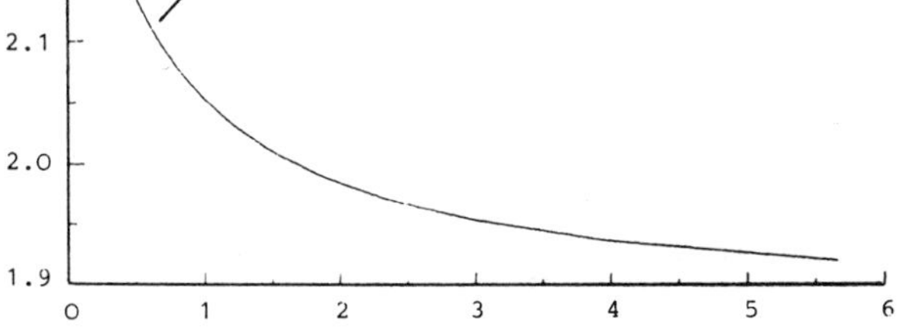

$\lambda_o(0,0,0,0) = 2.36502...$

$\lim_{x\to\infty} \lambda_o(0,x,0,0) = 1.87510...$ ($= \lambda_o(0,0,0,0)$ w.r.t. $P = \binom{1001}{0110}$)

The figure strongly suggests, for $k = 0$, the subsequent result which actually implies Proposition 5:

PROPOSITION 6. <u>For every</u> $i = 1,\dots,p$ <u>and</u> $k = 0,1,\dots,$ <u>the function</u>

(32) $$f(v_i) = \lambda_k(P, (v_1,\dots,v_p))$$

<u>is decreasing</u>.

Viewing both results in conjunction, we can say that $\lambda_k(\mathcal{L})$ is decreasing in $\mathcal{L}$ if we define the ordering $\mathcal{L} < \mathcal{L}'$ by $P < P'$ and $v \leq v'$.

Certainly, the above results are not all that can be said about eigenvalue inequalities for Pólya operators. Let us close with a conjecture, based on calculations of $\lambda_o(P,v)$ for $P = \binom{11010}{10010}$ and $v = (2,x,y,0,0)$ for 100 values of $(x,y) \in [1,10]$, as well as on some further examples.

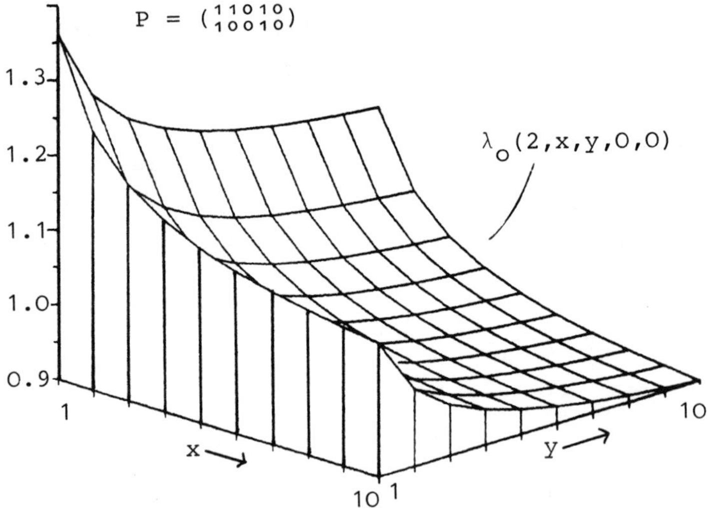

CONJECTURE. $\lambda_o(P,v)$ <u>is a convex function of</u> $v \in [0,\infty)$.

APPENDIX: PROOFS

Only the results of Section 4 will be proved here, for further proofs, cf. [4].

In the sequel, f denotes a polynomial of degree $d < p$. If $0 \leq j \leq d$, we write $s_j(f,0)$ for the sign of $f^{(j)}(x)$, $x \in (0,\varepsilon)$, where $\varepsilon > 0$ is such that $f^{(j)}$ has no zero in $(0,\varepsilon)$. The sign $s_j(f,1)$ is defined analogously in $(1-\varepsilon,1)$. The number of sign changes in the sequence $(s_0(f,x),\ldots,s_j(f,x))$ will be denoted by $V_j(f,x)$ $(x \in \{0,1\},\ 0 \leq j \leq d)$.

Let the coefficients of f be given by $f(x) = \sum_{k=0}^{d} a_k x^k$ and let $a_{k_1},\ldots,a_{k_r}$ be the nonzero coefficients, where $k_1 < \ldots < k_r = d$. Then we have

$$
(33)\qquad
\begin{aligned}
s_0(f,0) &= \ldots = s_{k_1}(f,0) = \operatorname{sgn}(a_{k_1}),\\
s_{k_1+1}(f,0) &= \ldots = s_{k_2}(f,0) = \operatorname{sgn}(a_{k_2}),\\
&\ \ \vdots \qquad\qquad\quad \vdots \qquad\qquad\quad \vdots\\
s_{k_{r-1}+1}(f,0) &= \ldots = s_{k_r}(f,0) = \operatorname{sgn}(a_{k_r}),
\end{aligned}
$$

so that

$$(34)\qquad V_d(f,0) \leq r-1.$$

Our aim is to show that in the situation of Propositions 1 and 3, inequality (34) becomes an equality.

Thus let $\mathcal{L}$ be a Pólya operator of order p, and let, for $x \in \{0,1\}$ and $0 \leq j \leq d$, $M_j(f,x)$ denote the number of boundary conditions δ of $\mathcal{L}$ such that: $\delta(f) = 0$, δ is a condition at the boundary point x, and the order of δ is $\leq j$. Furthermore, let $M_j(f) = M_j(f,0) + M_j(f,1)$. Note that $M_j(f) = M_j$ holds if f satisfies all boundary conditions of order $\leq j$, where M_j are the numbers defined in Section 2. We need the following result:

LEMMA. ([2], Lemma 3.6) <u>Let</u> $n_j(f)$ <u>be the number of zeros of</u> $f^{(j)}(x)$, $x \in (0,1)$, <u>where</u> $f \in C^j[0,1]$. <u>Then</u>

$$(35)\qquad n_j(f) = n_0(f) + M_{j-1}(f) - j$$

<u>implies</u>

$$(36)\qquad \begin{aligned} V_j(f,0) &= j - M_{j-1}(f,0), \\ V_j(f,1) &= M_{j-1}(f,1). \end{aligned}$$

Note that

$$(37)\qquad n_j(f) \geq n_o(f) + M_{j-1}(f) - j$$

holds for all polynomials f, by Rolle's Theorem ([2], Lemma 3.4).

From now on, let $1 \leq i \leq p$ and $f = G_i$ be a basic polynomial of $\mathcal{L}$. Then we have

$$(38)\qquad M_{d-1}(f) = M_d(f) = M_d - 1.$$

The first equality holds since $f^{(d)}(x) \equiv \text{const} \neq 0$, hence no Rolle type boundary condition $\delta(f) = 0$ of order d can be satisfied. The second one is obvious from the fact that $\delta_j(G_i) = \delta_{ij}$ $(j = 1,\dots,p)$.

Since the basic polynomials are strictly positive in (0,1), $n_o(f) = n_d(f) = 0$. Hence (37) and (38) yield for $j = d$

$$(39)\qquad 0 \geq M_{d-1}(f) - d = M_d - 1 - d \geq (d+1) - 1 - d = 0,$$

that is, equality (35) holds for $f = G_i$ and $j = d$. (The last inequality in (39) is Pólya's condition.)

Incidentally, since $M_d = d + 1$, we have proved that (15), i.e. $M_{j-1} > j$ for $j = 1,\dots,p-1$, implies $d = p - 1$. This yields the if-part of Proposition 2. For the converse, assume that $M_{j-1} = j$ for some $j < p$. Then the Pólya scheme P of $\mathcal{L}$ can be factored as $P = P_1 \circ P_2$ where P_1, resp. P_2, is the Pólya scheme consisting of the first j, resp. last $p - j$, columns of P. Then it is easy to see that j of the basic polynomials of $\mathcal{L}$ are also basic polynomials of a Pólya operator having the Pólya scheme P_1. Thus their degree is $\leq j - 1 < p - 1$, proving Proposition 2.

The proof of the Corollary to Proposition 2 will be left to the reader.

To prove Propositions 1 and 3 we now additionally assume that $\mathcal{L}$ is a standard Pólya operator. Then each of the $M_d(f,0)$ boundary conditions at $x = 0$ satisfied by $f = G_i$ renders one of the coefficients of f to be zero:

$$0 = \delta_k(f) = G_i^{(j_k)}(0) = j_k!\,a_{j_k},$$

thus the number r of nonzero coefficients has the bound

$$\begin{aligned} r &\leq (d+1) - M_d(f,0) \\ &= (d+1) - M_{d-1}(f,0) = V_d(f,0) + 1 \end{aligned}$$

by the above Lemma. Thus we have proved that equality holds in (34):

(40) $$r = V_d(f,0) + 1 = d + 1 - M_d(f,0).$$

Hence $(s_0(G_i,0),\ldots,s_d(G_i,0))$ has exactly $r-1$ sign changes, and by (33) the same holds for $(a_{k_1},\ldots,a_{k_r})$. This proves Proposition 1.

By the definition of G_i, the indexing order (10) implies that

(41) $$M_d(G_i,0) = M_{d,0} - \begin{cases} 1 & \text{if } i \leq q, \\ 0 & \text{if } i > q, \end{cases}$$

hence Proposition 3 is also a consequence of (40).

REFERENCES

1. S. S. Cheng, Isoperimetric eigenvalue problems of even order differential equations. Pac. J. Math. 99 (1982), 303 - 315.

2. A. Clausing, Pólya operators I: Total positivity. Math. Ann. 267 (1984), 37 - 59.

3. A. Clausing, Pólya operators II: Complete concavity. Math. Ann. 267 (1984), 61 - 81.

4. A. Clausing, Pólya operators. (In preparation)

5. U. Elias, Eigenvalue problems for the equation $Ly + \lambda p(x)y = 0$. J. Diff. Equ. 29 (1978), 28 - 57.

6. L. Euler, Specimen de usu observationum in mathesi pura. Opera omnia, Ser. 1, Vol. 2, 459 - 492, Teubner-Verlag, Leipzig, Berlin 1915.

7. U. Grenander, Mathematical experiments on the computer. Academic Press, New York 1982.

8. M. A. Neumark, Lineare Differentialoperatoren. Akademie-Verlag, Berlin 1963

9. G. Pólya, Bemerkungen zur Interpolation und zur Näherungstheorie der Balkenbiegung. Z. Angew. Math. Mech. 11 (1931), 445 - 449.

10. G. Pólya, Mathematics and plausible reasoning. Vol. 1: Induction and analogy in mathematics. Princeton University Press, Princeton, New Jersey 1954.

11. J. Schröder, Operator inequalities. Academic Press, New York 1980.

12. L. L. Schumaker, Spline functions: Basic Theory. J. Wiley and Sons, New York 1981.

Achim Clausing, Institut für Mathematische Statistik, Universität Münster, D-4400 Münster, West Germany

International Series of
Numerical Mathematics, Vol. 80

P-ESTIMATES FOR ULTRAPRODUCTS OF BANACH LATTICES AND REFLEXIVITY

Franziska Fehér and Elke Klaukien

Abstract. This paper is concerned with the relationship between reflexivity, superreflexivity, and indices of a Banach lattice in the sense of Grobler/Dodds. It is shown that the indices are strictly lying between 1 and ∞ if and only if the Banach lattice is superreflexive. The proof using ultrapower techniques, is based upon upper and lower p-estimates for ultraproducts of Banach lattices.

1. REFLEXIVITY, P-ESTIMATES, AND INDEX CONDITION

Reflexivity of Banach lattices was intensively studied by many authors, for instance by W.F. Eberlein, R.C. James, B. Maurey and others. In their papers, reflexivity is characterised by abstract topological conditions like weak*-compactness of the unit ball or by the infinite tree property.

On the other hand, since the Riesz representation theorem, it is well known that a Lebesgue space L^p is reflexive, if and only if $1 < p < \infty$. In [4] J.J. Grobler studied the reflexivity within the more general framework of Banach function spaces. In his paper he introduced the notion of upper and lower indices of a Banach function space L and showed that L is reflexive, if these indices are strictly lying between 1 and ∞. The notion of indices was generalised by P. Dodds [2] even to the case of abstract Banach lattices. Actually, let L denote a Banach lattice with norm $\| \ \|$, and let $1 \leq p \leq \infty$. Then he defined ([2],[4]):

a) The Banach lattice L has the ℓ^p-*decomposition property* if and only if $\{\|f_n\|\}_{n=1}^{\infty} \in \ell^p$ for any positive, disjoint, order-bounded sequence $\{f_n\}_{n=1}^{\infty}$ in L. The number

This paper is in final form and no version of it will be submitted for publication elsewhere.

$$\sigma(L) := \inf\{p \geq 1;\ L \text{ has } \ell^p\text{-decomposition property}\}$$

is called the *upper index* of L.

The Banach lattice L has the *ℓ^p-composition property*, if and only if for any positive, disjoint sequence $\{f_n\}_{n=1}^{\infty}$ in L with $\|f_n\| \leq 1$, $n = 1,2,\ldots$, and for any positive sequence $\{a_n\}_{n=1}^{\infty}$ in ℓ^p one has

$$\sup_{k\in\mathbb{N}} \|\sum_{k=1}^{n} a_k f_k\| < \infty .$$

The number

$$s(L) := \sup\{p \geq 1;\ L \text{ has } \ell^p\text{-composition property}\}$$

is called the *lower index* of L.

It can be shown that these properties are equivalent to the lower and upper p-estimates, respectively (see [2],[8]). Namely, a Banach lattice L is said to satisfy a *lower p-estimate* (*upper p-estimate*), if and only if there exists a constant $K > 0$ ($M > 0$) such that the inequalities

$$(\sum_{k=1}^{n} \|f_k\|^p)^{1/p} \leq K \|\sum_{k=1}^{n} |f_k|\| ,$$

$$\|\sup_{1\leq k\leq n} |f_k|\| \leq M(\sum_{k=1}^{n} \|f_k\|^p)^{1/p} ,$$

respectively, hold for any finite sequence $f_1, f_2, \ldots, f_n \in L$.

Then it follows that

$$\sigma(L) = \inf\{p \geq 1;\ L \text{ satisfies a lower p-estimate}\},$$

$$s(L) = \sup\{p \geq 1;\ L \text{ satisfies an upper p-estimate}\}.$$

In [2] P. Dodds proved that L is reflexive if $s(L) > 1$ and

$\sigma(L) < \infty$. J. Lindenstrauss/ L. Tzafriri [6] proved that this condition even implies the existence of an equivalent uniformly convex norm, i.e. that the space is superreflexive.

In this paper we give a different proof of the last fact using ultraproducts of Banach lattices. Then we show that the converse also holds, and hence superreflexivity is equivalent to the index condition. Finally, we give an example of a reflexive Banach lattice, which does not satisfy the index condition.

2. ULTRAPRODUCTS OF BANACH LATTICES

In the sequel, let $I \neq \emptyset$ be some index set and $\underline{U}$ an ultrafilter of subsets of I. For any compact Hausdorff space K and any family $(x_i)_{i \in I}$ of elements of K, the $\underline{U}$-limit is defined as the uniquely existing element $x \in K$ such that

$$\{i \in I;\ x_i \in V\} \in \underline{U}$$

for each neighbourhood V of x. The notion is $x = \lim_{\underline{U}} x_i$. The following is readily proved (for $x_i \in \mathbb{R}$):

a) <u>If</u> $f : \mathbb{R} \to \mathbb{R}$ <u>is continuous</u>, <u>then</u> $f(x) = \lim_{\underline{U}} f(x_i)$.

b) If $\{i \in I;\ x_i \geq 0\} \in \underline{U}$, <u>then</u> $x \geq 0$.

By means of the $\underline{U}$-limit the ultraproduct of Banach lattices can be defined as follows (see S. Heinrich [5]):

DEFINITION 2.1. Let $(L_i)_{i \in I}$ be a family of Banach lattices L_i. Then

a) $\ell^\infty(I, L_i) := \{(f_i)_{i \in I};\ \text{each } f_i \in L_i \text{ and } \sup_{i \in I} \|f_i\|_{L_i} < \infty\}$

b) $N_{\underline{U}} := \{(f_i)_{i \in I} \in \ell^\infty(I, L_i);\ \lim_{\underline{U}} \|f_i\|_{L_i} = 0\}$

c) The *ultraproduct* $(L_i)_{\underline{U}}$ is the quotient space

$$(L_i)_{\underline{U}} := \ell^\infty(I, L_i)/N_{\underline{U}}$$

equipped with the quotient norm $\|(f_i)_{\underline{U}}\|_{\underline{U}} = \lim_{\underline{U}} \|f_i\|_{L_i}$ and the order relation

$$(f_i)_{\underline{U}} \geq (g_i)_{\underline{U}} \quad \textit{iff}$$

$$f_i \geq g_i + h_i \quad (i \in I) \textit{ with } (h_i)_{i \in I} \in N_{\underline{U}}.$$

d) If $L_i = L$ for all $i \in I$, then $(L)_{\underline{U}}$ is called the *ultrapower* of L.

Among experts of abstract Banach space theory it is more or less known - although not explicitely appearing in literature - that properties like p-convex, p-concave, type, cotype ... carry over from Banach lattices to their ultrapowers. For our purpose we need the p-estimates for ultraproducts. Therefore, we prove explicitely:

THEOREM 2.1. Let $1 \leq p < \infty$. If there exists a constant $K > 0$ such that

$$\{i \in I;\ L_i \text{ satisfies a lower p-estimate with } K\} \in \underline{U},$$

then the ultraproduct $(L_i)_{\underline{U}}$ also satisfies a lower p-estimate with K. An analogous statement holds for upper p-estimates.

Proof. Let $(f_i^1)_{\underline{U}}, (f_i^2)_{\underline{U}}, \ldots, (f_i^n)_{\underline{U}} \in (L_i)_{\underline{U}}$ be any finite sequence in the ultraproduct. Then

$$\sum_{k=1}^{n} \|(f_i^k)_{\underline{U}}\|_{\underline{U}}^p = \sum_{k=1}^{n} (\lim_{\underline{U}} \|f_i^k\|_{L_i})^p = \lim_{\underline{U}} \sum_{k=1}^{n} \|f_i^k\|_{L_i}^p$$

$$\leq K^p \lim_{\underline{U}} \Big\| \sum_{k=1}^{n} |f_i^k| \Big\|_{L_i}^p = K^p \Big\| \sum_{k=1}^{n} |(f_i^k)_{\underline{U}}| \Big\|_{\underline{U}}^p.$$

For the converse of Theorem 2.1 we need the following notation: An ultrafilter $\underline{U}$ is *countably complete*, if and only if the intersection of any decreasing sequence of elements of $\underline{U}$ belongs to $\underline{U}$.

THEOREM 2.2. Let $1 \leqslant p < \infty$ and assume that the ultraproduct $(L_i)_{\underline{U}}$ satisfies a lower p-estimate with constant $K > 0$. If, in addition, $\underline{U}$ is countably complete, then the set

$$I_K := \{i \in I;\ L_i \textit{ satisfies a lower } p\textit{-estimate with } K\}$$

belongs to $\underline{U}$.
An analogous statement holds for upper p-estimates.

For the Proof we first assume that there exists a $K' > K$ such that

$$I_{K'} := \{i \in I;\ L_i \text{ satisfies a lower p-estimate with } K'\} \notin \underline{U}.$$

Then the complement $J := I \setminus I_{K'}$ must belong to $\underline{U}$. But, without loss of generality,

$$J = \{i \in I;\ \exists n_i \in \mathbb{N} \text{ (minimal) } \exists f_i^1, f_i^2, \dots, f_i^{n_i} \in L_i$$

$$\text{with } \|f_i^k\|_{L_i} \leqslant 1 \text{ such that}$$

$$\Big(\sum_{k=1}^{n_i} \|f_i^k\|_{L_i}^p\Big)^{1/p} > K' \Big\| \sum_{k=1}^{n_i} |f_i^k| \Big\|_{L_i}\} \in \underline{U}.$$

Now denote by $N := \sup_{i \in J} n_i$. If N is finite, then for $k \leqslant n$ and $i \in I$ let

$$g_{ik} := \begin{cases} f_i^k & (i \in J,\ k \leqslant n_i) \\ 0 & (\text{elsewhere}) \end{cases} .$$

Since $\|g_i^k\|_{L_i} \leqslant 1$, one has $(g_i^k)_{\underline{U}} \in (L_i)_{\underline{U}}$ for $k = 1, 2, \dots, N$. Moreover, because of the assumption $J \in \underline{U}$, it follows that

$$\Big(\sum_{k=1}^{N} \|(g_i^k)_{\underline{U}}\|_{\underline{U}}^p\Big)^{1/p} = \lim_{\underline{U}} \Big(\sum_{k=1}^{N} \|g_i^k\|_{L_i}^p\Big)^{1/p} =$$

$$= \lim_{\underline{U}} \left(\sum_{k=1}^{n_i} \| f_i^k \|_{L_i}^p \right)^{1/p} \geq K' \lim_{\underline{U}} \| \sum_{k=1}^{n_i} |f_i^k| \|_{L_i} =$$

$$= K' \| \sum_{k=1}^{N} |(g_i^k)_{\underline{U}}| \|_{\underline{U}} > K \| \sum_{k=1}^{N} |(g_i^k)_{\underline{U}}| \|_{\underline{U}} .$$

This is a contradiction to the lower p-estimate property of $(L_i)_{\underline{U}}$.

In the second case, i.e. if $N = \sup_{i \in J} n_i = \infty$, denote by

$$I_n := \{ i \in J;\ n_i \leq n \} \qquad (n \in \mathbb{N}).$$

If there exists a number $n \in \mathbb{N}$, such that $I_n \in \underline{U}$, then the contradiction is consructed as in the first case ($N < \infty$). If $I_n \notin \underline{U}$ for all $n \in \mathbb{N}$, then $I \setminus I_n \in \underline{U}$ for all n. Hence $J_n := J \cap (I \setminus I_n) \in \underline{U}$ for all $n \in \mathbb{N}$. But evidently, $J_1 \supseteq J_2 \supseteq \ldots$ and $\bigcap_{n=1}^{\infty} J_n = \emptyset$ in contradiction to $\underline{U}$ being countably complete.

Therefore $I_{K'} \in \underline{U}$ for all $K' > K$. But then $U_n := I_{k+1/n} \in \underline{U}$ for all $n \in \mathbb{N}$ and $U_1 \supseteq U_2 \supseteq \ldots$. $\underline{U}$ being countably complete, this implies $I_k = \bigcap_{n=1}^{\infty} U_n \in \underline{U}$.

In particular, if all $L_i = L$, i.e. for ultrapowers, we have the

COROLLARY 2.1. $\sigma((L)_{\underline{U}}) = \sigma(L)$ *and*

$$s((L)_{\underline{U}}) = s(L) .$$

Indeed, the inequalities $\sigma((L)_{\underline{U}}) \leq \sigma(L)$ and $s((L)_{\underline{U}}) \geq s(L)$ immediately follow from Theorem 2.1. The converse inequalities can be obtained by using the (order) isometric embedding of L into $(L)_{\underline{U}}$ via the mapping $x \to (x_i)_{\underline{U}}$ with $x_i := x$ for all $i \in I$.

REMARK. If $K_p(L)$ and $M_p(L)$ denote the smallest constants K and M, respectively, such that L satisfies a lower (upper) p-estimate with K (M), then for ultraproducts it can be shown that

$$\sigma((L_i)_{\underline{U}}) \leqslant \sup_{i \in U} \sigma(L_i)$$

for some $U \in \underline{U}$ such that the right hand side is finite and
$\sup_{i \in U} \sup_{p > \sigma(L_i)} K_p(L_i) < \infty$.

An analogous inequality holds for the lower index of $(L_i)_{\underline{U}}$.
The (quite straight forward) prove is omitted.

3. SUPERREFLEXIVITY

Now we consider the stronger property of superreflexivity. For this purpose we recall the following definitions:

DEFINITION 3.1. Let L and M denote two Banach spaces.

a) The Banach space M is *finitely representable* in L, if and only if for any $\varepsilon > 0$ and for any finite dimensional subspace M' of M there exists an isomorphism $T : M' \to L'$ such that $\|T\| \|T^{-1}\| \leqslant 1 + \varepsilon$.

b) The Banach space L is called *superreflexive*, if and only if any Banach space M which is finitely representable in L, is reflexive.

The property "superreflexive" is stronger than reflexivity (see e.g. [1]).

THEOREM 3.1. A Banach lattice L is superreflexive, if and only if $s(L) > 1$ and $\sigma(L) < \infty$.

The proof of necessity of the index condition is based upon a lemma of B. Beauzamy / B. Maurey on monotone basic sequences. We first consider the upper index $\sigma(L)$. Let $\{f_n\}_{n=1}^{\infty}$ be a positive, disjoint sequence in L. Moreover, let $r, s \in \mathbb{N}$ with $r \leqslant s$, and let $a_1, a_2, \ldots, a_s \in \mathbb{R}$ be some real numbers. Then the elements $a_n f_n$, $n = 1, 2, \ldots, s$, also are disjoint elements of L. Hence (see e.g. [7, p.69]

$$|\sum_{n=1}^{r} a_n f_n| = \sup_{1\leq n\leq r} |a_n f_n| \leq \sup_{1\leq n\leq s} |a_n f_n|$$

$$= |\sum_{n=1}^{s} a_n f_n| .$$

Since the norm of L is monotone, this yields

$$\| \sum_{n=1}^{r} a_n f_n \| \leq \| \sum_{n=1}^{s} a_n f_n \| ,$$

that is to say, the disjoint sequence $\{f_n\}_{n=1}^{\infty}$ is a monotone basic sequence in L.
On the other hand, the Banach lattice L is assumed to be superreflexive. By a result of B. Beauzamy / B. Maurey [1] therefore there exists a constant $c \geq 1$ and a number p with $1 < p < \infty$ such that

$$(\sum_{n=1}^{k} \| g_n \|^p)^{1/p} \leq C \| \sum_{n=1}^{k} g_n \|$$

for any finite monotone basic sequence $\{g_n\}_{n=1}^{k}$ in L.
A fortiori, this inequality holds for any finite, positive, disjoint sequence in L. That is to say, the Banach lattice L satisfies a lower p-estimate (comp. [6]), and therefore $\sigma(L) \leq p < \infty$.

The second statement, namely $s(L) > 1$, now follows by duality arguments. Indeed, the dual space L* is also superreflexive. Therefore, by applying the above arguments to L* instead of L, one has that $\sigma(L^*) < \infty$. But

$$\frac{1}{s(L)} = 1 - \frac{1}{\sigma(L^*)}$$

(see e.g. [2],[4],[8]), yielding $s(L) > 1$.

<u>Proof of sufficiency.</u> If $s(L) > 1$ and $\sigma(L) < \infty$, then by Corollary 2.1 one has that $s((L)_{\underline{U}}) > 1$ and $\sigma((L)_{\underline{U}}) < \infty$ for each ul-

trapower of L. From this index condition it follows [2] that each ultrapower $(L)_{\mathcal{U}}$ is reflexive, and this implies the superreflexivity of L, as S. Heinrich [5] has shown.

Finally, consider the Banach lattice $L := \prod_{n=1}^{\infty} \ell_1^n$ with the partial order

$$\{f_n\}_{n=1}^{\infty} \leqslant \{g_n\}_{n=1}^{\infty} :\Longleftrightarrow f_n \leqslant g_n \text{ in } \ell_1^n \qquad (n \in \mathbb{N})$$

and the norm

$$\| \{f_n\}_{n=1}^{\infty} \| := \left(\sum_{n=1}^{\infty} \| f_n \|_1^2 \right)^{1/2}.$$

This space is known to be reflexive [1]. On the other hand, L is not superreflexive [1], and therefore - by Theorem 3.1. - cannot satisfy the index condition. Actually, it can be shown that $s(L) = 1$ and $\sigma(L) \leqslant 2$.

ACKNOWLEDGEMENT: The authors would like to thank P. Dodds, J.J. Grobler, and A. Shep for the hint to ultrapower techniques.

REFERENCES

1. B. Beauzamy, Introduction to Banach Spaces and their Geometry. North Holland Publ. Comp., Amsterdam - New York - Oxford, 1982.

2. P. Dodds, Indices for Banach lattices. Proc. Netherl. Acad. Sci. A 80 (1977), 73-86.

3. F. Fehér, Indices of Banach function spaces and spaces of fundamental type. J. Approx. Theory 37 (1983), 12-28.

4. J.J. Grobler, Indices for Banach function spaces. Math. Z. 145 (1975), 99-109.

5. S. Heinrich, Ultraproducts in Banach space theory, J. Reine Angew. Math. 313 (1980), 72-104.

6. J. Lindenstrauss - L. Tzafriri, Classical Banach Spaces II, Function Spaces. Springer, Berlin - Heidelberg - New York, 1979.

7. W.A.J. Luxemburg - A.C. Zaanen, Riesz Spaces I. North Holland Publ. Comp., Amsterdam - London, 1971.

8. V.K. Vietsch, Abstract kernel operators and compact operators. Thesis, Leiden University, 1979.

Franziska Fehér and Elke Klaukien, Lehrstuhl A für Mathematik, RWTH Aachen, D-5100 Aachen, West Germany

Functional Equations and Inequalities

Walke slopes with farmstead

International Series of
Numerical Mathematics, Vol. 80

ON THE STABILITY OF A FUNCTIONAL EQUATION ARISING IN PROBABILISTIC NORMED SPACES

Claudi Alsina

Abstract. In this paper we solve for a given $\varepsilon > 0$ the inequality

$$d_L(\tau(F(j/a),F(j/b)),F(j/a+b)) \le \varepsilon ,$$

where d_L is the modified Lévy metric in the space Δ^+ of probability distribution functions of non-negative random variables, F is in Δ^+, a,b in $(0,\infty)$ are arbitrary and τ is a binary operation on Δ^+ to be found.

Let Δ^+ be the space of probability distribution functions of non-negative random variables, i.e.,

$$\Delta^+ = \{F \mid F: [-\infty,+\infty] \to [0,1],\ F(0) = 0,\ F(+\infty) = 1,\ F \text{ is non-decreasing and left-continuous on } [-\infty,+\infty)\},$$

and let ε_0 be the distribution function in Δ^+ defined by $\varepsilon_0(x) = 0$ for $x \le 0$ and $\varepsilon_0(x) = 1$ for $x > 0$. If j denotes the identity function on $[-\infty,+\infty]$, F is in Δ^+ and $a > 0$ then we denote by $F(j/a)$ the function in Δ^+ defined by $F(j/a)(x) = F(x/a)$.

Motivated by a problem on probabilistic normed spaces, D.M. Mouchtari and A.N. Šerstnev proved in [5] that if τ is a triangle function, i.e., $(\Delta^+,\tau,\varepsilon_0,\le)$ is an Abelian ordered semigroup with ε_0 as a unit, then the unique solution of the functional equation

$$\tau(F(j/a),F(j/b)) = F(j/a+b) , \tag{1}$$

where F is an arbitrary function in Δ^+ and a, b are any numbers

---

This paper is in final form and no version of it will be submitted for publication elsewhere.

in $(0,+\infty)$, is the operation τ_M defined by

$$\tau_M(F,G)(x) = \sup_{u+v=x} \mathrm{Min}(F(u),G(v)) .$$

A very simple proof of this result was given by C. Alsina and B. Schweizer in [1] by using the duality theorem established in [2].

Our aim here is to study the stability of (1) by solving for a given ε in $(0,1)$ the inequality

$$(2) \qquad d_L(\tau(F(j/a),F(j/b)),F(j/a+b)) \le \varepsilon ,$$

where F is any function in Δ^+, a and b are arbitrary positive numbers, τ is a continuous triangle function to be characterized and d_L is the modified Lévy metric introduced by D.A. Sibley in [8]:

$$d_L(F,G) = \inf \{h \mid G(x) \le F(x+h)+h \text{ and } F(x) \le G(x+h)+h \text{ for } x \text{ in } (0,1/h)\} .$$

We recall from [7] that given $\varepsilon > 0$ and F, G in Δ^+ then

$$d_L(F,G) \le \varepsilon ,$$

if and only if for any x in $(0,1/\varepsilon)$ we have

$$F(x) \le G(x+\varepsilon)+\varepsilon \quad \text{and} \quad G(x) \le F(x+\varepsilon)+\varepsilon .$$

The metric space (Δ^+,d_L) is compact and complete and convergence with respect to d_L is equivalent to the weak convergence of distribution functions. This metric modifies the classical Lévy metric in Δ^+ (see [3], [4], [6], [7]) and plays a fundamental role in the theory of probabilistic metric spaces ([7]). In the set of binary operations on Δ^+ we will consider the metric

$$\hat{d}_L(\tau,\tau') = \sup \{d_L(\tau(F,G),\tau'(F,G)) \mid F,G \in \Delta^+\} .$$

In order to solve (2) we will need several lemmas. First of all we note that since τ_M satisfies (1) obviously τ_M is a solution of (2). We will see immediately that, in fact, any operation ε-close to τ_M, according to $\hat{d}_L$, is also a solution of (2).

LEMMA 1. *Let τ be a binary operation on Δ^+ such that $\hat{d}_L(\tau,\tau_M) \le \varepsilon$. Then (2) holds.*

Proof. For any F in Δ^+ and for any $a,b>0$ we have

$$\begin{aligned} &d_L(\tau(F(j/a),F(j/b)),F(j/a+b)) \\ &\qquad \le d_L(\tau(F(j/a),F(j/b)),\tau_M(F(j/a),F(j/b))) \\ &\qquad\quad + d_L(\tau_M(F(j/a),F(j/b)),F(j/a+b)) \\ &\qquad = d_L(\tau(F(j/a),F(j/b)),\tau_M(F(j/a),F(j/b))) \\ &\qquad \le \hat{d}_L(\tau,\tau_M) \le \varepsilon . \end{aligned}$$

The following example shows how to find a large family of operations satisfying the previous lemma.

EXAMPLE 1. Let L be a continuous two-place function from $\mathbb{R}^+ \times \mathbb{R}^+$ onto $\mathbb{R}^+$ which is non-decreasing in each place and satisfies

(i) If $x<u$ and $y<v$ then $L(x,y)<L(u,v)$,

and

(ii) For all x and y in $\mathbb{R}^+$: $|L(x,y)-x-y| \le \varepsilon$,

then the binary operation $\tau_{M,L}$ defined by

$$\tau_{M,L}(F,G)(x) = \sup\{\mathrm{Min}\,(F(u),G(v)) \mid L(u,v) = x\}$$

satisfies

$$\hat{d}_L(\tau_{M,L},\tau_M) \le \varepsilon . \tag{3}$$

In order to show (3) we want to prove for any x in $(0,1/\varepsilon)$ the inequalities

$$\tau_{M,L}(F,G)(x) \le \tau_M(F,G)(x+\varepsilon)+\varepsilon \tag{4}$$

and

$$\tau_M(F,G)(x) \le \tau_{M,L}(F,G)(x+\varepsilon)+\varepsilon . \tag{5}$$

First we note that (4) is equivalent to prove that for any u_o, v_o such that $L(u_o,v_o)=x$ it is

$$\mathrm{Min}\,(F(u_o),G(v_o)) \le \sup_{u+v=x+\varepsilon} \mathrm{Min}\,(F(u),G(v)) + \varepsilon .$$

To this end, if $L(u_o,v_o)=x$ and (ii) holds, then we have

$$u_o + v_o - \varepsilon \le L(u_o, v_o) = x ,$$

i.e., $u_o + v_o \le x+\varepsilon$. Consider

$$u_1 = u_o + \frac{x+\varepsilon-u_o-v_o}{2} \quad \text{and} \quad v_1 = v_o + \frac{x+\varepsilon-u_o-v_o}{2}$$

Then $u_1 \ge u_o$, $v_1 \ge v_o$ and $u_1+v_1 = x+\varepsilon$, i.e.,

$$\begin{aligned} \text{Min}\,(F(u_o),G(v_o)) &\le \text{Min}\,(F(u_1),G(v_1)) \\ &\le \sup_{u+v=x+\varepsilon} \text{Min}(F(u),G(v)) \\ &\le \sup_{u+v=x+\varepsilon} \text{Min}(F(u),G(v)) + \varepsilon \end{aligned}$$

i.e., (4) follows. Now we proceed to show (5), i.e., for any $u_2+v_2 = x$

$$\text{Min}\,(F(u_2),G(v_2)) \le \sup_{L(u,v)=x+\varepsilon} \text{Min}(F(u),G(v)) + \varepsilon .$$

By (ii) again, $L(u_2,v_2)-\varepsilon \le u_2+v_2 = x$, i.e., $L(u_2,v_2) \le x+\varepsilon$. Since the function $f(t) = L(u_2+t,v_2+t)$ is a continuous strictly increasing mapping from $\mathbb{R}^+$ onto $\mathbb{R}^+$ and $f(0) = L(u_2,v_2) \le x+\varepsilon$, there will exist t_o in $\mathbb{R}^+$ such that $f(t_o) = L(u_2+t_o,v_2+t_o) = x+\varepsilon$. Then,

$$\begin{aligned} \text{Min}\,(F(u_2),G(v_2)) &\le \text{Min}\,(F(u_2+t_o),G(v_2+t_o)) \\ &\le \sup_{L(u,v)=x+\varepsilon} \text{Min}(F(u),G(v)) \\ &\le \sup_{L(u,v)=x+\varepsilon} \text{Min}(F(u),G(v)) + \varepsilon . \end{aligned}$$

Now we will prove a crucial result in this paper. Let $\Delta^+_{ic} = \{F \mid F \in \Delta^+,\ F \text{ is strictly increasing and continuous on } (0,\infty)\}$. It is a well known fact that Δ^+_{ic} is dense in Δ^+ with respect to the metric topology defined by d_L.

LEMMA 2. *If a binary operation on* Δ^+ *which is non-decreasing in each place satisfies* (2), *then*

(6) $$d_L(\tau(F,G),\tau_M(F,G)) \le \varepsilon$$

for all F *and* G *in* Δ^+_{ic}.

Proof. Assume that a non-decreasing binary operation τ on Δ^+ satisfies

$$d_L(\tau(H(j/a),H(j/b)),H(j/a+b)) \leq \varepsilon$$

for all $a,b>0$ and for all H in Δ^+, i.e., for any x in $(0,1/\varepsilon)$ we have

$$\tau(H(j/a),H(j/b))(x) \leq H(x+\varepsilon/a+b)+\varepsilon \tag{7}$$

and

$$H(x/a+b) \leq \tau(H(j/a),H(j/b))(x+\varepsilon)+\varepsilon\ . \tag{8}$$

Our aim here is to show that given any couple of functions F and G in Δ^+_{ic} we have

$$d_L(\tau(F,G),\tau_M(F,G)) \leq \varepsilon$$

or, equivalently, that for such F, G in Δ^+_{ic} and for any x in $(0,1/\varepsilon)$:

$$\tau(F,G)(x) \leq \tau_M(F,G)(x+\varepsilon)+\varepsilon \tag{9}$$

and

$$\tau_M(F,G)(x) \leq \tau(F,G)(x+\varepsilon)+\varepsilon\ . \tag{10}$$

To this end we define f from $[0,1]$ into $\mathbb{R}$ by

$$f(\lambda) = F(\lambda(x+\varepsilon)) - G((1-\lambda)(x+\varepsilon))\ .$$

Since F and G are continuous f is also continuous, and since F and G are strictly increasing on $(0,\infty)$ we have $f(0) = -G(x+\varepsilon) < 0$ and $f(1) = F(x+\varepsilon) > 0$. Thus there exists λ_o in $(0,1)$ with $f(\lambda_o) = 0$, i.e.,

$$F(\lambda_o(x+\varepsilon)) = G((1-\lambda_o)(x+\varepsilon))\ . \tag{11}$$

Let U be the distribution function in Δ^+_{ic} defined by

$$U(t) = \operatorname{Max}\left(F\left(\frac{\lambda_o t}{1-\lambda_o}\right), G(t)\right)\ .$$

The function U satisfies

$$U \geq G \qquad \text{and} \qquad U\left(\frac{1-\lambda_o}{\lambda_o}\, j\right) \geq F\ , \tag{12}$$

and by (11) we also have:

(13) $$U\left(\frac{x+\varepsilon}{\frac{\lambda_o}{1-\lambda_o}+1}\right) = U((1-\lambda_o)(x+\varepsilon)) = F(\lambda_o(x+\varepsilon)) = G((1-\lambda_o)(x+\varepsilon)) .$$

Using (12), (7) and (13) we obtain the following inequalities:

$$\begin{aligned}\tau(F,G)(x) &\le \tau\left(U\left(\frac{1-\lambda_o}{\lambda_o}\,j\right),U(j/1)\right)(x) \\ &\le U\left(\frac{x+\varepsilon}{\frac{\lambda_o}{1-\lambda_o}+1}\right) + \varepsilon \\ &= \mathrm{Min}\,(F(\lambda_o(x+\varepsilon)),G((1-\lambda_o)(x+\varepsilon))) + \varepsilon \\ &\le \sup_{u+v=x+\varepsilon} \mathrm{Min}\,(F(u),G(v)) + \varepsilon \\ &= \tau_M(F,G)(x+\varepsilon) + \varepsilon ,\end{aligned}$$

i.e., (9) holds.

Now let g be the function from [0,1] into $\mathbb{R}$ defined by

$$g(\mu) = F(\mu x) - G((1-\mu)x) .$$

Since F and G are in Δ^+_{ic}, g is continuous, $g(0) = -G(x) < 0 < F(x) = g(1)$, and there exists μ_o in (0,1) with $g(\mu_o) = 0$, i.e.,

$$F(\mu_o x) = G((1-\mu_o)x) .$$

We will show now that

(14) $$\sup_{u+v=x} \mathrm{Min}\,(F(u),G(v)) = F(\mu_o x) = G((1-\mu_o)x) .$$

This is equivalent to proving that

$$\mathrm{Min}\,(F(u),G(v)) \le F(\mu_o x) = G((1-\mu_o)x)$$

for all $u,v > 0$ such that $u+v = x$. If there would exist $u_1,v_1 > 0$ such that $u_1+v_1 = x$, but

$$\mathrm{Min}\,(F(u_1),G(v_1)) > F(\mu_o x) = G((1-\mu_o)x) ,$$

this would imply

$$F(u_1) > F(\mu_o x) \quad \text{and} \quad G(v_1) > G((1-\mu_o)x) .$$

Since F and G are strictly increasing on $(0,\infty)$, the above inequalities would yield $u_1 > \mu_o x$ and $v_1 > (1-\mu_o)x$ and from this $x = u_1+v_1 > \mu_o x+(1-\mu_o)x = x$ which is a contradiction. Thus (14) holds.

Next define the distribution function

$$V(t) = \mathrm{Min}\left(F\left(\frac{\mu_o}{1-\mu_o}\, t\right), G(t)\right) .$$

Then,

$$V \le G \quad \text{and} \quad V\left(\frac{1-\mu_o}{\mu_o}\, j\right) \le F . \tag{15}$$

Moreover, by (14) we have

$$\begin{aligned} \tau_M(F,G)(x) &= \sup_{u+v=x} \mathrm{Min}(F(u),G(v)) = F(\mu_o x) \\ &= G((1-\mu_o)x) = V((1-\mu_o)x) . \end{aligned} \tag{16}$$

Finally we have by (16), (14) and (8):

$$\begin{aligned} \tau_M(F,G)(x) &= V((1-\mu_o)x) = V\left(\frac{x}{\frac{\mu_o}{1-\mu_o}+1}\right) \\ &\le \tau\left(V\left(\frac{j}{\frac{\mu_o}{1-\mu_o}}\right), V(j/1)\right)(x+\varepsilon) + \varepsilon \\ &= \tau\left(V\left(\frac{1-\mu_o}{\mu_o}\, j\right), V\right)(x+\varepsilon) + \varepsilon \\ &\le \tau(F,G)(x+\varepsilon) + \varepsilon , \end{aligned}$$

whence (10) holds.

Now we will extend (6) to Δ^+, i.e.,

LEMMA 3. <u>Let τ be a continuous binary operation on Δ^+ such that</u>

$$d_L(\tau(F,G),\tau_M(F,G)) \le \varepsilon \tag{17}$$

<u>for all F and G in Δ^+_{ic}. Then (17) holds for all F and G in Δ^+ and consequently $\hat{d}_L(\tau,\tau_M) \le \varepsilon$.</u>

<u>Proof.</u> Since Δ^+_{ic} is dense in Δ^+, given F, G in Δ^+, there exist sequences (F_n) and (G_n) in Δ^+_{ic} such that $F = w\text{-}\lim_{n\to\infty} F_n$ and

$G = w\text{-}\lim_{n\to\infty} G_n$. Since τ and τ_M are continuous and τ satisfies (17) we will have

$$\lim_{n\to\infty} d_L(\tau(F_n,G_n),\tau(F,G)) = 0 , \tag{18}$$

$$\lim_{n\to\infty} d_L(\tau_M(F_n,G_n),\tau_M(F,G)) = 0 \tag{19}$$

and

$$d_L(\tau(F_n,G_n),\tau_M(F_n,G_n)) \le \varepsilon . \tag{20}$$

Since d_L is a metric we also have by (20):

$$\begin{aligned} d_L(\tau(F,G),\tau_M(F,G)) &\le d_L(\tau(F,G),\tau(F_n,G_n)) \\ &\quad + d_L(\tau(F_n,G_n),\tau_M(F_n,G_n)) \\ &\quad + d_L(\tau_M(F_n,G_n),\tau_M(F,G)) \\ &\le \varepsilon + d_L(\tau(F_n,G_n),\tau(F,G)) \\ &\quad + d_L(\tau_M(F_n,G_n),\tau_M(F,G)) , \end{aligned}$$

whence by (18) and (19) it follows that for F, G in Δ^+

$$d_L(\tau(F,G),\tau_M(F,G)) \le \varepsilon .$$

Thus $\hat{d}_L(\tau,\tau_M) \le \varepsilon$.

All previous lemmas yield the general solution of our problem:

THEOREM. *Let τ be a continuous nondecreasing binary operation on Δ^+ and let $\varepsilon > 0$ be given. Then,*

$$d_L(\tau(F(j/a),F(j/b)),F(j/a+b)) \le \varepsilon$$

for all F in Δ^+ and for all $a,b > 0$, if and only if

$$\hat{d}_L(\tau,\tau_M) \le \varepsilon .$$

REFERENCES

1. C. Alsina and B. Schweizer, On a theorem of Mouchtari and Šerstnev. Note di Matematica Vol. I (1981), 19-24.

2. M.J. Frank and B. Schweizer, On the duality of generalized

infimal and supremal convolutions. Rendiconti di Matematica 12 (1979), 1-23.

3. M. Fréchet, Recherches théoriques modernes sur le calcul des probabilités. Premier livre: Généralités sur les probabilités (1936); Elements aléatoires. Gauthier-Villars, Paris, deuxième ed., 1950.

4. B.V. Gnedenko and A.N. Kolmogorov, Limit Distributions for Sums of Independent Random Variables. Addison-Wesley, Reading, MA, 1954.

5. D.H. Mouchtari and A.N. Šerstnev, Les fonctions du triangle pour les espaces normés aléatoires. In: E.F. Beckenbach (ed.), General Inequalities 1 (Proceedings of the First International Conference on General Inequalities, Oberwolfach), pp.255-260, Birkhäuser Verlag, Basel, 1978.

6. B. Schweizer, Multiplications on the space of probability distribution functions. Aeq. Math. 12 (1975), 156-183.

7. B. Schweizer and A. Sklar, Probabilistic Metric Spaces. Elsevier North-Holland, New York, 1983.

8. A.N. Šerstnev, Random Normed Spaces: Problems of completeness. Kazan. Gos. Univ. Ucen. Zap 122 (1962), 3-20.

9. D.A. Sibley, A metric for weak convergence of distribution functions. Rocky Mountain J. Math. 1 (1971), 427-430.

Claudi Alsina, Departament de Matemàtiques i Estadîstica (ETSAB), Universitat Politècnica de Catalunya, Avda. Diagonal, 649, 08028 Barcelona, Spain

International Series of
Numerical Mathematics, Vol. 80

SOME ITERATIVE FUNCTIONAL INEQUALITIES AND SCHRÖDER'S EQUATION

B. Choczewski and M. Stopa

Abstract. Iterative functional inequalities of third and second order with constant coefficients are considered in connection with the Schröder functional equation $\phi \circ f = s\phi$.

Let f be a self-mapping of a real interval $I = [0,a)$, $a > 0$, and let f^n denote the n-th iterate of the function f. Continuous solutions Φ of the iterative functional inequality

$$\Phi(f^3(x)) + b_2\Phi(f^2(x)) + b_1\Phi(f(x)) + b_0\Phi(x) \le 0 \tag{1}$$

have been described in [3] for the case where $b_i \in \mathbb{R}$ and f is continuous and strictly increasing in I, $0 < f(x) < x$ in $(0,a)$. The set of solutions of (1) essentially depends on the roots of the characteristic equation

$$r^3 + b_2r^2 + b_1r + b_o = 0 . \tag{2}$$

Let s be a real root of (2). We denote the two other roots by p and q (they may be complex).

In this note we are going to prove a theorem that shows connections between solutions of inequality (1) and those of the Schröder equation

$$\phi(f(x)) = s\phi(x) . \tag{3}$$

The theorem is patterned by one from [1] concerning an iterative functional inequality of second order. Our main tool will be the following lemma on linear recurrences, which is due to D.C.Russell [2].

This paper is in final form and no version of it will be submitted for publication elsewhere.

LEMMA. *Let* $K_i \in \mathbb{R}$ $(i = 1,\dots,m;\ m > 1)$ *and* $K_1 + \dots + K_m = 1$. *A bounded sequence* (a_n) *which satisfies the inequality*

$$a_{n+m} \le \sum_{i=1}^{m} K_i a_{n+m-i}, \quad n = 0,1,\dots,$$

is convergent if and only if the polynomial

$$p(z) = 1 + (1-K_1)z + \dots + (1-K_1-\dots-K_{m-1})z^{m-1}, \quad z \in C,$$

does not possess any zero in the set $\{z \in C:\ |z| = 1,\ z \neq 1\}$.

Our result reads

THEOREM. *Let* $f\colon I \to I$ *and* $b_1, b_2, b_3 \in \mathbb{R}$ *be given. Assume that* $s \in \mathbb{R}$, p *and* q *are roots of equation* (2) *such that*

$$0 < p+q < s \quad \text{and} \quad |p| \neq s \quad \text{if } p \text{ and } q\ (= \bar{p}) \text{ are complex}.$$

If $\Phi\colon I \to \mathbb{R}$ *is nonnegative and satisfies inequality* (1) *and*

$$\Phi(f^2(x)) - (p+q)\Phi(f(x)) + pq\Phi(x) \ge 0, \tag{4}$$

then there exists the limit

$$\phi(x) = \lim_{n\to\infty} s^{-n}\Phi(f^n(x)), \quad x \in I. \tag{5}$$

The function ϕ *satisfies the Schröder equation* (3) *in* I.

Proof. Obviously the function ϕ given by (5) satisfies (3). Let us put, for a fixed $x \in (0,a)$,

$$a_n = s^{-n}\Phi(f^n(x)),$$

where $\Phi \ge 0$ is a solution of inequalities (1) and (4). We have $b_2 = -(p+q+s)$, $b_1 = pq+ps+qs$, $b_0 = -pqs$. Replace x by $f^n(x)$ in (1) and use the formulas for the b's to get

$$s^3 a_{n+3} - (p+q+s)s^2 a_{n+2} + (pq+ps+qs)s a_{n+1} - pqs a_n \le 0.$$

With the notations $K_1 = 1 + (p+q)/s$, $K_2 = -pqs^{-2} + (p+q)/s$, $K_3 = pqs^{-2}$ the above inequality changes over into

$$a_{n+3} \le K_1 a_{n+2} + K_2 a_{n+1} + K_3 a_n, \tag{6}$$

where $K_1 + K_2 + K_3 = 1$.

The polynomial

$$p(z) = 1 + (1-K_1)z + (1-K_1-K_2)z^2 = 1 - \frac{p+q}{s} z + \frac{pq}{s^2} z^2$$

has the roots s/p and s/q which do not belong to the unit circle. In virtue of the lemma, the boundedness of (a_n) implies its convergence.

To prove the boundedness we first derive from (6) by induction the inequality

$$a_{n+1} \le (K_1-1)a_n - K_3 a_{n-1} + B, \quad n \in \mathbb{N}, \tag{7}$$

where

$$B := a_2 - (K_1-1)a_1 + K_3 a_0 \doteq s^{-2}(\Phi(f^2(x)) - (p+q)\Phi(f(x)) + pq\Phi(x)).$$

By (4) we have $B \ge 0$. Since $K_3 > 0$ and $a_{n-1} \ge 0$, inequality (7) yields

$$a_{n+1} \le ca_n + B, \quad n \in \mathbb{N}, \tag{8}$$

where $c = K_1 - 1 = (p+q)/s$ so that $0 < c < 1$. Iterating (8) we obtain

$$a_{n+1} \le c^n a_1 + B(1 + c + \ldots + c^{n-1}), \quad n \in \mathbb{N}.$$

The sequence occurring on the right hand side of the above inequality converges. Thus the sequence (a_n) is bounded, as its terms are nonnegative. This completes the proof.

REFERENCES

1. D. Brydak and B. Choczewski, Continuous solutions of a functional inequality of second order. Demonstratio Math. 9 (1976), 221-228.

2. D.C. Russell, On bounded sequences satisfying a linear inequality. Proc. Edinb. Math. Soc. (2) 19 (1974), 11-16.

3. M. Stopa, A linear functional inequality of third order. Zeszyty Naukowe AGH w Krakowie, Opuscula Mathematica 3 (1986) (to appear).

B. Choczewski and M. Stopa, Institute of Mathematics, Academy of Mining and Metallurgy, al. Mickiewicza 30, 30-059 Kraków, Poland

International Series of
Numerical Mathematics, Vol. 80

ON AN INEQUALITY OF P.W. CHOLEWA

István S. Fenyö

Abstract. Let G be an Abelian group and X a Banach space. Consider a mapping $f\colon G \to X$ for which

$$\|f(x+y) + f(x-y) - 2f(x) - 2f(y)\| < \delta$$

holds $(x,y \in G)$. Then there exists a uniquely defined solution $g\colon G \to X$ of the functional equation

$$g(x+y) + g(x-y) - 2g(x) - 2g(y) = 0 \quad (x,y \in G)$$

such that

$$\|f(x) - g(x)\| < k\delta ,$$

where

$$k = (1/3)\delta + (p/3\delta) , \qquad p = \|f(0)\| ,$$

k is the best possible constant.

Let G be an Abelian group and X a Banach space. P.W. Cholewa [1] proved the following theorem:

THEOREM (Cholewa). *If the function* $f\colon G \to X$ *fulfills the condition*

(1) $$\|f(x+y) + f(x-y) - 2f(x) - 2f(y)\| < \delta \quad \textit{for all } x,y \in G$$

$(\delta > 0)$, *then there exists a uniquely defined solution* $g\colon G \to X$ *of the functional equation*

(2) $$g(x+y) + g(x-y) - 2g(x) - 2g(y) = 0 \quad (x,y \in G)$$

such that

(3) $$\|f(x) - g(x)\| < \frac{1}{2}\delta \quad (x \in G) ,$$

and the factor 1/2 *cannot be replaced by a smaller one*.

This paper is in final form and no version of it will be submitted for publication elsewhere.

Obviously, if f is exposed to a certain condition, then the constant 1/2 on the right hand side of (3) can be reduced. An important type of condition for f is an initial condition. In what follows we establish how the best constant for an estimate of type (3) for $\|f-g\|$ depends on the initial value of f. More precisely, we prove the following

THEOREM. Let $f: G \to X$ be a function for which (1) is valid. Then there exists a uniquely defined solution $g: G \to X$ which satisfies (2), for which

$$\|f(x) - g(x)\| < k\delta \quad (x \in G) \tag{4}$$

holds, where $k = \frac{1}{3}\delta + \frac{1}{3}\frac{p}{\delta}$; $p = \|f(0)\|$. The constant k in (4) cannot be replaced by a smaller one.

Proof. Put

$$d(x,y) := f(x+y) + f(x-y) - 2f(x) - 2f(y) \quad (x,y \in G). \tag{5}$$

Setting $x = y$, we get

$$f(2x) - 4f(x) + f(0) = d(x,x), \tag{6}$$

and substituting $y = 0$ into (5), we have

$$f(0) = -\frac{d(x,0)}{2} \quad (x \in G). \tag{7}$$

This means, if we consider (5) as a functional equation with respect to f, then a necessary condition for (5) to be solvable is that $d(x,0)$ should be a constant (independent of x).

(6) and (7) implies

$$f(2x) - 4f(x) = d(x,x) + \frac{d(x,0)}{2} \quad (x \in G). \tag{8}$$

Now substituting 2x for x in (8) and eliminating f(2x) from the resulting equation and from (6), we get

$$\frac{f(2^2x)}{4^2} - f(x) = \frac{1}{4}\left[d(x,x) + \frac{d(2x,2x)}{4} + \frac{d(x,0)}{2} + \frac{d(x,0)}{2^3}\right].$$

By induction it follows that

$$\frac{f(2^nx)}{4^n} - f(x) = \frac{1}{4}\sum_{k=0}^{n-1}\frac{d(2^kx,2^kx)}{4^k} + \frac{d(x,0)}{8}\sum_{k=0}^{n-1}\frac{1}{4^k} \quad (n = 1,2,3,\ldots). \tag{9}$$

As, by (1), $\|d(2^k x, 2^k x)\| < \delta$ for all $k = 0,1,2,\ldots$ and $x \in G$, the right side of (9) converges to a certain element of X for $n \to \infty$ (depending on $x \in G$), therefore, the left hand side also has a limit if $n \to \infty$. This means that

$$g(x) = \lim_{n\to\infty} \frac{f(2^n x)}{4^n}$$

exists for all $x \in G$, and hence the limit of both sides of (9) leads to the relation

$$g(x) - f(x) = \frac{1}{4} \sum_{k=0}^{\infty} \frac{d(2^k x, 2^k x)}{4^k} + \frac{d(x,0)}{8} \sum_{k=0}^{\infty} \frac{1}{4^k}$$

or, considering (7), to

$$(10) \qquad g(x) - f(x) = \frac{1}{4} \sum_{k=0}^{\infty} \frac{d(2^k x, 2^k x)}{4^k} - \frac{f(0)}{3},$$

from which

$$\|g(x) - f(x)\| < \frac{\delta}{3} + \frac{1}{3}\frac{p}{\delta}\delta = k\delta$$

follows by using the notation $\|f(0)\| = p$.

We have now to prove that g satisfies the functional equation (2). Indeed, by definition of g and considering (5), we have

$$g(x+y) - g(x-y) = \lim_{n\to\infty} \left[\frac{f(2^n x + 2^n y)}{4^n} + \frac{f(2^n x - 2^n y)}{4^n}\right]$$

$$= 2 \lim_{n\to\infty} \frac{f(2^n x)}{4^n} + 2 \lim_{n\to\infty} \frac{f(2^n y)}{4^n} + \lim_{n\to\infty} \frac{d(x,y)}{4^n}$$

$$= 2g(x) + 2g(y) .$$

In order to prove the uniqueness of g, we have to show that the unique solution of (2), bounded on G, is the identically vanishing function. If we put $d(x,y) = 0$ $(x,y \in G)$, then we have to write g instead of f in (5) and so, by (9) (putting $d = 0$, $f = g$),

$$(11) \qquad \frac{g(2^n x)}{4^n} - g(x) = 0 \qquad (x \in G) .$$

By the assumption that g is bounded on G, we have, considering the limit of (11) for $n \to \infty$, $g(x) = 0$ for all $x \in G$. Now using the assertion (b) of the Theorem in [2], we have the uniqueness of g. (10) shows that k is the best possible constant in (4).

This completes the proof.

If f is not subject to any initial condition, then by (7) follows that $p/\delta \leq 1/2$ and $k = 1/3 + 1/6 = 1/2$; this is exactly the statement of Cholewa. If, e.g., $f(0) = 0$, then $k = 1/3$. We see that k is between 1/3 and 1/2.

REFERENCES

1. P.W. Cholewa, Remarks on the stability of functional equations. Aeq. Math. 27 (1984), 76-86.

2. I. Fenyö, Osservazioni su alcuni teoremi di D.H. Hyers. Ist. Lombardo Accad. Sci. Lett. Rendiconti 114 (1980), 235-242.

István S. Fenyö, Mathematical Research Institute of the Hungarian Academy of Sciences, H-1053 Budapest, Realtanoda u. 13-15, Hungary

International Series of
Numerical Mathematics, Vol. 80

SUBADDITIVE MULTIFUNCTIONS AND HYERS-ULAM STABILITY

Zbigniew Gajda and Roman Ger

Abstract. A multifunction F from an Abelian semigroup (S,+) into the family of all nonempty closed convex subsets of a Banach space $(X, \|\cdot\|)$ is called subadditive provided that $F(x+y) \subset F(x) + F(y)$ for all $x,y \in S$. We show that if all the values of a subadditive multifunction F are uniformly bounded then F admits an additive selection, i.e. a homomorphism $a: S \to X$ such that $a(x) \in F(x)$ for all $x \in S$. An abstract version of this result is also presented. The subject is motivated by (and strictly related to) the Hyers-Ulam stability problem.

1. INTRODUCTION

The classical Hyers-Ulam stability problem for the celebrated Cauchy functional equation reads as follows: Given an Abelian semigroup (S,+), a Banach space $(X, \|\cdot\|)$ and a nonnegative real number ε, assume that a map $f: S \to Y$ satisfies the inequality

$$(1) \qquad \|f(x+y) - f(x) - f(y)\| \leq \varepsilon , \qquad x,y \in S \ ;$$

does there exist an additive mapping $a: S \to X$ such that

$$\|a(x) - f(x)\| \leq \varepsilon , \qquad x \in S \ ?$$

It is well-known that the answer to this question is positive (see Hyers [2]). Observe that inequality (1) may equivalently be expressed in the following way: The Cauchy difference of f, i.e. the two place function $S \times S \ni (x,y) \mapsto f(x+y) - f(x) - f(y)$, has all its values inside the closed ball $\bar{B}(0,\varepsilon)$ centred at zero and having radius ε. Replace the ball $\bar{B}(0,\varepsilon)$ by an arbitrarily given nonempty set $A \subset X$ and assume that

---

This paper is in final form and no version of it will be submitted for publication elsewhere.

(2) $$f(x+y) - f(x) - f(y) \in A \quad \text{for all } x,y \in S .$$

Now, consider a multifunction $F: S \to 2^X$ given by the formula

(3) $$F(x) := f(x) + A , \quad x \in S ,$$

and note that in view of (2) one has

$$F(x+y) = f(x+y) + A \subset (f(x) + f(y) + A) + A = F(x) + F(y)$$

for all x, y in S. In other words, relation (2) implies that the multifunction F associated with f is <u>subadditive</u>, i.e.

(4) $$F(x+y) \subset F(x) + F(y) , \quad x,y \in S ,$$

(see W. Smajdor [5]). Now, the stability question is nothing else but the question about the existence of an <u>additive selection</u> of F: Does there exist an additive function $a: S \to X$ such that

$$a(x) \in F(x) , \quad x \in S ?$$

More generally, one may ask whether <u>any</u> subadditive multifunction (not necessarily of the form (3)) admits an additive selection. Without any further assumptions the answer to this question is negative. The following counter-example has been proposed to us by Z. Páles (Debrecen); the authors wish to thank him for that at this place.

EXAMPLE. Consider the semigroup $([0,\infty),+)$ (non-negative real numbers with the usual addition), the Banach space $(\mathbb{R},+;\cdot)$ (the real line) and the multifunction $F: [0,\infty) \to 2^{\mathbb{R}}$ given by the formula $F(x) := [x^2,\infty)$, $x \in [0,\infty)$. Evidently, F is subadditive, i.e. F satisfies "inequality" (4); nevertheless, F does not admit any additive selection. Indeed, otherwise, there would exist an additive function $a: [0,\infty) \to \mathbb{R}$ such that $a(x) \in [x^2,\infty)$, $x \in [0,\infty)$. This forces a to be linear: $a(x) = cx$, $x \in [0,\infty)$ (see J. Aczél [1], for instance). Consequently, one would have $x^2 \le cx$ for all $x \in [0,\infty)$, a contradiction.

Noteworthy is the fact that all the values of the multifunction considered here are closed and convex.

2. EXISTENCE AND UNIQUENESS

The following results give a sufficient condition for a subadditive multifunction to admit an additive selection.

THEOREM 1. *Let* $(S,+)$ *be an Abelian semigroup and let* ccl (X) *denote the collection of all nonempty closed convex subsets of a Banach space* $(X,\|\cdot\|)$. *Suppose that a multifunction* $F: S \to \mathrm{ccl}\,(X)$ *is subadditive. If*

(5) $$\sup \{\mathrm{diam}\, F(x) : x \in S\} < \infty ,$$

then there exists exactly one additive selection of F.

Proof. Subadditivity of F jointly with the convexity of values of F imply easily that $F(2x) \subset 2F(x)$ for any x in S. Inductively, one may show that

$$\frac{1}{2^{n+1}} F(2^{n+1}x) \subset \frac{1}{2^n} F(2^n x) =: F_n(x) , \quad x \in S , \quad n \in \mathbb{N}_0 := \mathbb{N} \cup \{0\} .$$

Therefore, for any $x \in S$, the collection $\{F_n(x) : n \in \mathbb{N}_0\}$ forms a descending countable family of nonempty closed subsets of the Banach space $(X,\|\cdot\|)$. Moreover,

(6) $$\lim_{n\to\infty} \mathrm{diam}\, F_n(x) = 0 \quad \text{for all} \quad x \in S .$$

In fact,

$$\mathrm{diam}\, F_n(x) = \sup_{u,v\in F_n(x)} \|u-v\| = \sup_{u,v\in F(2^n x)} \left\|\frac{1}{2^n} u - \frac{1}{2^n} v\right\|$$

$$= \frac{1}{2^n} \sup_{u,v\in F(2^n x)} \|u-v\| = \frac{1}{2^n} \mathrm{diam}\, F(2^n x) \underset{n\to\infty}{\to} 0$$

because, by means of (5), the sequence $(\mathrm{diam}\, F(2^n x))_{n\in \mathbb{N}_0}$ is bounded. According to Cantor's classical theorem the intersection

$$G(x) := \bigcap_{n\in \mathbb{N}_0} F_n(x)$$

is a singleton. Let a(x) be the only element of the set G(x). We shall show that the map $S \ni x \mapsto a(x)$ yields an additive selection of F. The appurtenance $a(x) \in F(x)$, $x \in S$, is trivial. To prove the additivity of a, fix arbitrarily x and y from S and note that (4) and the commutativity of the binary law + imply easily the inclusions

$$F_n(x+y) = \frac{1}{2^n} F(2^n(x+y)) \subset \frac{1}{2^n} F(2^n x) + \frac{1}{2^n} F(2^n y) = F_n(x) + F_n(y)$$

for all $n \in \mathbb{N}$. Hence

$$a(x+y) \in \bigcap_{n \in \mathbb{N}_0} (F_n(x) + F_n(y)) .$$

On the other hand, since $a(x) + a(y)$ also belongs to $F_n(x) + F_n(y)$ for any $n \in \mathbb{N}_0$, we have

$$\begin{aligned} \|a(x+y) - a(x) - a(y)\| &\leq \operatorname{diam} (F_n(x) + F_n(y)) \\ &\leq \operatorname{diam} F_n(x) + \operatorname{diam} F_n(y) \end{aligned}$$

for every $n \in \mathbb{N}_0$, yielding the additivity of a because of (6).

Finally, to show the uniqueness, assume that $a_i: S \to X$, $i \in \{1,2\}$, are two additive selections of F and fix an $x \in S$. Then

$$2^n a_i(x) = a_i(2^n x) \in F(2^n x) , \quad n \in \mathbb{N}_0 , \quad i \in \{1,2\} ,$$

whence

$$a_i(x) \in F_n(x) , \quad n \in \mathbb{N}_0 , \quad i \in \{1,2\} .$$

Thus, on account of (6), we get $a_1(x) = a_2(x)$ which was to be proved.

COROLLARY 1. Let $(S,+)$ be an Abelian semigroup and let $(X, \|\cdot\|)$ be a Banach space. If a bounded set $B \subset X$ and a function $f: S \to X$ are given such that

(7) $$f(x+y) - f(x) - f(y) \in B \quad \text{for all} \quad x,y \in S ,$$

then there exists exactly one additive function $a: S \to X$ such that

$$f(x) - a(x) \in \operatorname{cl} \operatorname{conv} B \quad \text{for all} \quad x \in S .$$

Proof. Write

(8) $$A := \operatorname{cl} \operatorname{conv} B$$

and define a multifunction $F: S \to \operatorname{ccl}(X)$ by (3). Since, obviously, relations (7) and (8) imply (2), we infer that F is subadditive. Moreover,

$$\sup \{\operatorname{diam} F(x): x \in S\} \leq \operatorname{diam} A \leq \operatorname{diam} \operatorname{cl} \operatorname{conv} B < \infty .$$

Consequently, (5) is satisfied and it remains to apply Theorem 1.

COROLLARY 2 (the classical Hyers' stability theorem). Let S and X have the same meaning as stated above and let $f\colon S \to X$ satisfy relation (1) with some $\varepsilon \geq 0$. Then there exists exactly one additive function $a\colon S \to X$ such that

$$\|a(x) - f(x)\| \leq \varepsilon$$

for all $x \in S$.

Proof. Take $B := \bar{B}(0,\varepsilon)$ in Corollary 1.

3. GENERALIZATIONS

Hyers' stability theorem (see [2] and Corollary 2) has been generalized in many directions (cf. Moszner [3] for an ample bibliography). A pretty general version has been presented by Jürg Rätz during the second conference on General Inequalities (see [4]). He replaced the Banach space $(X,\|\cdot\|)$ by a sequentially complete linear topological space over the field $\mathbb{Q}$ of all rationals and postulated relation (2) with a $\mathbb{Q}$-convex bounded and symmetric set $A \subset X$ containing the origin (a set $A \subset X$ is termed $\mathbb{Q}$-convex provided that jointly with any two points $x,y \in A$ the set $\{\lambda x + (1-\lambda)y\colon \lambda \in [0,1] \cap \mathbb{Q}\}$ is contained in A).

Our aim now is to obtain an abstract analogue of Theorem 1 which, in particular, would allow to deduce a stability theorem at the Rätz level of generality. This causes two difficulties to overcome:

(a) to introduce an appropriate substitute of the notion of diameter;

(b) to get an analogue of Cantor's classical theorem on the intersection of a descending sequence of closed sets in a complete metric space.

So, given a linear topological space X over $\mathbb{Q}$ assume that U is a $\mathbb{Q}$-balanced neighbourhood of zero, i.e. such that $\lambda U \subset U$ for any $\lambda \in [-1,1] \cap \mathbb{Q}$. For $A \subset X$ arbitrarily fixed let

$$\mathbb{Q}_U(A) := \{\lambda \in \mathbb{Q} \cap (0,\infty)\colon A - A \subset \lambda U\} .$$

The number

$$\operatorname{diam}_U A := \begin{cases} \infty & \text{if } Q_U(A) = \emptyset , \\ \inf Q_U(A) & \text{otherwise} \end{cases}$$

is called the *diameter of* A *relative to* U.

It is easy to check that in the case where X is a normed linear space the usual diameter diam A of a set $A \subset X$ coincides with the number $\operatorname{diam}_{B(0,1)} A$ - the diameter of A relative to the open unit ball $B(0,1) \subset X$. The following lemma collects some simple arithmetic properties of the notion just introduced.

LEMMA. *Let* U *and* V *be two* Q-*balanced neighbourhoods of zero in a linear topological space* X *over* Q *and let* $A, B \subset X$ *be arbitrarily given sets. Then*

(i) $\lambda \in Q \cap (0,\infty)$ *implies* $\operatorname{diam}_U \lambda A = \lambda \operatorname{diam}_U A$;

(ii) $V + V \subset U$ *implies* $\operatorname{diam}_U (A+B) \le \max(\operatorname{diam}_V A, \operatorname{diam}_V B)$;

(iii) $\operatorname{diam}_U A < 1$ *implies* $A - A \subset U$.

Proof. Routine; to give a flavour let us prove assertion (ii) for instance. Let $r := \max(\operatorname{diam}_V A, \operatorname{diam}_V B)$. Without any loss of generality we may assume that r is finite. Fix arbitrarily $\varepsilon > 0$ and choose $\lambda \in Q \cap (0, r+\varepsilon)$ such that

$$A - A \subset \lambda V \quad \text{and} \quad B - B \subset \lambda V .$$

Then

$$(A+B) - (A+B) = (A-A) + (B-B) \subset \lambda V + \lambda V \subset \lambda U ,$$

whence

$$\operatorname{diam}_U (A+B) \le \lambda < r+\varepsilon ;$$

now, (ii) results since ε is arbitrary.

From now on the symbol seq cl A will stand for the sequential closure of a set $A \subset X$.

THEOREM 2. *Let* X *be a sequentially complete linear topological space over* Q *and let* $\mathcal{B}_0$ *be any base of* Q-*balanced neighbourhoods of the origin in* X. *Assume that* $\{A_n : n \in \mathbb{N}_0\}$ *is a descending* family of nonempty subsets of X and such that for any member U of $\mathcal{B}_o$ one has

(9) $$\operatorname{diam}_U A_n < 1$$

for all but a finite number of indices $n \in \mathbb{N}_o$. Then

(10) $$\bigcap_{n \in \mathbb{N}_o} \operatorname{seq\,cl} A_n \neq \emptyset .$$

Proof. Fix a $U \in \mathcal{B}_o$ and choose arbitrarily an $x_n \in A_n$, $n \in \mathbb{N}_o$. Obviously, for any two positive integers $m > n$ we have

$$x_m - x_n \in A_m - A_n \subset A_n - A_n .$$

On the other hand, point (iii) of the lemma jointly with (9) imply that $A_n - A_n \subset U$ for all $n \in \mathbb{N}_o$ large enough, say, for all $n \geq n_o$. Consequently,

$$x_m - x_n \in U \quad \text{for all} \quad m > n \geq n_o ,$$

which says that $(x_n)_{n \in \mathbb{N}_o}$ is a Cauchy sequence. Now, the sequential completeness of X states that there exists an $x \in X$ such that $x = \lim x_n$. Evidently, $x \in \operatorname{seq\,cl} A_n$ for all $n \in \mathbb{N}_o$; hence inequality (10) has been proved.

We are now in a position to prove an abstract analogue of Theorem 1; namely, we have the following

THEOREM 3. Let $(S,+)$ be an Abelian semigroup and let X be a sequentially complete linear topological space over the field $\mathbb{Q}$. Assume that $F\colon S \to 2^X \setminus \{\emptyset\}$ is a subadditive multifunction whose values are all $\mathbb{Q}$-convex and sequentially closed. If

(11) $$\sup \{\operatorname{diam}_U F(x)\colon x \in S\} < \infty$$

for all elements U from some base of $\mathbb{Q}$-balanced neighbourhoods of the origin in X, then F admits an additive selection; if, moreover, X is a T_1-space, then such a selection is unique.

Proof. Similarly as in the proof of Theorem 1 we infer that, for any $x \in S$, the sets $F_n(x) := \frac{1}{2^n} F(2^n x)$, $n \in \mathbb{N}_o$, form a descending family. Fix arbitrarily a base $\mathcal{B}_o$ of $\mathbb{Q}$-balanced neighbourhoods of zero and take a $U \in \mathcal{B}_0$. Put

$$c := \sup \{\operatorname{diam}_U F(x) : x \in S\} ;$$

c is finite because of (11). Note that for any $x \in S$ and any $n \in \mathbb{N}$ one has

$$\operatorname{diam}_U F_n(x) = \operatorname{diam}_U \frac{1}{2^n} F(2^n x) = \frac{1}{2^n} \operatorname{diam}_U F(2^n x) \le \frac{c}{2^n}$$

(see point (i) of the lemma). Therefore, there exists an $n_0(U) \in \mathbb{N}_0$ such that

(12) $$\operatorname{diam}_U F_n(x) < 1 \quad \text{for all } x \in S \text{ and all } n \ge n_0(U) .$$

With the aid of Theorem 2 we get the inequality

$$G(x) := \bigcap_{n \in \mathbb{N}_0} F_n(X) \neq \emptyset$$

valid for all x in S. Choose arbitrarily $g(x) \in G(x)$, $x \in S$. In particular,

(13) $$g(x) \in F(x) \quad \text{for all} \quad x \in S .$$

Moreover, since all the multifunctions F_n, $n \in \mathbb{N}_0$, are subadditive, for any x, y in S, we get

$$g(x+y) \in F_n(x+y) \subset F_n(x) + F_n(y) , \quad n \in \mathbb{N}_0 .$$

On the other hand, the sum $g(x) + g(y)$ also belongs to $F_n(x) + F_n(y)$, $n \in \mathbb{N}_0$. Fix a $U \in \mathcal{B}_0$ again and take a $V \in \mathcal{B}_0$ such that $V + V \subset U$. Relation (12) ensures that there exists a positive integer $n_0(V)$ such that

$$\operatorname{diam}_V F_n(x) < 1 \quad \text{for all } x \in S \text{ and all } n \ge n_0(V) .$$

Hence, on account of point (ii) of the lemma, we obtain

$$\operatorname{diam}_U (F_n(x) + F_n(y)) \le \max (\operatorname{diam}_V F_n(x), \operatorname{diam}_V F_n(y)) < 1$$

provided that $n \ge n_0(V)$. Applying the lemma again (point (iii), we get

$$g(x+y) - (g(x) + g(y)) \in (F_n(x) + F_n(y)) - (F_n(x) + F_n(y)) \subset U ,$$

whence, because of the unrestricted choice of $U \in \mathcal{B}_0$, we deduce that

$$g(x+y) - g(x) - g(y) \in \bigcap_{U \in \mathcal{B}_0} U = \operatorname{cl} \{0\} .$$

Let X_1 denote the subspace of X complementary to cl {0}, i.e.

$$X = X_1 \oplus \mathrm{cl}\,\{0\} \quad ,$$

and let, for any $x \in S$, the symbol $a(x) = \mathrm{proj}_{X_1}\, g(x)$ stand for the projection of g(x) onto the X_1-axis. Then, for all x, y in S,

$$a(x+y) - a(x) - a(y) = \mathrm{proj}_{X_1}\,(g(x+y) - g(x) - g(y)) \in \mathrm{proj}_{X_1}\,\mathrm{cl}\,\{0\} = \{0\}$$

which states that the map $S \ni x \to a(x) \in X$ is additive. Moreover, for all x in S, we have

$$a(x) \in g(x) + \mathrm{cl}\,\{0\} = \mathrm{cl}\,\{g(x)\}$$

as well as

$$g(x) \in F(x) = \mathrm{seq}\,\mathrm{cl}\,F(x)$$

because of (13) and the fact that (by assumption) the values of F are sequentially closed. Therefore (see Rätz's paper [4])

$$a(x) \in F(x) \quad \text{for all } x \in S \ ,$$

as required.

To prove the uniqueness of the additive selection a assume that X is a T_1-space and suppose that a_i, $i \in \{1,2\}$, are two additive selections of F. Fix an $x \in S$ arbitrarily and note that

$$2^n a_i(x) = a_i(2^n x) \in F(2^n x) \quad \text{for all } n \in \mathbb{N}_0 \text{ and } i \in \{1,2\} \ .$$

Applying (12) and point (iii) of the lemma again, we get the relation

$$a_1(x) - a_2(x) \in U$$

valid for any neighbourhood U from $\mathcal{B}_0$. Consequently,

$$a_1(x) - a_2(x) \in \bigcap_{U \in \mathcal{B}_0} U = \mathrm{cl}\,\{0\} = \{0\} \ .$$

This completes the proof.

COROLLARY 3. <u>Let (S,+) be an Abelian semigroup and let X be a sequentially complete linear topological space over the field $\mathbb{Q}$. Assume $B \subset X$ to be a set whose $\mathbb{Q}$-convex hull $\mathrm{conv}_{\mathbb{Q}} B$ is bounded. If $f: S \to X$ is a function such that relation (7) holds true, then there exists an additive function $a: S \to X$ such that</u>

$$f(x) - a(x) \in \text{seq cl conv}_{\mathbb{Q}} B \quad \underline{\text{for all}} \quad x \in S \ ;$$

if, moreover, X *is a* T_1*-space then such a function* a *is unique.*

Proof. Write $A := \text{seq cl conv}_{\mathbb{Q}} B$ and define a multifunction $F: S \to 2^X \setminus \{\emptyset\}$ by (3). One may easily check that F is a subadditive multifunction whose values are all $\mathbb{Q}$-convex and sequentially closed. Let U be an arbitrary $\mathbb{Q}$-balanced neighbourhood of origin in X. Then

$$\sup \{\text{diam}_U\, F(x) : x \in S\} = \text{diam}_U\, A < \infty$$

because the (assumed) boundedness of the set $\text{conv}_{\mathbb{Q}} B$ is not spoiled by the seq cl operation and, obviously, any bounded set has a finite relative diameter. It remains to apply Theorem 3.

COROLLARY 4 (Rätz's stability theorem). *Let* S *and* X *have the same meaning as stated above and let* $f: S \to X$ *satisfy relation* (7) *with some* $\mathbb{Q}$*-convex and bounded set* $B \subset X$*. Then there exists an additive function* $a: S \to X$ *such that*

$$f(x) - a(x) \in \text{seq cl}\ B$$

for all $x \in S$; *if, moreover,* X *is a* T_1*-space then such a function* a *is unique.*

4. TWO REMARKS

REMARK 1. Actually, Rätz [4] has adopted some assumptions on the binary law + in S which are essentially weaker than commutativity; for the sake of brevity we have simply assumed that the semigroup $(S,+)$ is Abelian.

REMARK 2. In general, the boundedness of B is not shared by its $\mathbb{Q}$-convex hull $\text{conv}_{\mathbb{Q}} B$. Therefore, we had to assume the boundedness of the latter in Corollary 3. Alternatively, one might assume that the space X is locally $\mathbb{Q}$-convex; in such a case the assumption of the boundedness of B itself would suffice to keep the assertion of Corollary 3 valid.

REFERENCES

1. J. Aczél, Lectures on functional equations and their applications. Academic Press, New York and London, 1966.

2. D.H. Hyers, On the stability of the linear functional equation. Proc. Nat. Acad. Sci. U.S.A. 27 (1941), 222-224.

3. Z. Moszner, Sur la stabilité de l'équation d'homomorphisme. Aequationes Math. 29 (1985), 290-306.

4. J. Rätz, On approximately additive mappings. In: E.F. Beckenbach (ed.), General Inequalities 2 (Proceedings of the Second International Conference on General Inequalities, Oberwolfach), pp.233-251, Birkhäuser Verlag, Basel - Boston - Stuttgart, 1980.

5. W. Smajdor, Subadditive and subquadratic set-valued functions. Uniwersytet Śląski, Katowice (to appear).

Zbigniew Gajda, Institute of Mathematics, Silesian University, Bankowa 14, 40-007 Katowice, Poland

Roman Ger, Institute of Mathematics, Silesian University, Bankowa 14, 40-007 Katowice, Poland

International Series of
Numerical Mathematics, Vol. 80

BEMERKUNGEN ZU EINEM EXISTENZ- UND EINDEUTIGKEITSPROBLEM VON W. WALTER AUS DEM GEBIET DER DIFFERENZENGLEICHUNGEN

Hans-Heinrich Kairies

Abstract. We discuss the following problem: Under which assumptions on $h : \mathbb{R}_+ \to \mathbb{R}$ has the difference equation $f(x+1) - f(x) = h(x)$, $x \in \mathbb{R}_+$, exactly one convex solution, normalized by $f(1) = 0$? The problem is completely solved in the case $\inf \{h(n+1) - h(n) \mid n \in \mathbb{N}\} = 0$.

1. EINLEITUNG

Wir behandeln in dieser Note Existenz- und Eindeutigkeitsaussagen für normierte Lösungen der Differenzengleichung

$$\text{(D)} \quad f(x+1) - f(x) = h(x) \quad \text{für alle } x \in \mathbb{R}_+ := (0,\infty) ,$$

wobei $f : \mathbb{R}_+ \to \mathbb{R}$, $h : \mathbb{R}_+ \to \mathbb{R}$. Ein wichtiges, wohlbekanntes Resultat aus diesem Problemkreis stammt von W. Krull [1]. Wir geben es hier in einer von M. Kuczma [2, p. 114] stammenden Version an.

SATZ A. h sei konkav, und es gelte $\lim_{x\to\infty} [h(x+1) - h(x)] = 0$. Dann existiert genau eine konvexe, durch $F(1) = 0$ normierte Lösung F von (D). Sie ist gegeben für $x \in \mathbb{R}_+$ durch

$$F(x) = (x-1)h(1) + \sum_{k=0}^{\infty} \{[h(k+1) - h(x+k)] + (x-1)[h(k+2) - h(k+1)]\}.$$

Im Zusammenhang damit wurde von Wolfgang Walter auf der 5. Internationalen Tagung über Allgemeine Ungleichungen die Frage gestellt: Welche Eigenschaften muß h besitzen, damit (D) genau eine konvexe, normierte Lösung hat? Es stellt sich heraus, daß die im Krullschen Satz aufgeführten Bedingungen an h keineswegs notwendig dafür sind, daß (D) genau eine konvexe, normierte Lösung besitzt. Für die Menge der uns interessierenden Lösungen von (D) führen

---

This paper is in final form and no version of it will be submitted for publication elsewhere.

wir eine Kurzbezeichnung ein:

$K(h) := \{f : \mathbb{R}_+ \to \mathbb{R} \mid f$ erfüllt (D), f ist konvex und $f(1) = 0\}$.

In 2. wird gezeigt, daß man die Eindeutigkeitsaussage aus Satz A auch dann erhält, wenn

"h sei konkav, und es gelte $\lim_{x\to\infty} [h(x+1)-h(x)] = 0$"

ersetzt wird durch die weitaus schwächere Forderung

$$\text{(I)} \quad \inf\{h(n+1)-h(n) \mid n \in \mathbb{N}\} = 0.$$

In 3. geben wir zunächst für die Menge der auf den Intervallen $(0,1]$ und $[n,n+1]$, $n \in \mathbb{N}$, affinen Störfunktionen h notwendige und hinreichende Bedingungen dafür an, daß card $K(h) = 1$ ist. Setzen wir dann noch voraus, daß $\lim_{x\to 0+} h(x) < h(1)$ ist, so kann in diesem Fall die Lösung des Problems von W. Walter sehr einfach formuliert werden: card $K(h) = 1$ gilt genau dann, wenn $(h(n))_{n\in\mathbb{N}}$ monoton wachsend ist und (I) erfüllt ist. Es ist damit klar, daß es Funktionen h gibt, die weder konvex noch konkav sind und doch card $K(h) = 1$ erzwingen. Anschließend wird die Lösung des Problems für die Menge derjenigen Störfunktionen h gegeben, die (I) erfüllen. Hier ist die Charakterisierung von card $K(h) = 1$ durch Eigenschaften von h nicht so einfach.

BEMERKUNG 1. a) Wenn $K(h)$ nichtleer ist, muß h notwendig monoton wachsend sein:
Mit $f \in K(h)$ und $s < t$ ergibt sich wegen der Konvexität von f

$$h(s) = f(s+1)-f(s) \leqq \frac{f(t+1)-f(s)}{t+1-s} \leqq f(t+1)-f(t) = h(t).$$

b) Es sei $f^*(x) := f(x) + (1-x)h(1)$. f ist genau dann eine Lösung von (D), wenn $f^*(x+1) - f^*(x) = h(x) - h(1)$ gilt. Ferner gilt $f(1) = 0$ genau dann, wenn $f^*(1) = 0$ ist, und f ist genau dann konvex, wenn f^* konvex ist. Es ist daher keine Einschränkung der Allgemeinheit, wenn $h(1) = 0$ angenommen wird. Das werden wir von nun an tun. Für ein $f \in K(h)$ gilt dann $f(1) = f(2) = 0$ und f ist monoton wachsend auf $[2,\infty)$.

2. EIN EINDEUTIGKEITSSATZ

Teil a) des folgenden Satzes enthält die schon in der Einleitung genannte Eindeutigkeitsaussage. In b) wird eine explizite Darstellung von F gegeben, falls $K(h) = \{F\}$ gilt.

SATZ 1. a) $h : \mathbb{R}_+ \to \mathbb{R}$ _erfülle_ (I). _Dann ist_ card $K(h) \leqq 1$.
b) $h : \mathbb{R}_+ \to \mathbb{R}$ _erfülle_ (I), _und es gelte_ $K(h) = \{F\}$. _Ferner sei_

$$H_n(x) := (x-1)h(n+1) - \sum_{k=0}^{n-1} [h(x+k) - h(1+k)].$$ _Dann gilt:_

(1) $F(x) = H_N(x)$ _für_ $x \in (1,2]$,

$F(x) = H_N(x+1) - h(x)$ _für_ $x \in (0,1]$ _und_

$F(x) = H_N(x-n) + \sum_{k=1}^{n} h(x-k)$ _für_ $x \in (n+1,n+2]$, $n \in \mathbb{N}$,

falls ein $N \in \mathbb{N}$ _existiert mit_ $h(N+1) - h(N) = 0$.

(2) $F(x) = \lim_{n\to\infty} H_n(x)$ _für_ $x \in \mathbb{R}_+$,

falls $\lim_{n\to\infty} [h(n+1) - h(n)] = 0$.

(3) $F(x) = \lim_{k\to\infty} H_{n_k}(x)$ _für_ $x \in (1,2]$,

$F(x) = \lim H_{n_k}(x+1) - h(x)$ _für_ $x \in (0,1]$ _und_

$F(x) = \lim H_{n_k}(x-n) + \sum_{k=1}^{n} h(x-k)$ _für_ $x \in (n+1,n+2]$, $n \in \mathbb{N}$,

falls eine Teilfolge $(n_k)_{k\in\mathbb{N}}$ _existiert mit_ $\lim_{k\to\infty} [h(n_k+1) - h(n_k)] = 0$.

Beweis. a) Nach Bemerkung 1 a) darf angenommen werden, daß $(h(n))$ monoton wachsend ist. Für $n \in \mathbb{N}$ sei $P_n = (n, f(n))$ und y_n beschreibe die Verbindungsgerade von P_n und P_{n+1}. Es gilt $y_n(x) = f(n) + h(n)(x-n)$, also $y_{n+1}(x) - y_n(x) = [h(n+1)-h(n)](x-n-1)$. Sind f_1 und f_2 Elemente von $K(h)$, so muß für $n+1 \leqq x \leqq n+2$ gelten: $|f_1(x) - f_2(x)| \leqq [h(n+1) - h(n)](x-n-1) \leqq h(n+1) - h(n)$. Nun ist $p := f_1 - f_2$ eine 1-periodische Funktion. Aus

$$|p(x)| \leqq h(n+1) - h(n) \text{ für alle } x \in \mathbb{R}_+ \text{ und } n \in \mathbb{N}$$

folgt aber wegen (I), daß p die Nullfunktion ist.

b) Es sei nun $K(h) = \{F\}$. Da F konvex auf $\mathbb{R}_+$ ist, gilt für $x \in (1,2]$ und $n \in \mathbb{N}$

$$F(n+1) - F(n) \leqq (x-1)^{-1}[F(n+x) - F(n+1)] \leqq F(n+2) - F(n+1), \text{ also}$$
$$(x-1)h(n) \leqq F(n+x) - F(n+1) \leqq (x-1)h(n+1), \text{ d.h.}$$

$$(x-1)h(n) \leq \sum_{k=0}^{n-1} [h(x+k) - h(1+k)] + F(x) \leq (x-1)h(n+1).$$

Mit der in der Behauptung des Satzes erklärten Funktion H_n kann die letzte Ungleichung auch in folgender Form geschrieben werden:

$$(4) \qquad -(x-1)[h(n+1) - h(n)] \leq F(x) - H_n(x) \leq 0.$$

Wegen der Infimumbedingung (I) muß wenigstens einer der drei in der Behauptung aufgeführten Fälle eintreten. Gibt es ein $N \in \mathbb{N}$ mit $h(N+1) - h(N) = 0$, so ist wegen (4) notwendig $F(x) = H_N(x)$ für $x \in (1,2]$. Fortsetzung dieser Lösung auf $(0,1]$ bzw. $(2,\infty)$ geschieht mit (D) und führt nach einfacher Rechnung auf die in (1) angegebene Darstellung.

Gilt $\lim\limits_{n\to\infty} [h(n+1) - h(n)] = 0$, so ist nach (4) notwendig $F(x) = \lim\limits_{n\to\infty} H_n(x)$ für $x \in (1,2]$. Wegen $H_n(x+1) - H_n(x) = h(x) + [h(n+1) - h(n+x)]$, $n \in \mathbb{N}$, folgt hier

$$\lim H_n(x+1) = \lim H_n(x) + h(x).$$

Daraus läßt sich entnehmen, daß $\lim H_n(x)$ für alle $x \in \mathbb{R}_+$ existiert und daß $\lim H_n$ die Differenzengleichung (D) erfüllt. Daher ist $F(x) = \lim H_n(x)$ für alle $x \in \mathbb{R}_+$.

Gibt es schließlich eine Folge $(n_k)_{k\in\mathbb{N}}$ natürlicher Zahlen mit $\lim\limits_{k\to\infty} [h(n_k+1) - h(n_k)] = 0$, so ist wegen (4) notwendig $F(x) = \lim\limits_{k\to\infty} H_{n_k}(x)$ für $x \in (1,2]$. Nun ist i. allg. $\lim H_{n_k}$ keine Lösung von (D). Daher erfolgt die Fortsetzung auf $\mathbb{R}_+$ wie im ersten Fall und liefert die in (3) angegebene Darstellung. -

BEMERKUNG 2. Gibt es ein $N \in \mathbb{N}$ mit $h(N+1) = h(N)$, so läßt sich aus der Darstellung (1) entnehmen, daß F auf $(0,n+1]$ bereits eindeutig durch die Werte von h auf dem endlichen Intervall $(0, \max\{n,N+1\}]$ bestimmt ist. Ferner ist dann F affin auf $[N,N+2]$. Nun sei angenommen: $h(n+1) \neq h(n)$ gilt für alle $n \in \mathbb{N}$. Dann werden im Fall $\lim[h(n+1) - h(n)] = 0$ schon zur Festlegung von F auf dem Intervall $(0,1]$ die Werte von h auf ganz $\mathbb{R}_+$ benötigt. Die in (2) gegebene Darstellung kann auch in die folgende mit der aus Satz A übereinstimmenden Form gebracht werden:

$$F(x) = \sum_{k=0}^{\infty} \{[h(k+1) - h(k+x)] + (x-1)[h(k+2) - h(k+1)]\}.$$

3. CHARAKTERISIERUNGEN VON card K(h) = 1.

Wir haben gezeigt, daß die Infimumbedingung (I) hinreicht, um card $K(h) \leq 1$ zu erzwingen. Wir geben nun eine Menge A von Funktionen h an, für die (I) - in leicht modifizierter Form - aus card $K(h) = 1$ folgt.

$A: = \{h : \mathbb{R}_+ \to \mathbb{R} \mid h \text{ affin auf } (0,1] \text{ und } [n,n+1],\ n \in \mathbb{N}\}.$

Für jedes $h \in A$ existiert die reelle Zahl $h(0): = \lim_{x \to 0+} h(x)$.

Ferner sei an die Vereinbarung aus der Bemerkung 1 b) erinnert: $h(1) = 0$.

SATZ 2. Es sei $h \in A$. Dann gilt:

a) Aus card $K(h) = 1$ folgt, daß h monoton wächst und daß

(I') $\inf \{h(n) - h(n-1) \mid n \in \mathbb{N}\} = 0$

erfüllt ist.

b) Aus $h(n-1) \leq h(n)$, $n \in \mathbb{N}$, und (I) folgt card $K(h) = 1$.

Beweis. a) Einem $h \in A$ wird ein $g : \mathbb{R}_+ \to \mathbb{R}$ zugeordnet, das durch folgende Vorschrift definiert ist:

$g(1) = 0,\ g(n+1) = h(1) + h(2) + \ldots + h(n)$ für $n \in \mathbb{N}$, $\lim_{x \to 0+} g(x) = -h(0)$.

Ferner sei g affin auf $(0,1]$ und auf $[n,n+1]$, $n \in \mathbb{N}$.

Es ist einfach nachzuprüfen, daß g eine normierte Lösung von (D) ist. Aus card $K(h) = 1$ folgt nach Bemerkung 1 a), daß h monoton wächst. Daraus ergibt sich aber die Konvexität der eben eingeführten Standardlösung g von (D):

Es ist nämlich

$$g(n+1) - 2g(n) + g(n-1) = h(n) - h(n-1) \geq 0 \quad \text{für } n \in \mathbb{N}.$$

Dabei ist $g(0): = \lim_{x \to 0+} g(x) = -h(0)$. Da $K(h)$ einelementig ist, folgt $K(h) = \{g\}$.

Wir nehmen nun an, daß (I') nicht gilt. Dann ist

(5) $\inf \{h(n) - h(n-1) \mid n \in \mathbb{N}\} =: 3a > 0.$

Nun werde eine 1-periodische Funktion $p : \mathbb{R}_+ \to \mathbb{R}$ erklärt durch

$p(x): = -ax$ für $x \in (0, \frac{1}{2})$, $p(x): = a(x-1)$ für $x \in [\frac{1}{2}, 1]$.

Ferner sei $G(x) := g(x) + p(x)$. Dann ist klar, daß G ebenfalls eine normierte Lösung von (D) ist, die auf den Intervallen $(0,1]$ sowie $[n,n+1]$, $n \in \mathbb{N}$, konvex ist. G ist sogar konvex auf ganz $\mathbb{R}_+$. Eine kurze Rechnung zeigt nämlich, daß für $n \in \mathbb{N}$ gilt

$$G'_+(n) - G'_-(n) = h(n) - a - [h(n-1) + a] = h(n) - h(n-1) - 2a \geq a > 0.$$

Daher wäre neben g auch noch G ein Element von $K(h)$, was unserer Voraussetzung card $K(h) = 1$ widerspricht. (5) ist damit widerlegt.

b) Wir hatten in a) schon gezeigt, daß aus $h(n-1) \leq h(n)$ für $n \in \mathbb{N}$ folgt: card $K(h) \geq 1$.
Nach Satz 1 a) liefert die Voraussetzung (I): card $K(h) \leq 1$. -

BEMERKUNG 3. a) Für die Funktionen $h \in A$ mit $h(0) < h(1)$ sind die Infimumbedingungen (I) und (I') äquivalent. In diesem Fall ist also eine einfache Charakterisierung möglich: card $K(h) = 1$ gilt genau dann, wenn $(h(n))$ monoton wachsend ist und (I) erfüllt ist.

b) Aus card $K(h) = 1$ folgt im allgemeinen Fall nicht notwendig (I). Man kann z. B. zeigen, daß die zu $h \in A$ mit $h(0) = 0$, $h(n) = n-1$ für $n \in \mathbb{N}$ gehörende Standardlösung g aus Satz 2 die einzige konvexe, normierte Lösung von (D) ist. Es ist aber
$\inf\{h(n+1) - h(n) \mid n \in \mathbb{N}\} = 1$.
Wir geben nun für die Menge der Störfunktionen h, welche (I) erfüllen, eine Charakterisierung von card $K(h) = 1$. Wir führen sie auf Satz 1 zurück und fassen uns kurz, indem wir auf die dort angegebenen Fallunterscheidungen (1), (2), (3) für die Darstellung der Lösung F verweisen.

SATZ 3. _Für die Menge der Funktionen_ $h : \mathbb{R}_+ \to \mathbb{R}$, _die_ (I) _erfüllen, gilt_ card $K(h) = 1$ _genau dann, wenn die gemäß der Fallunterscheidung in Satz 1 b) festgelegte Funktion_ F _auf_ $\mathbb{R}_+$ _konvex ist._

Beweis. Wenn card $K(h) = 1$, also $K(h) = \{F\}$ ist, so besitzt F nach Satz 1 b) die dort angegebene Darstellung. F ist nach Definition von $K(h)$ konvex.

Andererseits erzwingt die vorausgesetzte Bedingung (I) nach Satz 1 a), daß card $K(h) \leqq 1$ ist. Ist nun F konvex, so gilt $F \in K(h)$, denn die durch (1) bzw. (2) bzw. (3) dargestellte Funktion F ist eine normierte Lösung von (D).

BEMERKUNG 4. a) Auch beim Krullschen Satz A handelt es sich um eine Aussage über Störfunktionen h, für die (I) gilt. Sind nämlich die Krullschen Bedingungen "h sei konkav, und es gelte $\lim_{x\to\infty} [h(x+1) - h(x)] = 0$" erfüllt, so ist h notwendig monoton wachsend und (I) folgt. Wir sind dann im 2. Fall aus Satz 1 b). Aus der Darstellung (2)

$$F(x) = \lim_{n\to\infty} H_n(x)$$

läßt sich unmittelbar entnehmen, daß F konvex ist. Der Konvergenzbeweis ist dann einfach (vergl. [2]).

b) Andererseits gilt auch für die durch $h(0) = -1$, $h(2n-1) = h(2n) = n-1$, $n \in \mathbb{N}$, definierte Funktion $h \in A$, daß die nach (1) zugeordnete Funktion F konvex ist. Diese stimmt mit der im Beweis von Satz 2 a) erklärten Funktion g überein. In diesem Fall ist h weder konkav noch konvex auf irgendeinem Teilintervall der Länge 3 von $\mathbb{R}_+$.

LITERATUR

1. W. Krull, Bemerkungen zur Differenzengleichung $g(x+1) - g(x) = \phi(x)$. Math. Nachr. 1 (1948), 365 - 376.

2. M. Kuczma, Functional equations in a single variable. Polish Scientific Publishers, Warszawa, 1968.

Hans-Heinrich Kairies, Institut für Mathematik, Technische Universität Clausthal, D-3392 Clausthal-Zellerfeld, West Germany

Inequalities for Differential Operators

Church at Oberwolfach-Walke

International Series of
Numerical Mathematics, Vol. 80

LINEAR AND NONLINEAR DISCRETE INEQUALITIES IN n INDEPENDENT VARIABLES

Ravi P. Agarwal

Abstract. We introduce a discrete analogue of Riemann's function and use it to study discrete Gronwall type inequalities in n independent variables. Next, we provide an estimate of Riemann's function and use it to obtain Wendroff type estimates.

1. INTRODUCTION

In the last few years discrete analogues of several integral and differential inequalities known in the continuous case for a single independent variable have been studied and applied, e.g., [2,4,9 and references therein]. In two or more independent variables for discrete Gronwall type inequalities only Wendroff type estimates are known [3,5-8,10-12,14,15]. In this paper we shall introduce discrete analogue of Riemann's function and use it to obtain estimates (best possible results for particular cases) for certain Gronwall type inequalities which are discrete analogues of several results established in [1,13]. As a consequence of this approach we are able to relax some of the conditions on the functions appearing in the inequalities which were required in our earlier work [5].

2. DISCRETE RIEMANN'S FUNCTION

Let $N = \{0,1,\ldots\}$ and the product $N \times \ldots \times N$ (n times) be denoted by N^n. For all $a > b \in N$ and any $f(t)$ defined on N, $\sum_{t=a}^{b} f(t) = 0$ and $\prod_{t=a}^{b} f(t) = 1$. A point $(x_1^i,\ldots,x_n^i)$ in N^n is

---

This paper is in final form and no version of it will be submitted for publication elsewhere.

denoted by x^i. For all $s,x \in N^n$, $0 \leq s < x$, i.e., $0 \leq s_i < x_i$ and any $f(p)$ defined on N^n, $\mathop{\mathbf{S}}_{p=s}^{x-1} f(p)$ represents the n-fold sum $\sum_{p_1=s_1}^{x_1-1} \cdots \sum_{p_n=s_n}^{x_n-1} f(p_1,\ldots,p_n)$ and $\Delta_p^n f(p)$ denotes $\Delta_{p_1}\cdots\Delta_{p_n}$ $f(p_1,\ldots,p_n)$, where Δ is the forward difference operator $\Delta f(t) = f(t+1) - f(t)$.

We need the following two elementary lemmas.

LEMMA 2.1 *Let* $\phi_1(t)$ *and* $\phi_2(t)$ *be defined on* N, *then for all* $t \in N$

$$\sum_{\ell=0}^{t-1} \phi_1(\ell)\Delta\phi_2(\ell) = \phi_1(\ell)\phi_2(\ell)\Big|_{\ell=0}^{t} - \sum_{\ell=0}^{t-1} \Delta\phi_1(\ell)\phi_2(\ell+1).$$

LEMMA 2.2 *Let* $g(x)$ *be defined on* N^n, *then the function* $v(s,x)$, $s \leq x-1$, $(s,x) \in N^n \times N^n$ *is a solution of*

$$(-1)^n \Delta_s^n v(s,x) = g(s)v(s+1,x) \tag{2.1}$$

$$v(s,x) = 1;\ s_i = x_i,\quad 1 \leq i \leq n \tag{2.2}$$

if and only if

$$v(s,x) = 1 + \mathop{\mathbf{S}}_{p=s}^{x-1} g(p)v(p+1,x). \tag{2.3}$$

LEMMA 2.3 *The problem* (2.1), (2.2) *or equivalently* (2.3) *has a unique solution* $v(s,x)$. *Further, if* $g(x) \geq 0$ *on* N^n, *then* $v(s,x) \geq 1$ *on* $N^n \times N^n$.

Proof. Define the iterates

$$v_0(s,x) = 1$$

$$v_{m+1}(s,x) = 1 + \mathop{\mathbf{S}}_{p=s}^{x-1} g(p)v_m(p+1,x);\ m = 0,1,\ldots\ . \tag{2.4}$$

Then, an easy induction provides

$$\left|v_m(s,x) - v_{m-1}(s,x)\right| \leq G^m \frac{1}{(m!)^n} \prod_{i=1}^{n} (x_i-s_i)^m$$

wehre $G = \max_{0\leq p\leq x-1} |g(p)|$.

Therefore, for $(s,x) \in N^n \times N^n$ it follows that

$$|v_0(s,x)| + \sum_{k=1}^{m} |v_k(s,x)-v_{k-1}(s,x)| \leq 1 + \sum_{k=1}^{m} G^k \frac{1}{k!} \left[\prod_{i=1}^{n} (x_i-s_i)\right]^k$$

$$\leq \exp\left[G \prod_{i=1}^{n} (x_i-s_i)\right]$$

and hence the sequence $\{v_m(s,x)\}$ converges to a solution $v(s,x)$ of (2.3). The uniqueness of $v(s,x)$ and the inequality $v(s,x) \geq 1$ on $N^n \times N^n$ (when $g(x) \geq 0$ on N^n) are obvious from (2.4).

LEMMA 2.4 *Let* $g(x)$ *and* $h(x)$ *be defined and nonnegative on* N^n *and the following inequality holds*

$$\Delta_x^n u(x) \leq g(x)u(x) + h(x) \tag{2.5}$$

where

$$u(x_1,\ldots,x_{i-1},0,x_{i+1},\ldots,x_n) = 0,\ 1 \leq i \leq n. \tag{2.6}$$

Then,

$$u(x) \leq \sum_{s=0}^{x-1} h(s)v(s+1,x) \tag{2.7}$$

where $v(s,x)$ *is the solution of* (2.1), (2.2).

Proof. From (2.1) and (2.5), we have

$$\sum_{s=0}^{x-1} v(s+1,x)\Delta_s^n u(s) - \sum_{s=0}^{x-1} u(s)(-1)^n \Delta_s^n v(s,x) \leq \sum_{s=0}^{x-1} h(s)v(s+1,x). \tag{2.8}$$

An application of Lemma 2.1 provides

$$\sum_{s=0}^{x-1} u(s)(-1)^n \Delta_s^n v(s,x)$$

$$= (-1)^n \sum_{s_1=0}^{x_1-1} \cdots \sum_{s_{n-1}=0}^{x_{n-1}-1} [u(s)\Delta_{s_1\ldots s_{n-1}}^{n-1} v(s,x) \Big|_{s_n=0}^{x_n}$$

$$- \sum_{s_n=0}^{x_n-1} \Delta_{s_n} u(s)\Delta_{s_1\ldots s_{n-1}}^{n-1} v(s_1,\ldots,s_{n-1},s_n+1,x)]. \tag{2.9}$$

Using (2.2) and (2.6), the right side of (2.9) reduces to

$$(-1)^{n+1} \sum_{s_n=0}^{x_n-1} \sum_{s_1=0}^{x_1-1} \cdots \sum_{s_{n-1}=0}^{x_{n-1}-1} \Delta_{s_n} u(s)\Delta_{s_1\ldots s_{n-1}}^{n-1} v(s_1,\ldots,s_{n-1},s_n+1,x).$$

Repeating the above arguments successively, we obtain

$$(-1)^{2n-1} \sum_{s_n=0}^{x_n-1} \cdots \sum_{s_2=0}^{x_2-1} [\Delta_{s_n\ldots s_2}^{n-1} u(s)v(s_1,s_2+1,\ldots,s_n+1,x) \Big|_{s_1=0}^{x_1}$$

$$- \sum_{s_1=0}^{x_1-1} \Delta_s^n u(s)v(s+1,x)]$$

which is same as

$$(-1)^{2n-1}\sum_{s_n=0}^{x_n-1}\cdots\sum_{s_2=0}^{x_2-1}\Delta^{n-1}_{s_n\cdots s_2}u(x_1,s_2,\ldots,s_n)$$

$$+\ (-1)^{2n}\sum_{s=0}^{x-1}\Delta^n_s\, u(s)v(s+1,x)$$

or

$$-u(x)+\sum_{s=0}^{x-1}\Delta^n_s\, u(s)v(s+1,x).$$

Substituting this in (2.8), the result (2.7) follows.

REMARK 2.1 For all $g(x)$ and $h(x)$, equality in (2.5) implies equality in (2.7) and hence $v(s,x)$ the solution of (2.1), (2.2) is a discrete analogue of Riemann's function.

COROLLARY 2.5 *Let* $g(x)$ *and* $h(x)$ *be as in Lemma* 2.4 *and* $\phi(x)$, $\psi(x)$ *be defined on* N^n *and satisfy*

$$\Delta^n_x\ \phi(x) \le g(x)\phi(x)+h(x)$$

$$\Delta^n_x\ \psi(x) \ge g(x)\psi(x)+h(x)$$

$$\phi(x_1,\ldots,x_{i-1},0,x_{i+1},\ldots,x_n) = \psi(x_1,\ldots,x_{i-1},0,x_{i+1},\ldots,x_n).$$

Then,

$$\phi(x) \le \psi(x).$$

LEMMA 2.6 *Let* $g(x)$ *be as in Lemma* 2.4 *and* $v(s,x)$ *be the solution of* (2.1), (2.2). *Let* $w(s,x)$ *be defined for all* $s \le x-1$, $(s,x)\in N^n\times N^n$ *and*

$$(-1)^n\ \Delta^n_s\ w(s,x) \ge g(s)w(s+1,x), \tag{2.10}$$

$$w(s,x) = 1;\quad s_i = x_i,\ 1\le i\le n. \tag{2.11}$$

Then,

$$v(s,x) \le w(s,x).$$

Proof. Let $r(s,x)$ be defined and nonnegative for all $s \le x-1$ $(s,x)\in N^n\times N^n$ so that

$$(-1)^n\ \Delta^n_s\ w(s,x) = g(s)w(s+1,x) + r(s,x). \tag{2.12}$$

Next, we define the iterates

$$w_0(s,x) = v(s,x)$$

$$w_{m+1}(s,x) = 1+\sum_{p=s}^{x-1} g(p)w_m(p+1,x)+\sum_{p=s}^{x-1} r(p,x);\ m=0,1,\ldots\ .$$

Obviously, $w_m(s,x) \geq v(s,x)$ for all $m \geq 1$, and as in Lemma 2.3 the sequence $\{w_m(s,x)\}$ converges to $w(s,x)$ which is the solution of (2.12), (2.11).

3. LINEAR INEQUALITIES

In what follows we shall assume that the functions which appear in the inequalities are real valued, nonnegative and defined on N^n.

THEOREM 3.1 Let for all $x \in N^n$, the following inequality be satisfied

$$\phi(x) \leq a(x) + b(x) \sum_{r=1}^{m} E^r(x,\phi) \tag{3.1}$$

where

$$E^r(x,\phi) = \mathop{S}_{x^1=0}^{x-1} f_{r1}(x^1) \mathop{S}_{x^2=0}^{x^1-1} f_{r2}(x^2) \dots \mathop{S}_{x^r=0}^{x^{r-1}-1} f_{rr}(x^r)\phi(x^r).$$

Then,

$$\phi(x) \leq a(x) + b(x) \mathop{S}_{s=0}^{x-1} [\sum_{r=1}^{m} \Delta_s^n E^r(s,a)]v(s+1,x) \tag{3.2}$$

where $v(s,x)$ is the solution of

$$(-1)^n \Delta_s^n v(s,x) = [\sum_{r=1}^{m} \Delta_s^n E^r(s,b)]v(s+1,x), \quad s \leq x-1$$

$$v(s,x) = 1; \quad s_i = x_i, \quad 1 \leq i \leq n.$$

Proof. Define a function $u(x)$ such that

$$u(x) = \sum_{r=1}^{m} E^r(x,\phi)$$

then, we have

$$\Delta_x^n u(x) = \sum_{r=1}^{m} \Delta_x^n E^r(x,\phi). \tag{3.3}$$

Since $\phi(x) \leq a(x) + b(x)u(x)$ and $u(x)$ is nondecreasing in x, from (3.3) we get successively

$$\begin{aligned}
\Delta_x^n u(x) &\leq \sum_{r=1}^{m} \Delta_x^n E^r(x,a+bu) \\
&= \sum_{r=1}^{m} \Delta_x^n E^r(x,a) + \sum_{r=1}^{m} \Delta_x^n E^r(x,bu) \\
&\leq \sum_{r=1}^{m} \Delta_x^n E^r(x,a) + [\sum_{r=1}^{m} \Delta_x^n E^r(x,b)]u(x).
\end{aligned}$$

Now, an application of Lemma 2.4 provides

$$u(x) \leq \mathbf{S}_{s=0}^{x-1} \Big[\sum_{r=1}^{m} \Delta_s^n E^r(s,b) \Big] v(s+1,x). \tag{3.4}$$

The result (3.2) follows from (3.4) and the inequality $\phi(x) \leq a(x) + b(x)u(x)$.

REMARK 3.1 For $m = 1$, (3.2) is best possible, i.e., equality in (3.1) implies equality in (3.2).

CONDITION (c). We say condition (c) is satisfied if for all $x \in N^n$ the inequality (3.1) holds, where

$$f_{ii}(x) = f_i(x), \ 1 \leq i \leq m;$$

$$f_{i+1,i}(x) = f_{i+2,i}(x) = \dots = f_{m,i}(x) = g_i(x), \ 1 \leq i \leq m-1.$$

In our next result, for all $x \in N^n$ we shall denote

$$r_1(x) = \max\Big\{ \sum_{r=1}^{m} b(x) f_r(x), \ g_i(x); \ 1 \leq i \leq m-1 \Big\}$$

$$r_j(x) = \max\Big\{0, \sum_{r=1}^{m-j+1} b(x) f_r(x) - g_{m-j+1}(x),$$

$$g_i(x) - g_{m-j+1}(x); \ 1 \leq i \leq m-j \Big\}, \ 2 \leq j \leq m.$$

THEOREM 3.2 *Let the condition* (c) *be satisfied. Then,*

$$\phi(x) \leq a(x) + b(x) P_j(x), \ 1 \leq j \leq m \tag{3.5}$$

where

$$P_1(x) = \mathbf{S}_{s=0}^{x-1} \Big[a(s) \sum_{i=1}^{m} f_i(s) \Big] v_1(s+1,x)$$

$$P_j(x) = \mathbf{S}_{s=0}^{x-1} \Big[a(s) \sum_{i=1}^{m-j+1} f_i(s) + g_{m-j+1}(s) P_{j-1}(s) \Big] v_j(s+1,x), \ 2 \leq j \leq m$$

and $v_j(s,x)$, $1 \leq j \leq m$ *are the solutions of*

$$(-1)^n \Delta_s^n v_j(s,x) = r_j(s) v_j(s+1,x), \ s \leq x-1$$

$$v_j(s,x) = 1; \ s_i = x_i, \ 1 \leq i \leq n, \ 1 \leq j \leq m.$$

Proof. The proof is similar to the continuous case [13].

REMARK 3.2 For $m = 1$, (3.5) is same as (3.2).

THEOREM 3.3 *Let for all* $x \in N^n$, *the following inequality be satisfied*

$$\phi(x) \le p_0(x) + \sum_{i=1}^{m} p_i(x) \mathop{\mathbf{S}}_{s=0}^{x-1} q_i(s)\phi(s). \tag{3.6}$$

Then,

$$\phi(x) \le F_m[p_0(x)] \tag{3.7}$$

where

$$F_i = D_i\, D_{i-1}\ldots D_0$$

$$D_0 w = w$$

$$D_j w = w + (F_{j-1}[p_j]) \mathop{\mathbf{S}}_{s=0}^{x-1} q_j(s) w(s) v_j(s+1,x)$$

and $v_j(s,x)$, $1 \le j \le m$ are the solutions of

$$(-1)^n \Delta_s^n v_j(s,x) = q_j(s) F_{j-1}[p_j(s)] v_j(s+1,x), \quad s \le x-1$$
$$v_j(s,x) = 1; \quad s_i = x_i, \quad 1 \le i \le n, \quad 1 \le j \le m. \tag{3.8$_j$}$$

Proof. The proof is by finite induction. For $m = 1$, we have from Theorem 3.1 that

$$\phi(x) \le p_0(x) + p_1(x) \mathop{\mathbf{S}}_{s=0}^{x-1} q_1(s) p_0(s) v_1(s+1,x)$$

where $v_1(s,x)$ is the solution of $(3.8)_1$, and hence (3.7) is true. Now, assume that the result is true for some k, wehre $1 < k \le m-1$, then for $k+1$, we are given

$$\phi(x) \le [p_0(x) + p_{k+1}(x) \mathop{\mathbf{S}}_{s=0}^{x-1} q_{k+1}(s)\phi(s)] + \sum_{i=1}^{k} p_i(x) \mathop{\mathbf{S}}_{s=0}^{x-1} q_i(s)\phi(s)$$

and from (3.7), we find

$$\phi(x) \le F_k[p_0(x) + p_{k+1}(x) \mathop{\mathbf{S}}_{s=0}^{x-1} q_{k+1}(s)\phi(s)].$$

Next, using the definition of F_k and the fact that $\mathop{\mathbf{S}}_{s=0}^{x-1} q_{k+1}(s)\phi(s)$ is nondecreasing for all $x \in N^n$, the above inequality can be written as

$$\phi(x) \le F_k[p_0(x)] + F_k[p_{k+1}(s) \mathop{\mathbf{S}}_{s=0}^{x-1} q_{k+1}(s)\phi(s)]$$

$$\le F_k[p_0(x)] + F_k[p_{k+1}(x)] \mathop{\mathbf{S}}_{s=0}^{x-1} q_{k+1}(s)\phi(s)$$

and again an application of Theorem 3.1 provides

$$\phi(x) \le F_k[p_0(x)] + F_k[p_{k+1}(x)] \mathop{S}_{s=0}^{x-1} q_{k+1}(s)F_k[p_0(s)]v_{k+1}(s,x)$$
$$= F_{k+1}[p_0(x)].$$

This completes the induction.

4. WENDROFF TYPE INEQUALITIES

THEOREM 4.1 *Let for all* $x \in N^n$, *the inequality* (3.1) *be satisfied. Further, let* (i) $a(x)$ *be nondecreasing and* (ii) $b(x) \ge 1$. *Then,*

$$\phi(x) \le a(x)b(x)v(0,x) \tag{4.1}$$

where $v(s,x)$ *is same as in Theorem* 3.1.

Proof. Using (i) and (ii) in (3.2), we find

$$\phi(x) \le a(x)b(x) \ [1 + \mathop{S}_{s=0}^{x-1} \ (\sum_{r=1}^{m} \Delta_s^n \ E^r(s,b))v(s+1,x)]$$

and hence

$$\phi(x) \le a(x)b(x)[1 + \mathop{S}_{s=0}^{x-1} \ (-1)^n \ \Delta_s^n \ v(s,x)]. \tag{4.2}$$

Now using the fact that $v(s,x) = 1;\ s_i = x_i,\ 1 \le i \le n$ it follows from (4.2) that

$$\phi(x) \le a(x)b(x)[1+(-1)^{2n-1} \sum_{s_1=0}^{x_1-1} \Delta_{s_1} v(s_1,0,\ldots,0,x)]$$
$$= a(x)b(x)[1+(-1)^{2n-1}(v(x_1,0,\ldots 0,x)-v(0,x))]$$
$$= a(x)b(x)v(0,x).$$

REMARK 4.1 Let $w(s,x)$ be any function defined for all $s \le x-1$, $(s,x) \in N^n \times N^n$ and

$$(-1)^n \ \Delta_s^n \ w(s,x) \ge [\sum_{r=1}^{m} \ \Delta_s^n \ E^r(s,b)]w(s+1,x)$$
$$w(s,x) = 1; \quad s_i = x_i, \quad 1 \le i \le m. \tag{4.3}$$

Then, from Lemma 2.6 it follows that in (4.1) $v(0,x)$ can be replaced by $w(0,x)$. However, finding $w(s,x)$ in advance which satisfies (4.3) seems to be quite difficult.

In our next result we shall provide an estimate for v(s,x) which is quite adequate for practical applications.

THEOREM 4.2 Let v(s,x) be same as in Theorem 3.1. Then,

$$v(s,x) \leq \prod_{p_1=s_1}^{x_1-1} [1 + \sum_{p_2=s_2}^{x_2-1} \cdots \sum_{p_s=s_n}^{x_n-1} (\sum_{r=1}^{m} \Delta_p^n E^r(p,b))]. \qquad (4.4)$$

Proof. From Lemma 2.3 the function $v(s,x) \geq 1$, and from Lemma 2.2

$$v(s,x) = 1 + \underset{p=s}{\overset{x-1}{S}} (\sum_{r=1}^{m} \Delta_p^n E^r(p,b))v(p+1,x).$$

Thus, $(-1)^i \Delta^i_{s_1\ldots s_i} v(s,x) \geq 0,\ 1 \leq i \leq n.$

Since

$$(-1)^n \Delta_{s_n} [\frac{\Delta^{n-1}_{s_1\ldots s_{n-1}} v(s,x)}{v(s_1+1,\ldots,s_{n-1}+1,s_n,x)}]$$

$$+(-1)^n \Delta^{n-1}_{s_1\ldots s_{n-1}} v(s,x)[\frac{1}{v(s_1+1,\ldots,s_{n-1}+1,s_n,x)} - \frac{1}{v(s+1,x)}]$$

$$= \sum_{r=1}^{m} \Delta_s^n E^r(s,b)$$

we have

$$(-1)^n \Delta_{s_n} [\frac{\Delta^{n-1}_{s_1\ldots s_{n-1}} v(s,x)}{v(s_1+1,\ldots,s_{n-1}+1,s_n,x)}] \leq \sum_{r=1}^{m} \Delta_s^n E^r(s,b). \qquad (4.5)$$

In (4.5) keeping $s_1,\ldots,s_{n-1}$ fixed and setting $s_n = p_n$ and summing over $p_n = s_n$ to x_n-1, to obtain

$$(-1)^{n+1} [\frac{\Delta^{n-1}_{s_1\ldots s_{n-1}} v(s,x)}{v(s_1+1,\ldots,s_{n-1}+1,s_n,x)}]$$

$$\leq \sum_{p_n=s_n}^{x_n-1} \sum_{r=1}^{m} \Delta^n_{s_1\ldots s_{n-1}\, p_n} E^r(s_1,\ldots,s_{n-1},p_n,b).$$

Repeating the above arguments successively with respect to $s_{n-1},\ldots,s_2$ we find

$$(-1)^{2n-1}\left[\frac{\Delta_{s_1} v(s,x)}{v(s_1+1,s_2,\ldots,s_n,x)}\right]$$

$$\leq \sum_{p_2=s_2}^{x_2-1} \cdots \sum_{p_n=s_n}^{x_n-1} \sum_{r=1}^{m} \Delta^n_{s_1 p_2 \ldots p_n} E^r(s_1,p_2,\ldots,p_n,b)$$

which is same as

$$v(s,x) \leq \left[1 + \sum_{p_2=s_2}^{x_2-1} \cdots \sum_{p_n=s_n}^{x_n-1} \sum_{r=1}^{m} \Delta^n_{s_1 p_2 \ldots p_n} E^r(s_1,p_2,\ldots,p_n,b)\right] \times$$

$$v(s_1+1,s_2,\ldots,s_n,x).$$

The above inequality easily provides (4.4).

COROLLARY 4.3 *Let* $v(s,x)$ *be same as in Theorem* 3.1. *Then,*

$$v(s,x) \leq \min_{1\leq i\leq n} \left\{ \prod_{p_i=s_i}^{x_i-1} \left[1 + \sum_{p_1=s_1}^{x_1-1} \cdots \sum_{p_{i-1}=s_{i-1}}^{x_{i-1}-1} \sum_{p_{i+1}=s_{i+1}}^{x_{i+1}-1} \cdots \sum_{p_n=s_n}^{x_n-1} \sum_{r=1}^{m} \Delta^n_p E^r(p,b)\right]\right\}.$$

THEOREM 4.4 *Let the conditiions of Theorem* 4.1 *be satisfied. Then,*

$$\phi(x) \leq a(x)b(x) \min_{1\leq i\leq n} \left\{ \prod_{p_i=0}^{x_i-1} \left[1+ \sum_{r=1}^{m} \Delta_{p_i} E^r(x_1,\ldots,x_{i-1},p_i,x_{i+1},\ldots,x_n,b)\right]\right\}$$

Proof. The proof follows from Theorem 4.1 and Corollary 4.3.

THEOREM 4.5 *Let the condition* (c) *be satisfied and* (i) $a(x)$ *is nondecreasing* (ii) $b(x) \geq 1$. *Then,*

$$\phi(x) \leq a(x)b(x) \min_{1\leq i\leq n} \left\{ \prod_{p_i=0}^{x_i-1} \left[1+ \sum_{p_1=0}^{x_1-1} \cdots \sum_{p_{i-1}=0}^{x_{i-1}-1} \sum_{p_{i+1}=0}^{x_{i+1}-1} \cdots \sum_{p_n=0}^{x_n-1} r_1(p)\right]\right\}.$$

Proof. The proof is similar to that of Theorem 4.4.

5. NONLINEAR INEQUALITIES

Our first result for the nonlinear case is connected with the inequality

$$\phi(x) \leq a(x)[c + \sum_{r=1}^{m} H^r(x,\phi)] \tag{5.1}$$

where

$$H^r(x,\phi) = \mathop{S}_{x^1=0}^{x-1} f_{r1}(x^1)\phi^{\alpha_{r1}}(x^1) \ldots \mathop{S}_{x^r=0}^{x^{r-1}-1} f_{rr}(x^r)\phi^{\alpha_{rr}}(x^r)$$

and α_{ri}; $1 \leq i \leq r$, $1 \leq r \leq m$ are nonnegative real numbers and the real number $c > 0$.

In the following result we shall denote $\alpha_r = \sum_{i=1}^{r} \alpha_{ri}$ and $\alpha = \max_{1 \leq r \leq m} \alpha_r$.

THEOREM 5.1 <u>Let for all</u> $x \in N^n$, <u>the inequality</u> (5.1) <u>be satisfied</u>. <u>Then</u>,

$$\phi(x) \leq ca(x) \min_{1 \leq i \leq n} \{ \prod_{p_i=0}^{x_i-1} [1+\Delta_{p_i} Q(x_1,\ldots,x_{i-1},p_i,x_{i+1},\ldots,x_n)] \}, \quad \text{if } \alpha = 1 \tag{5.2}$$

$$\phi(x) \leq a(x)[c^{1-\alpha} + (1-\alpha)Q(x)]^{1/1-\alpha}, \quad \text{if } \alpha \neq 1 \tag{5.3}$$

<u>where</u>

$$Q(x) = \sum_{r=1}^{m} H^r(x,a)c^{\alpha_r-\alpha}$$

<u>and when</u> $\alpha > 1$, <u>we assume</u>, $c^{1-\alpha} + (1-\alpha)Q(x) > 0$.

<u>Proof</u>. The inequality (5.1) can be written as $\phi(x) \leq a(x)u(x)$, where

$$u(x) = c + \sum_{r=1}^{m} H^r(x,\phi).$$

Thus, on using the nondecreasing nature of u(x), we find

$$\Delta_x^n u(x) \leq \sum_{r=1}^{m} \Delta_x^n H^r(x,a)[u(x)]^{\alpha_r}.$$

Since $u(x) \geq c$, we get

$$\Delta_x^n u(x) \le \sum_{r=1}^{m} \Delta_x^n H^r(x,a)c^{\alpha_r-\alpha}[u(x)]^\alpha$$

$$= \Delta_x^n Q(x)[u(x)]^\alpha.$$

On using the nondecreasing nature of u(x), we obtain

$$\Delta_{x_n}\left[\frac{\Delta^{n-1}_{x_1\ldots x_{n-1}} u(x)}{u^\alpha(x)}\right] \le \Delta_x^n Q(x). \tag{5.4}$$

In (5.4) setting $x_n = p_n$ and summing over $p_n = 0$ to x_n-1, we find on using $\Delta^{n-1}_{x_1\ldots x_{n-1}} u(x_1,\ldots,x_{n-1},0) = \Delta^{n-1}_{x_1\ldots x_{n-1}} Q(x_1,\ldots,x_{n-1},0) = 0$ that

$$\frac{\Delta^{n-1}_{x_1\ldots x_{n-1}} u(x)}{u^\alpha(x)} \le \Delta^{n-1}_{x_1\ldots x_{n-1}} Q(x).$$

Repeating the above arguments successively with respect to $x_{n-1},\ldots,x_{i+1},\ x_{i-1},\ldots,x_1$ we get

$$\frac{\Delta_{x_i} u(x)}{u^\alpha(x)} \le \Delta_{x_i} Q(x). \tag{5.5}$$

If $\alpha = 1$, the result (5.2) immediately follows from (5.5) and the fact that $u(x_1,\ldots,x_{i-1},0,x_{i+1},\ldots,x_n) = c$.

If $\alpha \ne 1$, we have

$$\frac{\Delta_{x_i}[u(x)]^{1-\alpha}}{1-\alpha} = \int_{x_i}^{x_i+1} \frac{du(x_1,\ldots,x_{i-1},s,x_{i+1},\ldots,x_n)}{u^\alpha(x_1,\ldots,x_{i-1},s,x_{i+1},\ldots,x_n)} \le \frac{\Delta_{x_i} u(x)}{u^\alpha(x)}$$

and from (5.5), we obtain

$$\frac{\Delta_{x_i}[u(x)]^{1-\alpha}}{1-\alpha} \le \Delta_{x_i} Q(x). \tag{5.6}$$

In (5.6) setting $x_i = p_i$ and summing over $p_i = 0$ to x_{i-1}, we get the required inequality (5.3).

DEFINITION. A continuous function $W : [0,\infty) \to (0,\infty)$ is said to belong to the class T if

(i) $W(u)$ is positive and nondecreasing for all $u \geq 0$

(ii) $\frac{1}{v} W(u) \leq W(\frac{u}{v})$ for all $u \geq 0$, $v \geq 1$.

THEOREM 5.2 *Let for all* $x \in N^n$, *the following inequality be satisfied*

$$\phi(x) \leq a(x) + \sum_{r=1}^{m} E^r(x,\phi) + \sum_{i=1}^{\ell} g_i(x) \mathop{\mathbf{S}}_{s=0}^{x-1} h_i(s) W_i(\phi(s))$$

where (i) $a(x) \geq 1$ *and nondecreasing* (ii) $g_i(x) \geq 1$, $1 \leq i \leq \ell$ (iii) $W_i \in T$, $1 \leq i \leq \ell$. *Then,*

$$\phi(x) \leq a(x)\psi(x)e(x) \prod_{i=1}^{\ell} F_i(x)$$

where

$$\psi(x) = \min_{1 \leq i \leq n} \left\{ \prod_{p_i=0}^{x_i-1} [1 + \sum_{r=1}^{m} \Delta_{p_i} E^r(x_1,\ldots,x_{i-1},p_i,x_{i+1},\ldots,x_n,e)] \right\}$$

$$e(x) = \prod_{i=1}^{\ell} g_i(x)$$

$$F_k(x) = G_k^{-1}[G_k(1) + \mathop{\mathbf{S}}_{s=0}^{x-1} h_k(s)\psi(s)e(s) \prod_{j=1}^{k-1} F_j(s)];\ F_0(x) = 1,\ 1 \leq k \leq \ell$$

$$G_k(u) = \int_{u_0}^{u} \frac{ds}{W_k(s)},\quad u \geq u_0 > 0$$

as long as

$$G_k(1) + \mathop{\mathbf{S}}_{s=0}^{x-1} h_k(s)\psi(s)e(s) \prod_{j=1}^{k-1} F_j(s) \in \mathrm{Dom}(G_k^{-1}),\ 1 \leq k \leq \ell.$$

Proof. The proof is similar to the continuous case [13].

THEOREM 5.3 *In addition to the hypotheses of Theorem* 5.2 *let* $g_i(x)$, $1 \leq i \leq \ell$ *be nondecreasing*. *Then,*

$$\phi(x) \leq a(x)\psi^*(x) \prod_{i=1}^{\ell} F_i^*(x)$$

where $\psi^*(x)$ *is same as* $\psi(x)$ *with* $e(x) = 1$, *and*

$$F_k^*(x) = g_k(x) G_k^{-1}[G_k(1) + \mathop{\mathbf{S}}_{s=0}^{x-1} h_k(s)\psi^*(s)g_k(s) \prod_{i=1}^{k-1} F_i^*(s)];\ F_0^*(x) = 1,\ 1 \leq k \leq \ell$$

as long as

$$G_k(1) + \mathop{S}_{s=0}^{x-1} h_k(s)\psi^*(s)g_k(s) \prod_{i=1}^{k-1} F_i^*(s) \in \mathrm{Dom}(G_k^{-1}),\ 1 \le k \le \ell.$$

Proof. The proof is similar to the continuous case [13].

THEOREM 5.4 *Let for all* $x \in N^n$, *the following inequality be satisfied*

$$\phi(x) \le a(x) + \sum_{r=1}^{m} E^r(x,\phi) + \sum_{i=1}^{\ell} E^i(x,W(\phi)) \tag{5.7}$$

where (i) $a(x) \ge 1$ *and nondecreasing* (ii) $W \in T$. *Then,*

$$\phi(x) \le a(x)\psi^*(x)G^{-1}[G(1) + \sum_{i=1}^{\ell} E^i(x,\psi^*)]$$

where ψ^* *is same as in Theorem* 5.3 *and*

$$G(1) + \sum_{i=1}^{\ell} E^i(x,\psi^*) \in \mathrm{Dom}(G^{-1}).$$

Proof. The proof is similar to the continuous case [13].

THEOREM 5.5 *Let for all* $x \in N^n$, *the inequality* (5.7) *be satisfied, where* (i) $a(x)$ *is positive and nondecreasing* (ii) W *is positive, continuous, nondecreasing and submultiplicative. Then,*

$$\phi(x) \le a(x)\psi^*(x)G^{-1}[G(1) + \sum_{i=1}^{\ell} E^i(x, \frac{W(a\psi^*)}{a})] \tag{5.8}$$

where ψ^* *is same as in Theorem* 5.3 *and*

$$G(1) + \sum_{i=1}^{\ell} E^i(x, \frac{W(a\psi^*)}{a}) \in \mathrm{Dom}(G^{-1}).$$

Proof. We apply Theorem 4.4 for inequality (5.12), to obtain

$$\phi(x) \le [a(x) + \sum_{i=1}^{\ell} E^i(x,W(\phi))]\psi^*(x)$$

or

$$\frac{\phi(x)}{a(x)\psi^*(x)} \le 1 + \sum_{i=1}^{\ell} E^i(x,W(\frac{\phi}{a\psi^*}\, a\psi^*)/a). \tag{5.9}$$

Let $u(x)$ be the right side of (5.9), then

$$\Delta_x^n u(x) = \sum_{i=1}^{\ell} \Delta_x^n E^i(x,W(\frac{\phi}{a\psi^*}\, a\psi^*)/a).$$

Now using the fact that W is nondecreasing and submultiplicative, we get

$$\frac{\Delta_x^n u(x)}{W(u(x))} \leq \sum_{i=1}^{\ell} \Delta_x^n E^i(x,W(a\psi^*)/a)$$

which implies that

$$u(x) \leq G^{-1}[G(1) + \sum_{i=1}^{\ell} E^i(x,W(a\psi^*)/a)]$$

and from this the inequality (5.8) follows.

As a final remark we note that the inequalities obtained in Sections 3-5 can be directly used to prove the boundedness, uniqueness and continuous dependence of the solutions of discrete versions of hyperbolic partial differential systems and hyperbolic integrodifferential equations of a more general type then those given in [3,5-8,10-12,14,15], sicne the arguments are simialr the details are not repeated here.

REFERENCES

1. R. P. Agarwal, On an integral inequality in n independent variables. J. Math. Anal. Appl. 85 (1982), 192-196.

2. R. P. Agarwal, Difference calculus with applications to difference equations. General Inequalities 4, ed. by W. Walter, Birkhauser Verlag Basel, 1984, 95-110.

3. R. P. Agarwal, Sharp estimates for the Wendroff discrete inequality in n-independent variables. An. st. Univ. Iasi 30 (1984), 65-68.

4. R. P. Agarwal and E. Thandapani, On some new discrete inequalities. Appl. Math. Comp. 7 (1980), 205-224.

5. R. P. Agarwal and E. Thandapani, On discrete inequalities in n independent variables. Rivista di Matematica della Univ. Parma (to appear).

6. R. P. Agarwal and S. J. Wilson, On discrete inequalities involving higher order partial differences. An. st. Univ. Iasi 30 (1984), 41-50.

7. P. R. Beesack, Lower bounds from discrete inequalities of Gollwitzer-Langenhop type. An. st. Univ. Iasi 30 (1984), 25-30.

8. B. G. Pachpatte and S. M. Singare, Discrete generalized Gronwall inequalities in three independent variables. Pacific J. Math. 82 (1979), 197-210.

9. J. Popenda, On the asymptotic behaviour of the solutions of an n-th order difference equation. Ann. Polon. Math. 44, (1984), 95-111.

10. S. M. Singare and B. G. Pachpatte, On some fundamental discrete inequalities of the Wendroff type. An. st. Univ. Iasi 26 (1980), 85-94.

11. S. M. Singare and B. G. Pachpatte, Wendroff type discrete inequalities and their applications. J. Mathl. Phyl. Sci. 13 (1979), 149-167.

12. E. Thandapani and R. P. Agarwal, Some new discrete inequalities in two independent variables. An. st. Univ. Iasi 27 (1981), 269-278.

13. E. Thandapani and R. P. Agarwal, On some new inequalities in n independent variables. J. Math. Anal. Appl. 86 (1982), 542-561.

14. Yeh, Cheh Chih, Discrete inequalities of the Gronwall-Bellman type in n independent variables. J. Math. Anal. Appl. 105 (1985), 322-332.

15. Yeh, Cheh Chih, Discrete inequalities of the Gronwall-Bellman type in n independent variables II. J. Math. Anal. Appl. 106 (1985), 282-285.

Ravi P. Agarwal, Department of Mathematics, National University of Singapore, 10 Kent Ridge Crescent, Singapore 0511.

International Series of
Numerical Mathematics, Vol. 80

EXTREMAL PROBLEMS FOR EIGENVALUES OF THE STURM-LIOUVILLE TYPE

Catherine Bandle

Abstract. Extremal problems are constructed which yield upper bounds for the eigenvalues of certain Sturm-Liouville problems. An isoperimetric inequality of Troesch [7] is extended. The method is based on rearrangement techniques and on the discussion of a nonlinear functional.

1. INTRODUCTION

Consider the eigenvalue problem

$$(1.1)\qquad (\sigma u')' + \lambda u = 0 \quad \text{in } (-a,a) \;, \quad (\sigma u')(\pm a) = 0 \;.$$

If σ is continuous and positive in $[-a,a]$, there exists a countable number of eigenvalues (cf. [8], p.337)

$$0 = \lambda_1 < \lambda_2 < \quad \dots \quad , \lambda_n \to \infty \quad \text{as} \quad n \to \infty \;.$$

In his study on sloshing frequencies of liquids [7], Troesch derived the following isoperimetric inequality:

$$(1.2)\qquad \lambda_2(\sigma) \le 3V/a^3 \quad \text{for all } \sigma \quad \text{with} \int_{-a}^{a} \sigma \, dx = 2V \;.$$

Equality holds for $\sigma(x) = 3V(a^2-x^2)/(2a^3)$. The aim of this paper is to extend this isoperimetric inequality to the case where σ ranges over the class

$$S(p,\sigma_o) := \left\{ \sigma \in C^o[-a,a] : 0 < \sigma(x) \le \sigma_o \text{ in } (-a,a), \int_{-a}^{a} \sigma^p \, dx = 2V \right\} .$$

Here $p \ge 1$, σ_o and V are positive numbers such that $a\sigma_o^p > V$. It turns out that there exists an extremal function $\sigma^* \in S(p,\sigma_o)$

This paper is in final form and no version of it will be submitted for publication elsewhere.

which maximizes $\lambda_2(\sigma)$. Using a result of B. Schwarz [5] we shall also be able to construct an extremal function for the higher eigenvalues $\lambda_n(\sigma)$.

According to the well-known variational characterization of the eigenvalues we have

$$(1.3) \qquad \lambda_n(\sigma) = \inf_{L_n} \max_{v \in L_n} \int_{-a}^{a} \sigma v'^2 \, dx \Big/ \int_{-a}^{a} v^2 \, dx$$

where $L_n \subset W^{1,2}(-a,a)$ denotes an n-dimensional linear subspace. Instead of looking at the eigenvalues of (1.1) we shall consider the slightly more general problem

$$(1.4) \quad \lambda_n(\sigma) \rightarrow \sup, \ \sigma \in S(p,\sigma_0), \ \lambda_n(\sigma) \text{ being defined in (1.3)} .$$

It is clear that in the case of non-differentiable functions σ , the eigenfunctions corresponding to $\lambda_n(\sigma)$ satisfy (1.1) in a weak sense, i.e.

$$(1.5) \quad \int_{-a}^{a} \sigma u' \eta' \, dx = \lambda \int_{-a}^{a} u \eta \, dx \quad \text{for all} \quad \eta \in C^\infty[-a,a] .$$

Problems of this type have been treated by different authors. Kuzanek [3] established the existence of an isoperimetric upper bound for $\lambda_n(\sigma)$ when σ is subject to

$$\int_{-a}^{a} \sqrt{1 + \sigma'^2} \, dx = \text{const} .$$

He derived an Euler equation for the extremal function.

A different approach for treating extremal problems has been proposed by Barnes [2] . He gives a necessary condition for the extremal function assuming certain regularity properties.

We shall use a direct approach. If $p = 1$, we can construct the solution explicitly as in [7] . For $p > 1$ we are led to the discussion of a nonlinear functional also considered by Ôtani [4] in a completely different context. It should be mentioned that our method applies also to the cases $0 < \sigma_1 \leq \sigma$ and $0 < \sigma_1 \leq \sigma \leq \sigma_0$. The main results have been announced in [1] and the case $0 < p < 1$ will be treated in a forthcoming paper.

2. PRELIMINARIES

From the variational characterization of λ_n we obtain immediately the

LEMMA 2.1. <u>If</u> $\sigma_1 \le \sigma_2$, <u>then</u> $\lambda_n(\sigma_1) \le \lambda_n(\sigma_2)$ <u>for all</u> n.

Let $\sigma^{(\varepsilon)}(x)$ be a class of functions in $C^\infty[-a,a]$ with the following properties:

(i) $0 < \varepsilon \le \sigma^{(\varepsilon)}(x) \le \sigma_0$ in $[-a,a]$

(ii) $\sigma \le \sigma^{(\varepsilon)}$ in $[-a,a]$

(iii) $\sigma^{(\varepsilon)} - \sigma \le \varepsilon$ in $[-a,a]$.

Put for short $\tilde{\lambda}_n(\varepsilon) := \lambda_n(\sigma^{(\varepsilon)})$.

LEMMA 2.2. $\lim\limits_{\varepsilon \to \infty} \tilde{\lambda}_n(\varepsilon) = \lambda_n(\sigma)$.

<u>Proof.</u> By Lemma 2.1 and (ii) we have

$$\lambda_n(\sigma) \le \tilde{\lambda}_n(\varepsilon) \; . \tag{2.1}$$

Let $\varphi_1, \dots, \varphi_n$ be the eigenfunctions corresponding to $\lambda_1(\sigma)$, ... $\lambda_n(\sigma)$ and $L_n = \text{span}\{\varphi_1, \dots, \varphi_n\}$. Then

$$\tilde{\lambda}_n(\varepsilon) \le \max_{v \in L_n} \left(\int_{-a}^{a} \sigma^{(\varepsilon)} v'^2 \, dx \right) / \int_{-a}^{a} v^2 \, dx$$

$$\le \max_{v \in L_n} \left(\int_{-a}^{a} \sigma v'^2 \, dx \right) / \int_{-a}^{a} v^2 \, dx + \varepsilon \max_{v \in L_n} \left(\int_{-a}^{a} v'^2 \, dx \right) / \int_{-a}^{a} v^2 \, dx$$

$$= \lambda_n(\sigma) + \varepsilon c \; ,$$

c being independent of $\sigma^{(\varepsilon)}$. This inequality together with (2.1) proves the assertion.

Let us now consider the following subclasses of $S(p,\sigma_0)$:

$$S_2^-(p,\sigma_0) := \{\sigma \in S(p,\sigma_0) : \sigma(x) = \sigma(-x), \; \sigma(x) \text{ non-decreasing in } (-a,0)\}$$

and more generally

$$S_n^-(p,\sigma_0) := \Big\{\sigma \in S(p,\sigma_0) : \sigma(x) = \sigma(x + \tfrac{2a}{n-1}), \; \sigma(-a + \tfrac{a}{n-1} - x) = \sigma(-a + \tfrac{a}{n-1} + x),$$
$$x \in (0, \tfrac{a}{n-1}), \; \sigma \text{ non-decreasing in } (-a, -a + \tfrac{a}{n-1}) \Big\} ,$$

cf. fig. 1

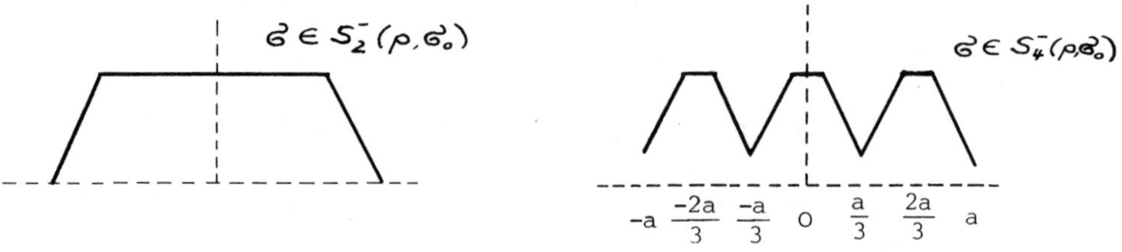

LEMMA 2.3. $\sup\limits_{\sigma \in S(p,\sigma_o)} \lambda_n(\sigma) = \sup\limits_{\sigma \in S_n^-(p,\sigma_o)} \lambda_n(\sigma)$.

<u>Proof.</u> Consider the family $\sigma^{(\varepsilon)}$ and the corresponding eigenvalues $\tilde{\lambda}_n(\varepsilon)$. Then $\tilde{\lambda}_n(\varepsilon)$ is the n-th eigenvalue of $(\sigma^{(\varepsilon)}u')' + \omega u = 0$ in $(-a,a)$, $u'(\pm a) = 0$. If we put $\sigma^{(\varepsilon)}u' = \psi$ it follows that $\tilde{\lambda}_n(\varepsilon)$ is the $(n-1)^{st}$ eigenvalue of

$$\psi'' + \frac{\omega}{\sigma^{(\varepsilon)}(x)}\psi = 0 \quad \text{in } (-a,a),\ \psi(\pm a) = 0 \ . \tag{2.2}$$

We now recall a theorem of B. Schwarz [5] regarding the frequencies of inhomogeneous strings. Set for short

$$\rho(x:\varepsilon) = (1/\sigma^{(\varepsilon)}(x)) \ .$$

DEFINITION 2.1. The <u>decreasing</u> <u>rearrangement</u> ρ^- of ρ of degree n-1 is defined as follows:

(i) ρ^- and ρ are equimeasurable over $(-a,a)$

(ii) $\rho^-(x+\frac{2ak}{n-1}) = \rho^-(x)$ for $-a \le x \le -a + \frac{2a}{n-1}$ and $k = 1,\dots,n-2$

(iii) ρ^- is non-increasing in $[-a,-a+\frac{a}{n-1}]$ and

$$\rho^-(-a+\frac{a}{n-1}-x) = \rho^-(-a+\frac{a}{n-1}+x) \text{ for all } x \in [0,\frac{a}{n-1}] \ .$$

According to [5] we have

$$\lambda_n(1/\rho^-) = \omega_{n-1}(\rho^-) \ge \omega_{n-1}(\rho) = \tilde{\lambda}_n(\varepsilon) \ . \tag{2.3}$$

Now observe that

$$\overline{\lim_{\varepsilon \to 0}}\,(1/\rho^-) =: \tilde{\sigma}(x)$$

exists and that $\tilde{\sigma}(x) \in S_n^-(p,\sigma_o)$. Hence, by (2.3) and Lemma 2.1 we have $\lambda_n(\sigma) \leq \lambda_n(\tilde{\sigma})$, which establishes the lemma.

LEMMA 2.4. _If_ $\sigma \in S_2^-(p,\sigma_o)$, _then_

$$\lambda_2(\sigma) = \inf_{v(0)=0} \int_0^a \sigma v'^2 \, dx \Big/ \int_0^a v^2 \, dx . \tag{2.4}$$

Proof. If $u(x)$ is a minimizer corresponding to $\lambda_2(\sigma)$, so is $u(-x)$. Then, in view of (1.3),

$$\lambda_2(\sigma) \leq \max_{c,d} \frac{c^2 a_{11} + 2cda_{12} + d^2 a_{22}}{c^2 b_{11} + 2cdb_{12} + d^2 b_{22}} = \max_{c,d} Q[c,d] ,$$

where

$$a_{11} = \int_{-a}^{a} \sigma u'^2(x)dx , \quad a_{12} = -\int_{-a}^{a} \sigma u'(x)u'(-x)dx , \quad a_{22} = \int_{-a}^{a} \sigma u'^2(-x)dx$$

$$b_{11} = \int_{-a}^{a} u^2(x)dx , \quad b_{12} = \int_{-a}^{a} u(x)u(-x)dx , \quad b_{22} = \int_{-a}^{a} u(-x)^2 dx .$$

An elementary consideration shows that $Q[c,d] \equiv \lambda_2(\sigma)$. Hence $v(x) = u(x) - u(-x)$ is also a minimizer and the conclusion is now obvious.

3. THE CASE $p = 1$

This case will be solved by choosing an appropriate trial function for (2.4).

Consider the functions

$$\hat{\sigma}(x;\alpha) := \begin{cases} \sigma_o & \text{in } [0,\alpha] \\ -A(a-x)^2 + B(a-x) & \text{in } [\alpha,a] \\ \hat{\sigma}(-x;\alpha) & \text{in } [-a,0] \end{cases}$$

and

$$\hat{v}(x;\alpha) := \begin{cases} \sin\sqrt{\hat{\lambda}_2/\sigma_o}\; x & \text{in } [0,\alpha] \\ bx + c & \text{in } [\alpha,a] \\ -\hat{v}(-x) & \text{in } [-a,0] \end{cases} .$$

A straightforward computation yields

LEMMA 3.1. *Let* $0 < \alpha < a$. *Then there exist* $A, B \in \mathbb{R}^+$ *and* $b, c \in \mathbb{R}$ *such that* $\hat{\sigma}$ *is continuous and* $\hat{v}$ *is the second eigenfunction of* (1.1) *with* $\sigma = \hat{\sigma} \cdot \hat{\lambda}_2$ *is the smallest positive solution of*

$$-\lambda(a-\alpha)/2 + \sigma_o/(a-\alpha) = \sqrt{\lambda\sigma_o}\ \mathrm{tg}(\sqrt{\lambda/\sigma_o}\,\alpha) .$$

Proof. If we insert $\hat{v}$ into (1.1) we find that the following conditions must hold :

$$\lambda = 2A \quad \text{and} \quad 2Aab - bB + \lambda c = 0 .$$

By the continuity of $\hat{\sigma}$ at α

$$-A(a-\alpha)^2 + B(a-\alpha) = \sigma_o .$$

By the continuity of $\hat{v}$ and $\hat{v}'$ at α, $b\alpha + c = \sin\sqrt{\lambda/\sigma_o}\,\alpha$ and $b = \sqrt{\lambda/\sigma_o}\cos\sqrt{\lambda/\sigma_o}\,\alpha$. The assertion is now obvious.

Moreover we have

LEMMA 3.2. *There exists* $\alpha^* \in [0,a]$ *such that* $\overset{*}{\sigma} := \hat{\sigma}(x;\alpha^*) \in S(1;\sigma_o)$.

Proof. Since $B > 2A(a-\alpha)$, we have $\hat{\sigma} \le \sigma_o$ for all α . It remains to show that

$$\int_{-a}^{a} \hat{\sigma}(x;\alpha^*)\,dx = 2V \quad \text{for some } \alpha^* .$$

The function

$$J(\alpha) = \int_0^a \hat{\sigma}(x;\alpha)\,dx = \sigma_o\alpha + \frac{B}{2}(a-\alpha)^2 - \frac{A}{3}(a-\alpha)^3$$

depends continuously on α . Since $2Aa^3/3 = J(0)$ and $J(a) = a\sigma_o > V$, there is always an α^* with the desired property.

REMARK. If $V < 2a\sigma_o/3$, we then set $\alpha^* = 0$ and Troesch's inequality (1.2) applies.

THEOREM 3.1. _If_ $V/a \leq \sigma_0 \leq 3V/(2a)$, _then_

$$\lambda_2(\sigma) \leq \lambda_2(\sigma^*) \quad \forall \ \sigma \in S(1,\sigma_0) . \tag{3.1}$$

Moreover, the extremal function is unique.

Proof. According to Lemma 2.3 we can restrict ourselves to functions $\sigma \in S_2^-(1,\sigma_0)$. Because of the hypothesis on σ_0 there is an $\alpha > 0$ such that $\sigma = \sigma_0$ in $(0,a]$. We shall distinguish two cases.

(i) $\alpha \geq \alpha^*$.

By Lemma 2.4 we have for $v = \hat{v}(x;\alpha^*)$,

$$\lambda_2(\sigma) \leq [\ \sigma_0 \int_0^{\alpha^*} \hat{v}'^2 \ dx + b^2 \int_{\alpha^*}^{a} \sigma \ dx\ ] \ / \int_0^a \hat{v}^2 \ dx = \lambda_2(\sigma^*) .$$

(ii) $\alpha < \alpha^*$.

In this case

$$\int_0^{\alpha^*} \sigma \ dx < \int_0^{\alpha^*} \sigma^* \ dx \qquad \text{and} \qquad \int_{\alpha^*}^{a} \sigma \ dx > \int_{\alpha^*}^{a} \sigma^* \ dx .$$

By the monotonicity of the eigenvalues we have $\lambda_2(\sigma^*) \leq \lambda_2(\sigma_0) = (\pi/2a)^2 \sigma_0$. Hence $\hat{v}$ is in $[0,a]$ an increasing, concave function. The mean value theorem yields

$$\int_0^a (\sigma-\sigma^*)\hat{v}'^2 dx = \hat{v}'^2(x_1) \int_0^{\alpha^*} (\sigma-\sigma^*) \ dx + \hat{v}'^2(x_2) \int_{\alpha^*}^{a} (\sigma-\sigma^*) \ dx \tag{3.2}$$

for some $x_1 \in [0,\alpha^*]$ and $x_2 \in [\alpha^*,a]$. Since $\hat{v}'^2(x_1) \geq \hat{v}'^2(x_2)$ the right-hand side is nonpositive. Consequently by Lemma 2.4

$$\lambda_2(\sigma) \leq (\int_0^a \sigma \hat{v}'^2 dx) \ / \int_0^a \hat{v}^2 dx \leq (\int_0^a \sigma^* \hat{v}'^2 dx) \ / \int_0^a \hat{v}^2 dx = \lambda_2(\sigma^*)$$

which proves (3.1) . To prove the uniqueness suppose that σ_1^* and σ_2^* are two extremal functions. Let w be the second eigenfunction corresponding to $\chi := (\sigma_1{}^* + \sigma_2{}^*)/2$. From the recursive characterization of the eigenvalues it follows that

$$\frac{1}{2}\,[\,\lambda_2(\sigma_1{}^*) + \lambda_2(\sigma_2{}^*)\,] \le \int_{-a}^{a} \chi\, w'^2\, dx \Big/ \int_{-a}^{a} w^2\, dx = \lambda_2(\chi) \;.$$

Because of the extremality of $\sigma_i{}^*$, $i=1,2$ the equality sign holds and w is an eigenfunction corresponding both to $\lambda_2(\sigma_1{}^*)$, $\lambda_2(\sigma_2{}^*)$. Hence

$$\int_{-a}^{a} (\sigma_1{}^* - \sigma_2{}^*)\, w'\, \eta'\, dx = 0 \qquad \text{for any } \eta \in C_0^\infty(-a,a) \;,$$

and finally, since $w' \neq 0$ in $(-a,a)$ we conclude that $\sigma_1{}^* \equiv \sigma_2{}^*$ in $(-a,a)$.

DEFINITION 3.1. Let $\sigma_n{}^*$ be the extremal function for $\lambda_2(\sigma)$ in $I := (\,-a\,,\,-a + 2a/(n-1)\,)$ with

$$0 < \sigma(x) \le \sigma_0 \quad \text{in } I \quad \text{and} \quad \int_I \sigma\, dx = 2V/(n-1) \;.$$

Extend $\sigma_n{}^*$ as a periodic function to the whole interval $(-a,a)$ (cf. Fig. 2) .

Fig. 2

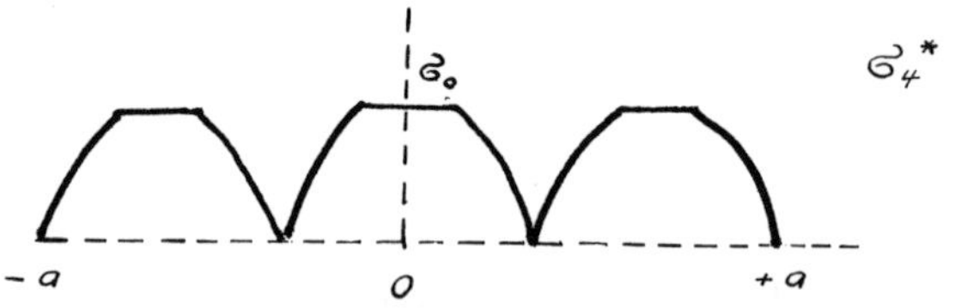

Lemma 2.3 combined with Theorem 3.1 implies

THEOREM 3.2. <u>Under the same assumptions as for Theorem 3.1 we have</u>

$$\lambda_n(\sigma) \le \lambda_n(\sigma_n{}^*) \qquad \text{for all } \sigma \in S(1,\sigma_0) \;.$$

Proof. We only remark that $\lambda_n(\sigma_n^*)$ corresponds to the second eigenvalue of (1.1) in $[-a, -a+2a/(n-1)]$ with σ replaced by σ_n^*.

REMARK. The unrestricted case $\sigma \in S(1,\infty)$ has already been discovered by Troesch [7].

4. THE CASE $p > 1$

4.1. Consider the functional

$$\Lambda[\alpha;v] := \left\{\sigma_0 \int_0^\alpha v'^2\,dx + \sigma_1(\alpha)\left[\int_\alpha^a |v'|^{2q}\,dx\right]^{1/q}\right\} \Big/ \int_0^a v^2\,dx ,$$

where

$$0 < \alpha < V\sigma_0^{-p}, \quad p^{-1}+q^{-1} = 1 \quad \text{and} \quad \sigma_1(\alpha) := [V - \alpha\sigma_0^{p}]^{1/p} .$$

Define

$$\Lambda(\alpha) := \inf_K \Lambda[\alpha;v] \quad \text{where} \quad K = \left\{v \in W^{1,2}(0,a) \cap W^{1,2q}(\alpha,a),\ v(0) = 0,\ v \not\equiv 0\right\}.$$

LEMMA 4.1. (i) *There exists a minimizer* $v_* \in K$ *such that* $\Lambda(\alpha) = \Lambda[\alpha;v_*]$.

(ii) v_* *is monotonic*.

Proof. The proof of (i) is standard and will therefore be omitted.

In order to establish (ii) we assume that v_* isn't monotonic. Then the function w determined by $w' = \{\max v_*', 0\}$, $w(0) = 0$ belongs to K, and we have $\Lambda[\alpha;w] < \Lambda(\alpha)$. This is impossible by the definition of $\Lambda(\alpha)$.

Set

$$F_1[v] := \sigma_0 \int_0^\alpha v'^2\,dx + \sigma_1\left[\int_\alpha^a |v'|^{2q}\,dx\right]^{1/q} \quad \text{and} \quad F_0[v] := \int_0^a v^2\,dx .$$

By Lagrange's multiplier rule there exists a real number ω such that the Fréchet derivatives F_i', i=0,1 satisfy

(4.1) $$F_1'[v_*] + \omega F_0'[v_*] = 0 .$$

(4.1) is equivalent to the Euler equation

(4.2) $$\sigma_o \int_0^{\alpha} v_*'\eta' \, dx + \sigma_1 \gamma \int_{\alpha}^{a} |v_*'|^{2q-2} v_*'\eta' \, dx - \Lambda(\alpha) \int_0^a v_* \eta \, dx = 0$$

$$\text{for all } \eta \in K , \qquad \gamma := [\int_{\alpha}^{a} |v_*'|^{2q} dx]^{1/q - 1} .$$

LEMMA 4.2. <u>The minimizer</u> v_* <u>has the following properties</u>:
(i) v_* <u>is a classical solution of</u>

(4.3)' $$\sigma_o v_*'' + \Lambda v_* = 0 \quad \text{in} \quad (0,\alpha) , \quad v_*(0) = 0$$

(4.3)'' $$\sigma_1 \gamma (|v_*'|^{2q-2} v_*')' + \Lambda v_* = 0 \quad \text{in} \quad (\alpha,a) , \quad v_*'(a) = 0$$

(4.3)''' $$\sigma_o v_*'(\alpha-0) = \sigma_1 \gamma \{v_*'(\alpha+0)\}^{2q-1} .$$

(ii) <u>Up to a factor</u>, v_* <u>is uniquely determined</u>.

<u>Proof.</u> First observe that $v_* \in C^0[0,a]$. (4.3)' is immediate if we restrict ourselves to functions η having their support in $[0,\alpha]$. In order to prove (4.3)'' we insert in (4.2) the function

$$\eta(\xi) = \begin{cases} 0 & \text{in } (0,\alpha) \\ \xi - \alpha & \text{in } [\alpha,x) \\ x - \alpha & \text{in } [x,a] \end{cases}$$

and obtain

$$\sigma_1 \gamma |v_*'|^{2q-2} v_*'(x) = - \Lambda \int_x^a v_* \, d\xi \quad \text{in} \quad (\alpha,a] .$$

This proves (4.3)'' .

If we integrate (4.2) by parts, we find, using (4.3)' and (4.3)" ,

$$\sigma_0 v_*'(\alpha-0)\,\eta(\alpha) - \sigma_1 \gamma |v_*'(\alpha+0)|^{2q-2} v_*'(\alpha+0)\,\eta(\alpha) = 0 \quad .$$

Since this identity must hold for all values of η at α , (4.3)" thus follows.

For the proof of (ii) consider the problem

$$(4.4)' \qquad \sigma_1 (z'^{2q-1})' + \Lambda z = 0 \quad \text{in} \quad (-a+\alpha, 0)$$

$$(4.4)'' \qquad z(-a+\alpha) = z'(0) = 0 \quad .$$

As in [4] it can be shown that there exists a unique positive solution. Observe that v_* can always be normalized such that $\gamma = 1$. Moreover for any $\nu \in \mathbb{R}^+$

$$w(x;\nu) := \nu^{-q/(q-1)} z[\nu(x-a)]$$

is a solution of (4.3)" with $\gamma = 1$, satisfying $w'(a;\nu) = 0$. Choose ν_0 such that $w(a;\nu_0) = v_*(a)$. Then (4.3)" yields

$$\sigma_1 [w'^{2q-1}(x;\nu_0) - v_*'^{2q-1}(x)] = \frac{\Lambda}{2} [v_*^2(x) - w^2(x;\nu_0)]$$

and consequently

$$(4.5) \qquad v_*(x) \equiv w(x;\nu_0) \ .$$

Next we show that ν_0 is uniquely determined. Notice that $\nu_0 < 1$. In $(0,\alpha)$ the minimizers are of the form $v_*(x) = A \sin\sqrt{\Lambda/\sigma_0}\, x$. Hence by (4.3)" and the continuity of v_*

$$A\sqrt{\Lambda/\sigma_0} \cos\sqrt{\Lambda/\sigma_0}\,\alpha = w'(\alpha;\nu_0)$$

and

$$A \sin\sqrt{\Lambda/\sigma_0}\,\alpha = w(\alpha;\nu_0) \quad ,$$

whence

$$(4.6)\qquad \nu_0 z'[\nu_0(\alpha-a)] = \sqrt{\Lambda/\sigma_o}\ \mathrm{ctg}\sqrt{\Lambda/\sigma_o}\,\alpha\ z[\nu_0(\alpha-a)] \quad .$$

If $\nu_0 \in [0,1]$ increases, the expression at the left-hand side of (4.6) increases whereas the one at the right-hand side decreases. Consequently there is a unique root ν_0 .

In the proof of the next lemma we shall need the

PROPOSITION 4.1. *For any* $\alpha_1 < V\sigma_o^{-p}$ *there exists a constant* $c_0 > 0$ *such that*
(i) $\Lambda(\alpha) \geq c_0$ *for all* $\alpha \leq \alpha_1$.
(ii) *For any* $\alpha_0 > 0$ *there is a constant* $c_1 > 0$ *such that all normalized minimizers* $v_*(x;\alpha)$, $\|v_*\|_{L^2} = 1$, *with* $\alpha \in [\alpha_0, \alpha_1]$ *satisfy* $|v_*'| \leq c_1$ *in* $[0,a]$.

Proof. From Hölder's inequality we get

$$\Lambda[v;\alpha] \geq \left\{\sigma_o \int_0^\alpha v'^2\,dx + \left(\frac{V-\alpha\sigma_o^p}{a-\alpha}\right)^{1/p} \int_\alpha^a v'^2\,dx\right\} \Big/ \int_0^a v^2\,dx$$

$$\geq \min\left\{\sigma_o\ ,\ \left(\frac{V-\alpha_1\sigma_o^p}{a}\right)^{1/p}\right\} \frac{\pi^2}{4a^2} \quad ,$$

which implies (i) .
In $(0,\alpha)$ we have $v_* = A\sin\sqrt{\Lambda/\sigma_o}\,x$, $A > 0$. By the concavity of v_*

$$v_*'(x;\alpha) \leq A\sqrt{\Lambda/\sigma_o} \quad .$$

Since $\sqrt{\Lambda/\sigma_o}\,\alpha \leq \pi/2$, it follows that

$$\sin\sqrt{\Lambda/\sigma_o}\,x \geq (2/\pi)\sqrt{\Lambda/\sigma_o}\,x$$

and therefore

$$1 = \|v_*\|^2_{L^2} \geq \int_0^{\alpha} (A\cdot(2/\pi)\sqrt{\Lambda/\sigma_0}\; x)^2\, dx = (4\Lambda\alpha^3/3\pi^2\sigma_0)\, A^2$$

which, together with (i) proves the assertion.

LEMMA 4.3. (i) $\Lambda(\alpha)$ <u>is continuous in</u> $(0,a)$.
(ii) <u>If</u> $v_*(x;\alpha)$ <u>denotes a minimizer such that</u> $\|v_*\|_{L^2} = 1$, <u>then for all</u> $\alpha_0 > 0$ <u>we have</u>

$$\lim_{\alpha\to\alpha_0} v_*(x;\alpha) = v_*(x;\alpha_0) \quad \underline{\text{uniformly in}} \quad [0,a] .$$

<u>Proof.</u> It suffices to prove (i) . The second assertion is then a direct consequence of the construction of v_* , given in Lemma 4.2 .
By definition

$$\Lambda(\alpha+\Delta\alpha) \leq \Lambda[\alpha+\Delta\alpha;\ v_*(x;\alpha)]$$

$$= \Lambda(\alpha) + \sigma_0 \int_{\alpha}^{\alpha+\Delta\alpha} v_*'^2(x;\alpha)dx + \sigma_1(\alpha+\Delta\alpha)\left\{\int_{\alpha+\Delta\alpha}^{a} v_*'^{2q}(x;\alpha)dx\right\}^{1/q}$$

$$- \sigma_1(\alpha)\left\{\int_{\alpha}^{a} v_*'^{2q}(x;\alpha)dx\right\}^{1/q} ,$$

whence

$$(4.7) \qquad \Lambda(\alpha+\Delta\alpha) \leq \Lambda(\alpha) + \eta(\Delta\alpha) , \quad \lim_{\Delta\alpha\to 0} \eta(\Delta\alpha) = 0 .$$

On the other hand

$$\Lambda(\alpha) \leq \Lambda[\alpha;\ v_*(x;\alpha+\Delta\alpha)]$$

$$= \Lambda(\alpha+\Delta\alpha) - \sigma_0 \int_{\alpha}^{\alpha+\Delta\alpha} v_*'^2(x;\alpha+\Delta\alpha)dx + \sigma_1(\alpha)\left\{\int_{\alpha}^{a} v_*'^{2q}(x;\alpha+\Delta\alpha)dx\right\}^{1/q}$$

$$- \sigma_1(\alpha+\Delta\alpha)\left\{\int_{\alpha+\Delta\alpha}^{a} v_*'^{2q}(x;\alpha+\Delta\alpha)dx\right\}^{1/q} .$$

From Proposition 4.1 (ii) we deduce that

$$\Lambda(\alpha) \leq \Lambda(\alpha+\Delta\alpha) + \omega(\Delta\alpha) , \quad \lim_{\Delta\alpha\to 0} \omega(\Delta\alpha) = 0 .$$

The assertion is now obvious.

4.2. Let $v_*(x;\alpha)$, $\alpha > 0$ be any positive minimizer. Define

$$\hat{\sigma}(x;\alpha) = \begin{cases} \sigma_o & \text{in } [0,\alpha] \\ \sigma_o v_*'^{2q/p}(x;\alpha) \,/\, v_*'^{2q/p}(\alpha+0;\alpha) & \text{in } [\alpha,a] \\ \hat{\sigma}(-x;\ ) & \text{in } [-a,0] \end{cases}$$

Note that $\hat{\sigma}(x;\alpha) \leq \sigma_o$. Moreover by Lemma 4.3 (ii) it depends continuously on α . By assumption we have

$$\int_0^a \hat{\sigma}^p(x;a)dx = \sigma_o^p\, a > V .$$

If

(4.8) $$\int_0^a \hat{\sigma}^p(x;\alpha)dx < V \quad \text{for} \quad \alpha \in (0,\varepsilon) ,$$

then there exists a α^* such that

$$\int_0^a \hat{\sigma}^p(x;\alpha^*)dx = V .$$

DEFINITION 4.1. If (4.8) holds, put $\sigma^*(x) := \hat{\sigma}(x;\alpha^*)$. Otherwise set $\alpha^*(x) := \omega v_*'^{2q/p}(x;0)$ in $(0,a)$, $\sigma^*(-x) = \sigma^*(x)$ where ω is determined such that σ^* belongs to $S(p,\sigma_o)$.

From the construction of σ^* we have immediately

THEOREM 4.1. <u>Let</u> σ^* <u>be defined as above and let</u> $v_* = v_*(x;\alpha^*)$ <u>for</u> $x \geq 0$. <u>Set</u> $v_*(x) = -v_*(-x)$ <u>for</u> $x \leq 0$. <u>Then</u> v_* <u>is a weak solution of</u>

$$(\sigma^* v_*')' + \Lambda(\alpha^*)v_* = 0 \quad \text{in } (-a,a) , \quad \sigma^* v_*'(\pm a) = 0 .$$

4.3. After the preparation of the previous sections we are now in position to construct an isoperimetric upper bound for $\lambda_2(\sigma)$, $\sigma \in S(p,\sigma_0)$.

THEOREM 4.2. (i) <u>For any</u> $\sigma \in S(p,\sigma_0)$ <u>we have</u>

$$\lambda_2(\sigma) \leq \lambda_2(\sigma^*) = \Lambda(\alpha^*) \quad .$$

(ii) σ^* <u>is the only function in</u> $S(p,\sigma_0)$ <u>with this property</u>.

<u>Proof.</u> According to Lemma 2.3 it suffices to establish (i) for symmetric functions. Let v_* be as in Theorem 4.1 . Then by Lemma 2.4 we have

$$\lambda_2(\sigma) \leq \int_0^a \sigma v_*'^2 \, dx \Big/ \int_0^a v_*^2 \, dx \quad . \tag{4.9}$$

Suppose that $\sigma \equiv \sigma_0$ in $[0,\alpha]$. We shall distinguish between two cases.

(i) $\alpha \geq \alpha^*$.

If we apply Hölder's inequality in (4.9) we find

$$\lambda_2(\sigma) \leq \left\{\sigma_0 \int_0^{\alpha^*} v_*'^2 dx + \sigma_1(\alpha^*)\left[\int_{\alpha^*}^a v_*'^{2q} dx\right]^{1/q}\right\} \Big/ \int_0^a v_*^2 dx = \Lambda(\alpha^*).$$

Since v_* vanishes only once, it is the second eigenfunction of (1.1) with σ replaced by σ^* . Hence by Theorem 4.1 $\Lambda(\alpha^*) = \lambda_2(\sigma^*)$.

(ii) $\alpha < \alpha^*$.

We shall show that

$$\int_0^a (\sigma - \sigma^*) \, v_*'^2 \, dx \leq 0 \quad . \tag{4.10}$$

For all σ we have

$$(4.11)\qquad \sigma^p - \sigma^{*p} \geq p\,(\sigma^*)^{p-1}\,(\sigma - \sigma^*)\ .$$

Let $I^+ = (x_1^+, x_2^+)$ be an interval where $\sigma - \sigma^* > 0$, and similarly we define I^-. From (4.11) it follows that

$$\frac{\sigma - \sigma^*}{\sigma^p - \sigma^{*p}} \leq \frac{1}{p\,(\sigma^*)^{p-1}} \qquad \text{in } I^+$$

$$\frac{\sigma - \sigma^*}{\sigma^p - \sigma^{*p}} \geq \frac{1}{p\,(\sigma^*)^{p-1}} \qquad \text{in } I^-\ .$$

Observe that all I^+ lie in $[\alpha^*, a]$, and that $v^{*\prime 2} = c(\sigma^*)^{p/q}$ in I^+. Hence

$$v_*'^2\,\frac{\sigma - \sigma^*}{\sigma^p - \sigma^{*p}} \leq \frac{c}{p} \qquad \text{in } I^+$$

and

$$(4.12)\qquad v_*'^2\,\frac{\sigma - \sigma^*}{\sigma^p - \sigma^{*p}} \geq \frac{c}{p} \qquad \text{in } I^- \cap [\alpha^*, a]\ .$$

Since $v_*'^2$ is non-increasing, (4.12) holds also in $I^- \cap [\alpha, \alpha^*]$. Whence

$$\int_0^a (\sigma - \sigma^*)\,v_*'^2 dx = \int_0^a (\sigma^p - \sigma^{*p})\,\frac{v_*'^2(\sigma - \sigma^*)}{\sigma^p - \sigma^{*p}}\,dx \leq \frac{c}{p}\int_0^a (\sigma^p - \sigma^{*p})\,dx = 0.$$

The assertion is now obvious.

In order to prove the second statement suppose that there are two extremal functions σ_1^* and σ_2^*. Then the function

$$\chi(x) := (\sigma_1^* + \sigma_2^*)\,/\,2$$

satisfies

$$\chi(x) \leq \sigma_o \quad \text{and} \quad \int_0^a \chi^p\,dx \leq V\ .$$

We can construct a function $\chi^* \geq \chi$ which belongs to $S(p,\sigma_o)$. By Lemma 2.1 we have

$$\lambda_2(\chi^*) \geq \lambda_2(\chi) \quad .$$

By Schwarz's result [5] , σ_1^* and σ_2^* and thus χ must be symmetric. Hence Lemma 2.4 applies and we have

$$\lambda_2(\chi) \geq \frac{1}{2} [\lambda_2(\sigma_1^*) + \lambda_2(\sigma_2^*)] \quad .$$

In view of the extremal property of σ_1^* and σ_2^* , we must have

$$\lambda_2(\chi^*) = \frac{1}{2} [\lambda_2(\sigma_1^*) + \lambda_2(\sigma_2^*)]$$

and, in addition, the eigenfunctions w corresponding to $\lambda_2(\chi^*)$ are simultaneously eigenfunctions corresponding to $\lambda_2(\sigma_i^*)$, i=1,2 . Consequently

$$\int_{-a}^{a} \sigma_i^* w' \eta' \, dx = \lambda_2(\chi^*) \int_{-a}^{a} w \eta \, dx \quad \forall \eta \in C_0^\infty(-a,a) \, , \quad \int_{-a}^{a} \eta \, dx = 0.$$

This implies

$$\int_{-a}^{a} (\sigma_1^* - \sigma_2^*) w' \eta' \, dx = 0 \quad \text{for all admissible } \eta \, ,$$

whence $\sigma_1^* \equiv \sigma_2^*$. Here we have used the fact that $w \equiv v_*$ has a non-vanishing derivative in $(-a,a)$.

Similarly as for $p = 1$ we define σ_n^* .

DEFINITION 4.2. Let σ_n^* be the extremal function for $\lambda_2(\sigma)$ in $[-a , -a + 2a/(n-1)] =: I$ with

$$0 < \sigma \leq \sigma_o \qquad \text{and} \qquad \int_I \sigma^p \, dx = 2V/(n-1) \quad .$$

Extend $\sigma_n{}^*$ as a periodic function in the whole interval (-a,a). Since $\sigma_n{}^*$ vanishes at the endpoints of I , the extended function is continuous. The same arguments as in Theorem 3.2 yield, together with Lemma 2.3 and Theorem 4.1

THEOREM 4.2. *Let* $\sigma_n{}^*$ *be defined as above. Then*

$$\lambda_n(\sigma) \le \lambda_n(\sigma_n{}^*) \qquad \text{for all } \sigma \in S(p,\sigma_o) \quad .$$

REFERENCES

1. C. Bandle and B. Zemp, Problèmes aux valeurs propres extrémaux du type de Sturm-Liouville. C. R. Acad. Sc. Paris 301 (1985), 895-897.

2. D.C. Barnes, Extremal problems for eigenvalues with applications to buckling, vibration and sloshing. SIAM J. Math. Anal. 16 (1985), 341-357.

3. J.F. Kuzanek, Existence and uniqueness of solutions to a fourth order nonlinear eigenvalue problem. SIAM J. Appl. Math. 27 (1974), 341-354.

4. M. Ôtani, Sur certaines équations différentielles ordinaires du second ordre associées aux inégalités du type Sobolev-Poincaré. C. R. Acad. Sc. Paris 296 (1983), 415-418.

5. B.Schwarz, On the extrema of frequencies of nonhomogeneous strings with equimeasurable density. J. Math. Mech. 10 (1961), 401-422.

6. I. Tadjbaksh and J.B. Keller, Strongest columns and isoperimetric inequality for eigenvalues. J. Appl. Mech. 29 (1962), 159-164.

7. B.A. Troesch, An isoperimetric sloshing problem. Comm. Pure Appl. Math. 18 (1965), 319-338.

8. Ph. Hartman, Ordinary Differential Equation. Birkhäuser-Verlag, Basel -Boston -Stuttgart, 2nd Edition, 1982.

Catherine Bandle, Mathematisches Institut, Universität Basel, Rheinsprung 21, CH-4051 Basel, Switzerland

International Series of
Numerical Mathematics, Vol. 80

THE HELP INEQUALITY IN THE REGULAR CASE

Christer Bennewitz

Abstract. In 1972 W.N. Everitt, generalizing a well known inequality by Hardy, Littlewood and Pólya, studied (0.1) in the case when a is a regular and b a singular point for the equation (0.2). This paper gives conditions close to being necessary and sufficient for the validity of (0.1) with a finite K in the case when both a and b are regular points.

0. INTRODUCTION

Starting in 1972 Everitt [8] and later others, see [5] - [7] and further references there, studied a generalization of the well-known Hardy-Littlewood-Pólya inequality

$$\left(\int_0^\infty |u'|^2\right)^2 \le 4 \int_0^\infty |u|^2 \int_0^\infty |u''|^2 .$$

In general the problem is to decide whether there is a finite K such that

$$\left\{\int_a^b (p|u'|^2+q|u|^2)\right\}^2 \le K^2 \int_a^b |u|^2 w \int_a^b |(pu')'-qu|^2/w \tag{0.1}$$

for any u making the right hand side finite. Here p, q and $w \ge 0$ are given real-valued functions so that it is natural that the differential equation

$$-(pu')' + qu = \lambda wu \tag{0.2}$$

plays an important role in the theory of (0.1). Everitt [8] discussed the case when (0.2) is regular at a, but has a singularity of "strong limit-point" type at b, and it was thought that no valid inequality was possible if both a and b were regular.

This paper is in final form and no version of it will be submitted for publication elsewhere.

In [3] I showed that $p(x) = w(x) = 1$, $q(x) = -1$ gives a counter-example to this conjecture if $b-a$ is an integer multiple n of π. The constant K is then a solution of a certain transcendental equation which gives $K \approx 2.48$ for $n = 1$ and then strictly decreasing with n to the limit 2 (which is the correct constant for the half-line case). In this paper we shall give conditions quite close to being necessary and sufficient for an inequality (0.1) to hold in the regular case. Along the way we also obtain new conditions, necessary within a fairly general class of coefficients p, for the original Everitt inequality.

In Section 1 we state the main result and then review some facts from the general theory of inequalities such as (0.1). In particular we note that for deciding whether a finite K exists (but not for the determination of its value) it is sufficient to investigate the behaviour of solutions of (0.2) for λ near 0 and ∞. In Section 2 we deduce necessary and sufficient conditions for the appropriate behaviour at 0. Finally, in Section 3 we deduce a sufficient condition for the appropriate behaviour at ∞. This condition is also necessary within a fairly wide class of co-efficients p, including all cases where p has a fixed sign in each of a right neighbourhood of a and a left neighbourhood of b. The condition is also appropriate for the behaviour at the regular endpoint of the original Everitt inequality. The whole of Section 3 depends heavily on the results of [4].

1. STATEMENTS OF RESULTS

We will always assume that the equation (0.2) is regular at a and our main concern is with the case when it is also regular at the other endpoint b. Thus we assume that $1/p$, q and w are in $L_{loc}[a,b)$ and mostly even in $L(a,b)$. We also assume that p and q are real-valued and that $w \geq 0$ does not vanish identically. There would be little difficulty in assuming, more generally, that $1/p$, q and w are measures with corresponding properties. See [1, Chapter 11] or [4, Section 1] for the interpretation of (0.2) in this case.

To describe our main result we need a few definitions. Let

$$P_a(x) = \sup_{a\le s\le t\le x} \left|\int_s^t 1/p\right| \quad \text{and} \quad P_b(x) = \sup_{x\le s\le t\le b} \left|\int_s^t 1/p\right|$$

(cf. [2], [4, (2.2)]). We shall need the property

$$\text{(1.1)} \quad \begin{cases} P_a(x) \sim \int_a^x 1/p \quad \text{or} \quad P_a(x) \sim -\int_a^x 1/p \quad \text{as } x \searrow a , \\ P_b(x) \sim \int_x^b 1/p \quad \text{or} \quad P_b(x) \sim -\int_x^b 1/p \quad \text{as } x \nearrow b . \end{cases}$$

Clearly (1.1) is automatically satisfied if there is a right neighbourhood of a and a left neighbourhood of b where p is (a.e.) of one sign. It may be, however, that (1.1) is satisfied also in cases when p oscillates wildly. An example, for $a = 0$, is given by $p(x) = \{1 + x^{-1/2} \sin(1/x)\}^{-1}$. If we put $W_a(x) = \int_a^x w$, $W_b(x) = \int_x^b w$ then W_a and W_b are monotone, but not necessarily strictly so. We may nevertheless consider "generalized" inverses W_a^{-1} and W_b^{-1}, defined in a right neighbourhood of 0, by setting

$$W_a^{-1}(x) = \inf \{y > a \mid W_a(y) \ge x\} , \quad W_b^{-1} = \sup \{y < b \mid W_b(y) \ge x\} .$$

Geometrically W_a^{-1} and W_b^{-1} are obtained by mirroring the graphs in $y = x$ and then deleting vertical line segments in such a way as to make the new functions left-continuous (cf. [4, Definition 2.1]). It follows that if we define

$$\text{(1.2)} \quad \begin{cases} S_a(x) = \overline{\lim_{u\downarrow 0}}\, P_a \circ W_a^{-1}(xu) \,/\, P_a \circ W_a^{-1}(u) , \\ S_b(x) = \overline{\lim_{u\downarrow 0}}\, P_b \circ W_b^{-1}(xu) \,/\, P_b \circ W_b^{-1}(x) , \end{cases}$$

then the functions S_a and S_b are increasing, defined for $x \ge 0$ (possibly with value ∞) and ≤ 1 for $x < 1$. It is clear that these functions are submultiplicative (e.g. $S_a(xy) \le S_a(x)S_a(y)$). It follows that in (0,1) either $S_a \equiv 1$ or $S_a(x) \to 0$ as $x \searrow 0$ and similarly for S_b (given x in (0,1) we have $S_a(x^n) \le (S_a(x))^n$ so as $n \to \infty$ follows $S_a(0+) = 0$ unless $S_a(x) = 1$). Our main result is the following

THEOREM. *For the existence of a finite K in (0.1) it is necessary that the origin is a double eigenvalue for (0.2) and some (non-separated) boundary conditions at a, b. Furthermore, if p satisfies (1.1) then it is necessary and sufficient for the existence of a finite K that in addition to the double eigenvalue property it holds that neither S_a nor S_b, as defined in (1.2), is $\equiv 1$ in the interval (0,1).*

Before we go into the proof we must review some facts from the general theory of (0.1). Everitt [8] showed that when b is in the strong limit-point case the inequality (0.1) is valid with $K = \max\,(1/|\cos v_1|, 1/|\cos v_2|)$ where v_1 is least in $[0,\pi/2]$ and v_2 largest in $[\pi/2,\pi]$ so that $-\mathrm{Im}\,(\lambda^2 m(\lambda)) \geq 0$ for all λ with $\arg\lambda = v_1$ or $\arg\lambda = v_2$. Thus, if either v_1 or v_2 equals $\pi/2$ then (0.1) is not valid for any finite K. Here $m(\lambda)$ is the Titchmarsh-Weyl m-function belonging to the Neumann boundary condition $pu'(a) = 0$ at a. In Bennewitz [3] was shown that in more general cases (including cases of higher order) the expression $-\mathrm{Im}\,(\lambda^2 m(\lambda))$ is replaced by a Hermitean form on the space of solutions of (0.2) which are square integrable in (a,b) with respect to the weight w, the conditions being otherwise the same. This space is one-dimensional in Everitt's case. In the present case when both a and b are regular for (0.2) the space is of course 2-dimensional and the form is given by $-\mathrm{Im}\,(\lambda^2 [u\overline{pu'}]_a^b)$ for any solution u of (0.2). It is easy to see that when λ is purely imaginary the form is positive definite so by a simple compactness argument it suffices to examine the behaviour near $\lambda = 0$ and $\lambda = \infty$ of this form to decide whether there is a finite K for which (0.1) holds. This we propose to do in the following sections.

2. BEHAVIOUR FOR SMALL λ

If the inequality (0.1) is to be valid in a regular case, the left member must vanish on the 2-dimensional solution-space of (0.2) for $\lambda = 0$. An integration by parts shows that this means that $[pu'\,\bar{u}]_a^b$ must vanish for these solutions which is easily seen to mean that they must satisfy the non-separated boundary conditions $u(a) = Du(b)$, $pu'(b) = Dpu'(a)$ for some real constant D. Thus, for a valid inequality the origin must be a double eigenvalue for some self-adjoint boundary conditions (see [3]). It was deduced by Everitt (personal communication) that the origin must be an eigenvalue both for the Dirichlet conditions $u(a) = u(b) = 0$ and the Neumann conditions $pu'(a) = pu'(b) = 0$. The two necessary conditions are actually the same. To see this, note that Everitt's conditions is obviously implied by the double eigenvalue condition. On the other hand, assuming Everitt's condition and letting θ, ϕ be the eigenfunctions for the Dirichlet and Neumann conditions, respectively, we have $p\theta'(b)/p\theta'(a) = \phi(a)/\phi(b) = D$ (say) since the Wronskian $p\theta'\phi - p\phi'\theta$ is constant. Thus ϕ and θ, and therefore all solutions, satisfy the non-separated boundary conditions.

The necessary condition just deduced is in fact sufficient to guarantee the correct behaviour of the form $-\text{Im}\,\{\lambda^2[u\overline{pu'}]_a^b\}$ for small λ. To see this, let ϕ and θ be solutions of (0.2) satisfying

$$\begin{cases} \phi(a,\lambda) = -1 \\ p\phi'(a,\lambda) = 0 \end{cases} \qquad \begin{cases} \theta(a,\lambda) = 0 \\ p\theta'(a,\lambda) = 1 . \end{cases}$$

Then $\phi_o(x) = \phi(x,0)$ and $\theta_o(x) = \theta(x,0)$ are the eigenfunctions for the Neumann and Dirichlet conditions, respectively. Using this it is easily seen that

$$[\phi\overline{p\phi'}]_a^b = -\lambda \int_a^b |\phi_o|^2 w + O(|\lambda|^2) ,$$

$$[\theta\overline{p\theta'}]_a^b = \lambda \int_a^b |\theta_o|^2 w + O(|\lambda|^2) ,$$

$$[\theta\overline{p\phi'}]_a^b = O(|\lambda|^2) \quad \text{and}$$

$$[\phi\overline{p\theta'}]_a^b = 2i\ \text{Im}\ \lambda \int_a^b \phi_o\overline{\theta_o} w + O(|\lambda|^2) ,$$

all for small $|\lambda|$. Now for every solution u of (0.2) we may write $u = s\phi + t\theta$, ϕ and θ being linearly independent. We therefore obtain

$$-\text{Im}\ \lambda^2[u\overline{pu'}]_a^b = A|s|^2 + B|t|^2 + 2\ \text{Re}\,(s\bar{t}C),$$

where

$$A = -\mathrm{Im}\ \{\lambda^2[\phi\overline{p\phi'}]_a^2\} = |\lambda|^2\ \mathrm{Im}\ \lambda \int_a^b |\phi_o|^2 w + O(|\lambda|^4)\ ,$$

$$B = -\mathrm{Im}\ \{\lambda^2[\theta\overline{p\theta'}]_a^2\} = -\mathrm{Im}\ \lambda^3 \int_a^b |\theta_o|^2 w + O(|\lambda|^4)\ ,$$

$$C = i[\lambda^2\phi\overline{p\theta'} - \bar{\lambda}^2 p\phi'\bar{\theta}]_a^b/2 = -\lambda^2\ \mathrm{Im}\ \lambda \int_a^b \phi_o\bar{\theta}_o w + O(|\lambda|^4)\ ,$$

valid for small $|\lambda|$. Clearly $A > 0$ for small λ in the upper half plane with fixed argument. Setting $\lambda = re^{iv}$ we also have, for small $r > 0$ and fixed v,

$$AB - |C|^2 = r^6 \sin^2 v\{(1 - 4\cos^2 v)\|\phi_o\|^2\|\theta_o\|^2 - |(\phi_o,\theta_o)|^2\} + O(r^7),$$

where $\|\cdot\|$ and $(\cdot,\cdot)$ denote the norm and scalar product, respectively, in the space $L^2_w(a,b)$ of functions with integrable square with respect to the weight w on the interval (a,b). For $v = \pi/2$ the bracketed expression is strictly positive, ϕ_o and θ_o being linearly independent. It is therefore clear that choosing v in a sufficiently small, fixed neighbourhood of $\pi/2$ we have $AB - |C|^2 > 0$ if r is small. In fact, we get a lower bound

$$K^2 \geq 4\{1 - |(\phi_o,\theta_o)|^2 / \|\phi_o\|^2\|\theta_o\|^2\}^{-1}$$

for the best constant in (0.1). Clearly this is least restrictive when ϕ_o and θ_o are orthogonal, in which case we get $K \geq 2$. This orthogonality occurs e.g. when the coefficients p, q and w are symmetric around $(a+b)/2$.

3. BEHAVIOUR FOR LARGE λ

It is now convenient to choose a basis for the solutions different from that in Section 2. First choose $m(\lambda)$ so that $\psi(x,\lambda) = \theta(x,\lambda) + m(\lambda)\phi(x,\lambda)$ satisfies $p\psi'(b,\lambda) = 0$, i.e. m is the Titchmarsh-Weyl function to the Neumann conditions at both ends. Furthermore let $\tilde{\theta}$ and $\tilde{\phi}$ be solutions with

$$\begin{cases} \tilde{\theta}(b,\lambda) = 0 \\ p\tilde{\theta}'(b,\lambda) = 1 \end{cases} \qquad \begin{cases} \tilde{\phi}(b,\lambda) = 1 \\ p\tilde{\phi}'(b,\lambda) = 0 \end{cases}$$

and choose $\tilde{m}(\lambda)$ so that $\tilde{\psi}(x,\lambda) = \tilde{\theta}(x,\lambda) + \tilde{m}(\lambda)\tilde{\phi}(x,\lambda)$ satisfies

$p\tilde{\psi}'(a,\lambda) = 0$. Since $p\psi'(a,\lambda) = 1$ the functions ψ and $\tilde{\psi}$ are linearly independent (and well-defined for non-real λ) so any solution u of (0.2) may be written $u = s\psi + t\tilde{\psi}$. Thus

$$-\mathrm{Im}\ \{\lambda^2[u\overline{pu'}]_a^b\} = A|s|^2 + B|t|^2 + 2\ \mathrm{Re}\,(s\bar{t}C)\ ,$$

where

$$A = -\mathrm{Im}\ \{\lambda^2[\psi\overline{p\psi'}]_a^b\} = -\mathrm{Im}\ \{\lambda^2 m(\lambda)\}\ ,$$

$$B = -\mathrm{Im}\ \{\lambda^2[\tilde{\psi}\overline{p\tilde{\psi}'}]_a^b\} = -\mathrm{Im}\ \{\lambda^2\tilde{m}(\lambda)\} \quad \text{and}$$

$$C = i[\lambda^2\psi\overline{p\tilde{\psi}'} - \bar{\lambda}^2 p\psi'\overline{\tilde{\psi}}]_a^b/2 = -\mathrm{Im}\ \{\lambda^2 M(\lambda)\},$$

where $M(\lambda) = \psi(b,\lambda) = p\tilde{\psi}'(x,\lambda)\psi(x,\lambda) - \tilde{\psi}(x,\lambda)p\psi'(x,\lambda) = -\tilde{\psi}(a,\lambda)$. By [4, Theorem 6.1] it follows that $\ln M(\lambda) \sim -\int_a^b \sqrt{-\lambda w/p}$ as $\lambda \to \infty$ along non-real rays, the root being the principal branch. In particular, $C = o(1)$ as $\lambda \to \infty$ along any non-real ray.

In the rest of the paper we assume that p has the property (1.1). (To avoid duplication of argument we shall always consider only the first alternative in (1.1) for P_a.) In order to complete the proof of the theorem we must briefly review some material from [4, Sections 1-4]. Firstly, it is possible to assign m-functions, with all the usual properties, to systems of equations defined on an interval $[0,c)$ and of the form

$$(3.1)\qquad \begin{cases} u(x) = u(0) + \int_0^x v\,dP_\infty \\ v(x) = v(0) - \lambda \int_0^x u \end{cases}$$

where P_∞ is a function of locally bounded variation in $[0,c)$. The Weyl circle at a non-real λ (for the Neumann boundary condition at 0) can have at most one real point k and this happens precisely if $dP_\infty(x) = k\delta$ where δ is the Dirac measure. Secondly, let f be the generalized inverse (in the sense already described) of the function $x \mapsto 1/(xW_a \circ P_a^{-1}(x))$, where P_a^{-1} is the generalized inverse of P_a. Then there are absolute constants $M,N > 0$ such that $N\,|\sin\arg\lambda|\,f(|\lambda|) \le |m(\lambda)| \le Mf(|\lambda|)/|\sin\arg\lambda|$. Thirdly, if $r_j \to \infty$ as $j \to \infty$ and $P_a \circ W_a^{-1}(xu_j)/P_a \circ W_a^{-1}(u_j) \to P_\infty(x)$ for $0 < x < 1$ and

$u_j = 1/(r_j f(r_j))$, then it follows that $m(r_j\mu)/f(r_j)$ (or $-m(-r_j\mu)/f(r_j)$ depending on which alternative holds in (1.1)) is asymptotically as $j\to\infty$ in the Weyl circle at $\lambda=\mu$ for (3.1) on the interval $[0,1)$. This holds uniformly for μ in any compact set not intersecting the real line. We can now prove a basic lemma.

LEMMA. $|m(\lambda)| = O(|\operatorname{Im} m(\lambda)|)$ _as_ $\lambda\to\infty$ _in any non-real sector_ (_i.e. a sector not intersecting the real axis_) _if_ $S_a \not\equiv 1$ _in the interval_ $(0,1)$.

The converse also holds, but we will actually prove a sharper statement later.

Proof. Assume that with $\lambda_j\to\infty$ in a fixed non-real sector holds $\operatorname{Im} m(\lambda_j)/m(\lambda_j)\to 0$ as $j\to\infty$ and set $r_j = |\lambda_j|$. For each r the function $P_r(x) = P_a\circ W_a^{-1}(xu)/P_a\circ W_a^{-1}(u)$, where $u = 1/(rf(r))$, is increasing and maps $[0,1]$ into itself. By the Helly theorem we may choose a subsequence of r_j so that P_r converges, pointwise and boundedly as $r\to\infty$ along this sequence, to some increasing function P_∞. We may also assume that λ_j/r_j has a (non-real) limit μ. Thus $m(\lambda_j)/f(r_j)$, which by assumption is asymptotically real (but non-zero), is (along a subsequence) asymptotically in the Weyl circle of (3.1) at μ which therefore contains a non-zero real point and so P_∞ has a jump at 0. But $S_a \ge P_\infty$, so S_a also jumps at 0 and is therefore $\equiv 1$. The lemma follows.

We wish to study the coefficient A along some ray in the upper half plane, so put $\lambda = r(t+i)$ with t real. Then $A = r^2((1-t^2)\operatorname{Im} m(\lambda) - 2t\operatorname{Re} m(\lambda))$. Assuming $S_a \not\equiv 1$ in $(0,1)$ the lemma provides a constant L such that $|m(\lambda)| \le L \operatorname{Im} m(\lambda)$ for large r if $|t|$ is, say, < 1. Hence $A \ge r^2(1-t^2-2tL)|m(\lambda)|/L$ for large r. Since always, in the case of equation (0.2), $\lambda m(\lambda)\to\infty$ as $\lambda\to\infty$ (see [4, Section 2]) it follows that $A\to+\infty$ as $r\to\infty$ if $|t|$ is sufficiently small. Similarly, if $S_b\not\equiv 1$ it follows in the same way that $B\to+\infty$ as $r\to\infty$ if $|t|$ is sufficiently small. Since $C = o(1)$ along any non-real ray it follows that there are rays in the interior of both the first and second quadrants along which the form

$-\mathrm{Im}\,\{\lambda^2[u\overline{pu'}]_a^b\}$ is eventually positive definite. The sufficiency part of the theorem now follows.

To prove the necessity part of the theorem, suppose $S_a \equiv 1$ in $(0,1)$. It is then clear that one may choose a sequence $r_j \to \infty$ such that $P_r(x)$, defined as above, for $0 < x < 1$ tends to 1 as $r \to \infty$ along this sequence. Therefore $m(r_j\mu)/f(r_j)$ is asymptotically in the Weyl circle of (3.1) at μ where dP_∞ is the Dirac measure and the interval $[0,1)$. This circle has center $1 + i/\mathrm{Im}\,\mu$ and radius $1/|\mathrm{Im}\,\mu|$. If $\lambda = r_j\mu$ we therefore have asymptotically that $A \le -2r_j^2 f(r_j)\,\{\mathrm{Im}\,\mu\ \mathrm{Re}\,\mu - |\mathrm{Im}\,\mu|\}$. Given any ray in the first quadrant it follows that one may choose μ on that ray with large absolute value so that $A < 0$ for $\lambda = r_j\mu$ and large j. Similarly one may show that if $S_b \equiv 1$, then $\underline{\lim}\, B < 0$ on all rays in the first or second quadrant, depending on which alternative holds in (1.1). This completes the proof of the theorem.

Note that in the original Everitt inequality the conditions on the solutions of (0.2) for λ near 0 are more subtle, because of the possible presence of a continuous spectrum. On the other hand, the condition for appropriate behaviour for large λ is simply that $S_a \not\equiv 1$, as follows from the reasoning above, provided that the relevant part of the condition (1.1) holds.

REFERENCES

1. F.V. Atkinson, Discrete and continuous boundary problems. Academic Press, New York, 1964.

2. F.V. Atkinson, The order of magnitude of the Titchmarsh-Weyl m-function. To appear.

3. C. Bennewitz, A general version of the Hardy-Littlewood-Pólya-Everitt (HELP) inequality. Proc. Roy. Soc. Edinburgh Sect. A 97 (1984), 9-20.

4. C. Bennewitz, Spectral asymptotics for Sturm-Liouville equations. To appear.

5. W.D. Evans and W.N. Everitt, A return to the Hardy-Littlewood integral inequality. Proc. Roy. Soc. London Ser. A 380 (1982), 447-486.

6. W.D. Evans, W.N. Everitt, W.K. Hayman and S. Ruscheweyh, On a class of integral inequalities of Hardy-Littlewood type. To appear.

7. W.D. Evans and A. Zettl, Norm inequalities involving derivatives. Proc. Roy. Soc. Edinburgh Sect. A 82 (1978), 51-70.

8. W.N. Everitt, On an extension to an integro-differential inequality of Hardy, Littlewood and Pólya. Proc. Roy. Soc. Edinburgh Sect. A 69 (1971/72), 295-333.

Added in proof: The estimate of $m(\lambda)$ at the bottom of page 343 is not true without further assumptions, see [4, section 2]. A sufficient such assumption is that $S_a \not\equiv 1$. Since the estimate is only used (in the lemma) under this assumption the results are true as stated.

Christer Bennewitz, Department of Mathematics, University of Uppsala, Thunbergsvägen 3, S-752 38 Uppsala, Sweden

International Series of
Numerical Mathematics, Vol. 80

ON ESTIMATING EIGENVALUES OF A SECOND ORDER LINEAR DIFFERENTIAL OPERATOR

Matts Essén

Abstract. For an eigenvalue problem on $I_0 = (0,1)$, we determine the maxima and infima of all eigenvalues when the coefficient p in the operator $-y'' + py$ is allowed to vary in the class of all integrable functions with $\int p_+ = B$ or $\int p_- = B$ where we integrate over the interval I_0.

0. INTRODUCTION

The first part of this paper gives one of the answers to a question of A. Ramm [9] (references will be given below). Let E be the class of nonnegative functions on $(0,1)$ with integral 1. If $p \in E$, we consider the eigenvalue problem

$$(0.1) \qquad -y'' + py = \lambda y \,, \quad t \in (0,1) \,, \quad y(0) = y(1) = 0 \,.$$

Let $\lambda_1(p)$ be the first positive eigenvalue. Find $\max_{p \in E} \lambda_1(p) = \Lambda(1)$ and $p_0 \in E$ such that $\lambda_1(p_0) = \Lambda(1)$!

This problem has been solved by several authors. What is special in my case is that the solution follows in a natural way from the theory developed in [5] and [6]: I use an optimization method based on a classical rearrangement inequality due to Hardy, Littlewood and Pólya.

REMARK. This first part was written in 1983 and the results were announced in [7]. When I sent a preprint to A. Ramm, he told me about the other solutions of Talenti [12] and Harrell [8a,b].

Let me also mention that my solution was one of the starting points of the work of Egnell [4], where higher-dimensional ver-

---

This paper is in final form and no version of it will be submitted for publication elsewhere.

sions of the problem of Ramm are solved.

The second part is devoted to the problem of finding upper and lower bounds for higher eigenvalues. One of the basic ideas in my proof is inspired by a paper of B. Schwarz [11] on rearrangements of the coefficient in another eigenvalue problem (details will be given below). Schwarz uses also the inequality of Hardy, Littlewood and Pólya mentioned above.

I. ON THE MAXIMUM OF THE FIRST EIGENVALUE

Let $E(B)$ be the class of integrable functions on $(0,1)$ which are such that $p \in E(B)$ if and only if $\int_0^1 p_+ = B$ where $p_+ = (|p| + p)/2$. We prove

THEOREM 1. *Let* $p \in E(B)$ *and let* $\lambda_1(p)$ *be the first eigenvalue of* (0.1). *Then*

$$\max_{p\in E(B)} \lambda_1(p) = (\pi + \sqrt{\pi^2 + 4B})^2/4 = \Lambda(B) = \Lambda .$$

The extremal coefficient p_0 *is defined by*

$$p_0(t) = \begin{cases} \Lambda, & \eta < t < 1-\eta, \\ 0, & 0 < t < \eta,\ 1-\eta < t < 1, \end{cases}$$

where $\eta = \pi/(2\sqrt{\Lambda})$.

REMARK. We have $\Lambda(1) \approx 11.7847$.

In the proof, we need certain concepts from [6]. If Φ is a given nonincreasing function in $L^\infty(0,1)$, we let $F = F(\Phi)$ be the class of measurable functions on $(0,1)$ which are such that if $f \in F$, then for all $s \in \underline{R}$, we have

$$|\{t \in (0,1): f(t) > s\}| = |\{t \in (0,1): \Phi(t) > s\}| .$$

Here $|\cdot|$ denotes Lebesgue measure on $\underline{R}$. If $n \geq 0$, we define

$$E_{B,n} = \{p \in E(B): 0 \leq p(t) \leq n \text{ for all } t \in [0,1]\} .$$

We first quote Theorem 2 in [6] which we state as

THEOREM A. *Let* Φ *be as above. We consider the equation*

$$(*) \qquad y'' + qy = 0, \quad y(0) = 1, \quad y'(0) = 0, \quad t \in (0,1),$$

where $q \in F(\Phi)$. We assume that when q varies in the class F, there exist solutions of (*) which are positive on $(0,1)$. Then the supremum of $y(1)$ when q varies in F is attained when q is the nondecreasing function in F . The supremum is assumed for this function only.

REMARK. In [6], Theorem 2 is stated for the case that q is nonnegative. It is clear from the remark in Section 3 in [6] that the theorem is correct also for coefficients which may change sign.

As a corollary, we have

THEOREM B. Let $\lambda > 0$ be given and consider (*) when $q \in E_{B,\lambda}$. We assume that when q varies in $E_{B,\lambda}$, there exist solutions of (*) which are positive on $(0,1)$. Then $\sup y(1)$ when q varies in $E_{B,\lambda}$ is assumed only when $q = q_0$ which is defined by

$$q_0(t) = \begin{cases} 0, & t \in (0,b) \\ \lambda, & t \in (b,1) . \end{cases}$$

Here b is chosen in such a way that $\int_0^1 q_0 = B$.

Proof of Theorem 1. First, we note that it suffices to consider nonnegative functions in $E(B)$. Let $\Lambda_n = \sup \lambda_1(p)$, $p \in E_{B,n}$. To show that Λ_n is attained, we choose a sequence $\{p_k\}_1^\infty$ in $E_{B,n}$ such that $\lambda_1(p_k) \to \Lambda_n$, $k \to \infty$. Without loss of generality, we can assume that the sequence $\{p_k\}_1^\infty$ is weak* convergent in $L^\infty(0,1)$ with limit p_0 . Let y_k be an eigenfunction belonging to $\lambda_1(p_k)$ which we normalize in such a way that $\max y_k(t) = 1$, $t \in [0,1]$. Let us assume that the maximum is attained at t_k . From the differential equation, we see that there exists a constant $C(n)$ such that $\|y_k^{(\nu)}\|_\infty \le C(n)$ for all k , $\nu = 1, 2$. By Ascoli's theorem (cf. [10], p. 179), there exists a subsequence (which we shall also call $\{y_k\}_1^\infty$) such that $y_k \to y_0$ and $y'_k \to y'_0$: the convergence is uniform on $[0,1]$. Letting $k \to \infty$ in the integrated differential equation, we find that $t_k \to a$ and that

$$y_0(t) = 1 + \int_a^t (p_0(s) - \Lambda_n) y_0(s)(t-s)\,ds .$$

Hence y_0 is an eigen-function for the extremal couple (p_0, Λ_n) in the class $E_{B,n}$.

In the argument which follows, we shall write $\Lambda_n = \lambda$. We first claim that $p_0 \le \lambda$ on $(0,1)$. Let us assume that the claim is false on $(a,1)$. Let p_0^* be the decreasing rearrangement of p_0 on the interval $(a,1)$ (for the definition, we refer to the introduction in [5]). Then there exists $\alpha > 0$ such that $p_0^*(t) > \lambda$, $t \in (a, a+\alpha)$. Consider the equation

$$z'' + (\lambda - p_0^*)z = 0 , \quad z(a) = 1 , \quad z'(a) = 0 , \quad t \in (a,1) .$$

We know that y_0 has a maximum at a. It follows that $p_0 \ne p_0^*$. According to Theorem A, we have $z(1) > y_0(1) = 0$.

Let us first consider the case when z has a local maximum at $b \in (a,1)$. We define

$$P(t) = \begin{cases} \lambda , & t \in (a,b) \\ p_0^*(t) , & t \in (b,1) \\ p_0(t) , & t \in (0,a) . \end{cases}$$

To see that $\int_0^1 P < \int_0^1 p_0 = B$, it is sufficient to note that

$$\int_a^1 p_0^* = \int_a^1 p_0 ,$$

$$\int_a^b (\lambda - p_0^*) = \int_a^b (-z''/z) = -\int_a^b (z'/z)^2 < 0 .$$

The function w defined by

$$w(t) = \begin{cases} z(b) y_0(t) , & t \in (0,a) \\ z(b) , & t \in (a,b) \\ z(t) , & t \in (b,1) \end{cases}$$

is a solution of the differential equation

$$w'' + (\lambda - P)w = 0 , \quad w(0) = 0 , \quad w(1) = z(1) > 0 .$$

Since $\int_0^1 P < B$, we can find $P_1 \in E_{B,n}$ such that $P_1 \ge P$ and $\int_0^1 (P_1 - P)$ is positive. A simple comparison shows that the solution w_1 of the equation

$$w_1'' + (\lambda - P_1)w_1 = 0 , \quad w_1(0) = 0 , \quad w_1'(0) = w'(0) ,$$

is a majorant of w on $[0,1]$ and in particular that $w_1(1) \geq w(1) > 0$. Hence there exists a first eigenvalue $\lambda' > \lambda = \Lambda_n$ of equation (0.1) with a coefficient in $E_{B,n}$. But Λ_n is the maximum of such eigenvalues. The contradiction shows that we must have $p_0 \leq \lambda$ on $(a,1)$. The same argument can be used on $(0,a)$. Thus there can not be a local maximum of z in $(0,a)$ or $(a,1)$.

In the remaining case when z is increasing on $(a,1)$, it is easy to find a first eigenvalue which is too big. We leave the details to the reader.

We know that $\lambda - p_0$ is nonnegative and that

$$y_0'' + (\lambda - p_0)y_0 = 0, \quad y_0(a) = 1, \quad y_0'(a) = 0, \quad t \in (a,1).$$

Let q_0 be an extremal coefficient of the type described in Theorem B for this equation on the interval $(a,1)$. This means that q_0 vanishes on an interval (a,c) and takes the value λ on the interval $(c,1)$: here c is chosen in such a way that $\int_a^1 q_0 = \int_a^1 (\lambda - p_0)$. If $\lambda - p_0 \neq q_0$, it follows from Theorem B that if y is a solution of the equation

$$y'' + q_0 y = 0, \quad y(a) = 1, \quad y'(a) = 0, \quad t \in (a,1).$$

then we have $y(1) > y_0(1) = 0$. Again, it is now easy to construct a first eigenvalue associated with the class $E_{B,n}$ which is too big. Thus we must have $\lambda - p_0 = q_0$.

Repeating this argument on the interval $(0,a)$, we find that the extremal p_0 for the whole interval in the class $E_{B,n}$ must have the form

$$p_0(t) = \begin{cases} \lambda, & t \in J \\ 0, & t \in (0,1) \setminus J, \end{cases}$$

where J is an interval in $(0,1)$ such that $\lambda|J| = B$. For all sufficiently large n, p_0 is independent of n and we have also found the extremal for the class $E_B = \lim E_{B,n}$, $n \to \infty$. To compute Λ, we let $J = (c, c + B/\Lambda)$ and find that

$$y_0(t) = \begin{cases} \cos\sqrt{\Lambda}(c - t), & t \in (0,c) \\ 1, & t \in J \\ \cos\sqrt{\Lambda}(t - c - B/\Lambda), & t \in (c + B/\Lambda, 1). \end{cases}$$

Since $y_0(0) = y_0(1) = 0$ and y_0 is positive in $(0,1)$, it follows that

$$\sqrt{\Lambda} - B/\sqrt{\Lambda} = \pi$$

which gives the expression for Λ in Theorem 1.

II. ON THE EXTREMA OF HIGHER EIGENVALUES

1. The main results

Let I be an interval on $\underline{R}$ and let

$$E(B:I) = \{p \in L^1(I): \int_I p_+ = B\}$$

$$F(B,I) = \{p \in L^1(I): \int_I p_- = B\},$$

where $p_+ = (|p| + p)/2$ and $p_- = (|p| - p)/2$.

For $p \in L^1(0,1)$, we consider the eigenvalue problem

$$(1.1) \qquad -y'' + py = \lambda y, \quad t \in (0,1), \ y(0) = y(1) = 0.$$

Let $\{\lambda_n(p)\}_1^\infty$ be the eigenvalues, arranged in increasing order. Our problem is to find

$$(1.2) \qquad \max_{p \in E(B)} \lambda_n(p) = \Lambda_n^+(B), \quad n = 1, 2, \ldots,$$

$$(1.2) \qquad \inf_{p \in F(B)} \lambda_n(p) = \Lambda_n^-(B), \quad n = 1, 2, \ldots,$$

where $E(B) = E(B, (0,1))$ and $F(B) = F(B, (0,1))$.

The question of determining $\Lambda_1^+(B)$ was discussed in the first part of the present paper and the answer was given in Theorem 1.

We quote a result of G. Talenti [12]:

THEOREM C. $\Lambda_1^-(B)$ <u>is the smallest eigenvalue of the problem</u>

$$(1.4) \qquad -y'' - B\,\delta_{1/2}\, y = \mu y, \quad t \in (0,1), \quad y(0) = y(1) = 0,$$

<u>where</u> $\delta_{1/2}$ <u>is the Dirac measure at</u> $x = 1/2$ <u>and</u> μ <u>is the unique root of the transcendental equation</u>

$$(1.5) \qquad (\sqrt{\mu}/2)^{-1}\tan(\sqrt{\mu}/2) = 4/B, \quad -\infty < \mu < \pi^2.$$

The coefficient in (1.4) is not in $L^1(0,1)$. From [12], we quote some of the remarks of Talenti:

The differential equation in (1.4) must be understood in the sense of distributions: it means that

$$\begin{cases} -y'' = \mu y\,, & t \in (0,1)\,, \quad t \neq 1/2\,, \\ y'(1/2+) - y'(1/2-) + By(1/2) = 0\,, & y(0) = y(1) = 0\,. \end{cases}$$

The left-hand member of (1.5) contains the function

$$(\sqrt{x}/2)^{-1}\tan(\sqrt{x}/2) = \sum_0^\infty 2(\pi^2(n+1/2)^2 - x/4)^{-1} = G(x)\,,$$

which is increasing, real-valued and convex on $(-\infty, \pi^2)$. We have $G(-\infty) = 0$, $G(0) = 1$ and $G(\pi^2-) = \infty$. Thus μ is negative when $B > 4$ and positive when $B < 4$.

Our main results are as follows:

THEOREM 2. $\Lambda_n^+(B) = (n^2/4)(\pi + \sqrt{\pi^2 + 4Bn^{-2}})^2$, $n = 2, 3, \ldots$. *For each* n, *the maximum is assumed if* $p = p_n \in E(B)$, *where* p_n *has period* $1/n$ *and*

$$p_n(t) = \begin{cases} \Lambda_n\,, & \eta_n < t < n^{-1} - \eta_n\,, \\ 0\,, & 0 < t < \eta_n\,, \quad n^{-1} - \eta_n < t < n^{-1}\,, \end{cases}$$

with $\Lambda_n = \Lambda_n^+(B)$ *and* $\eta_n = \pi/(2\sqrt{\Lambda_n^+(B)})$.

COROLLARY 1. *Let* λ_n *be the* n^{th} *eigenvalue of (1.1) where* $p \in E(B)$. *Then for* $n = 2, 3, \ldots,$ *we have*

$$\text{(1.6)} \qquad \sqrt{\lambda_n} \leq (n/2)(\pi + \sqrt{\pi^2 + 4Bn^{-2}}) = n\pi + B(\pi n)^{-1} - B^2(\pi n)^{-3} + \underline{O}(B^3/n^5)\,.$$

REMARK: The uniqueness of the extremal coefficient is discussed at the end of Section 3.

THEOREM 3. $\Lambda_n^-(B) = \mu_n$, $n = 2, 3, \ldots,$ *where* μ_n *is the unique root of the equation*

$$\text{(1.7)} \qquad G(\sqrt{\mu}/2n) = 4n^2/B\,, \quad \text{or equivalently,} \quad \tan(\sqrt{\mu}/2n) = 2n\sqrt{\mu}/B.$$

The infimum is assumed when $p = -p_n^- = -Bn^{-1}\sum_0^{n-1}\delta_i$, *where* δ_i *is the Dirac measure supported by* $n^{-1}(i+1/2)$, $i = 0, 1, \ldots, n-1$.

COROLLARY 2. *Let* λ_n *be the* n^{th} *eigenvalue of (1.1) where* $p \in F(B)$. *Then*

(1.8) $$\sqrt{\lambda_n} \geq n\pi - B(\pi n)^{-1} - B^2(n\pi)^{-3} + \underline{O}(B^3/n^5) .$$

Corollaries 1 and 2 give bounds for eigenvalues when the coefficient p is such that $\|p\|_1 = B$. There are classical results of this type (cf. e.g. Theorem 9 in Ch. 10 in [3]).

REMARK. One of the extremal configurations for the nth eigenvalue for problem (1.1) is in $I_n = (0,1/n)$ also extremal for the first eigenvalue over the classes $E(B/n,I_n)$ or $F(B/n,I_n)$ for the problem

(1.9) $$-y'' + py = \lambda y , \quad t \in I_n , \quad y(0) = y(1/n) = 0 .$$

Let $\Phi: [0,\infty) \to [0,1]$ be a given decreasing functions with $\Phi(0) = 1$ and $\Phi(\infty) = 0$. Let $F(\Phi)$ be the class of all measurable functions p on $(0,1)$ which are such that

$$|\{t \in (0,1): p(t) \geq s\}| = \Phi(s) , \quad 0 \leq s < \infty .$$

In [11], B. Schwarz considers the equation

(1.10) $$y'' + \lambda p y = 0 , \quad t \in (0,1) , \quad y(0) = y(1) = 0 ,$$

and the problem of determining $p \in F(\Phi)$ which gives the supremum or the infimum of the nth eigenvalue for (1.10). The first step in our method for determining $\Lambda_n^+(B)$ and $\Lambda_n^-(B)$ is inspired by the method used by Schwarz in his discussion of the second eigenvalue.

We are grateful to C. Bandle for telling us about this paper of B. Schwarz.

The proof of Theorem 2 is given in Sections 3 and 4. The basic idea should be clear from the discussion of the second eigenvalue in Section 3. The general case is treated in Section 4.

The proof of Theorem 3 is given in Section 5.

The methods of the present paper can also be used to prove analogues of Theorems 2 and 3 when the L^1-norm in the definitions of the classes $E(B)$ and $F(B)$ are replaced by L^r-norms, $r > 1$. This is discussed in a remark by H. Egnell at the end of the paper.

2. Preliminaries

We shall need the following classical results.

A) For each fixed $p \in L^1(0,1)$, the eigenvalues of (1) form a strictly increasing sequence $\{\lambda_n\}_1^\infty$ with $\lim_{n\to\infty} \lambda_n = \infty$. The eigenfunction y_n belonging to the eigenvalue λ_n has exactly $n-1$ zeros in the interval $(0,1)$ and is uniquely determined up to a constant factor (cf. Theorem 5 in Ch. 10 in [3]).

B) If $AC(I)$ is the class of absolutely continuous functions on the interval I, we define

$$X = \{y \in L^2(0,1) : y \in AC((0,1)),\ y' \in L^2(0,1),\ y(0) = y(1) = 0\} .$$

To the boundary value problem in (1.1), we associate the Rayleigh quotient

$$R[y,p] = \int_0^1 ((y')^2 + py^2) / \int_0^1 y^2 ,$$

which is defined when the denominator is positive.

For each $p \in L^1(0,1)$, $R[y,p]$ has a finite lower bound when $y \in X \setminus \{0\}$. Hence Poincaré's principle tells us that (cf. Theorem 3.2 in [2])

$$\lambda_n(p) = \min_{\{L_n\}} \max_{y \in L_n} R[y,p] , \quad n = 1, 2, \ldots ,$$

where L_n is an n-dimensional subspace of X and the minimum is taken over all such subspaces.

A direct consequence is a monotonicity principle: if $p_1 \le p_2$, then

$$\lambda_n(p_1) \le \lambda_n(p_2) , \quad n = 1, 2, \ldots .$$

It follows that when we study the maximum $\Lambda_n^+(B)$, it suffices to study nonnegative coefficients $p \in E(B)$. Similarly, when studying the infimum $\Lambda_n^-(B)$, we can confine ourselves to nonpositive coefficients $p \in F(B)$.

We shall say that p is extremal in $E(B,I)$ if p is the extremal coefficient for the first eigenvalue in the sense of Theorem 1 with $(0,1)$ replaced by I. We shall also say that p is extremal for $F(B,I)$ if p is the extremal coefficient for the first eigenvalue in the sense of Theorem C with $(0,1)$ replaced by I. We note that in the second case, the extremal coefficient is not in the class.

If $p \in L^1(I)$ where $I = (a,b)$ is an interval, $\lambda(p) = \lambda(p,I)$ denotes the first eigenvalue of the problem

$$-y'' + py = \lambda y, \quad t \in I, \quad y(a) = y(b) = 0.$$

The first eigenfunction, normalized so that it has maximum 1 over I, is denoted by $y(p,I)$.

The following construction will be used several times in our proofs. Let $p_1 : I_1 \to \underline{R}$ and $p_2 : I_2 \to \underline{R}$ be such that if y_1 and y_2 are nonnegative eigenfunctions belonging to the first eigenvalue λ of the two problems

$$-y_i'' + p_i y_i = \lambda y_i, \quad t \in I_i, \quad y_i(t) = 0, \quad t \in \partial I_i, \quad i = 1, 2,$$

then y_1 and y_2 have vanishing derivatives at the midpoints of I_1 and I_2, respectively. We shall also assume that y_1 and y_2 take the same value at these two midpoints. A typical example is when the coefficients p_1 and p_2 are extremal in the sense of Theorem 1 in I_1 and I_2.

Forgetting changes depending on possible translations of the intervals I_1 and I_2, we define y to be the combination of y_1 and y_2 in the interval $I = (a,b)$ of length $(|I_1|+|I_2|)/2$ if we have

$$y(t) = \begin{cases} y_1(t), & a < t < a + |I_1|/2, \\ y_2(t), & a + |I_1|/2 \le t < b. \end{cases}$$

Then y is the first eigenfunction of the problem

$$-y'' + py = \lambda y, \quad t \in I, \quad y(a) = y(b) = 0, \tag{2.1}$$

$$p(t) = \begin{cases} p_1(t), & a < t < a + |I_1|/2, \\ p_2(t), & a + |I_1|/2 < t < b. \end{cases}$$

If p_1 and p_2 are symmetric with respect to the midpoints of I_1 and I_2, we have

$$\int_I p = \left(\int_{I_1} p_1 + \int_{I_2} p_2\right)/2. \tag{2.2}$$

What we have done is simply to paste together the left-hand

half of y_1 with the right-hand half of y_2 in such a way that the new function y will be differentiable at $a+|I_1|/2$.

3. The maximum of the second eigenvalue

Without loss of generality, we confine ourselves to studying nonnegative coefficients $p \in E(B)$ (cf. Section 2).

Let $p \in E(B)$, let $\lambda = \lambda_2(p)$ be the second eigenvalue and let y be an eigenfunction belonging to λ, i.e. y is a solution of (1.1) and y has exactly one zero a in $(0,1)$ (cf. A)). We have

$$\lambda = \lambda(p,(0,a)) = \lambda(p,(a,1)) . \tag{3.1}$$

If $\int_0^a p_+ = B_1$ and $\int_a^1 p_+ = B_2$, we let p_1 and p_2 be the extremals in $E(B_1,(0,a))$ and $E(B_2,(a,1))$, respectively, with first eigenvalues $\lambda(p_1)$ and $\lambda(p_2)$. If $\lambda(p_2) > \lambda(p_1)$, say, we consider the function $\psi(s) = \lambda(sp_2)$, $s \in [0,1]$. Since p is nonnegative, it follows from the monotonicity principle for the first eigenvalue on the interval $(0,a)$ that

$$\psi(0) \leq \lambda \leq \lambda(p_1) \leq \psi(1) .$$

By continuity, there exists $s_0 \in [0,1]$ such that $\psi(s_0) = \lambda(p_1)$. Defining

$$\tilde{p} = \begin{cases} p_1 , & t \in (0,a) \\ s_0 p_2 , & t \in (a,1) , \end{cases}$$

we see that

$$\lambda(\tilde{p}) = \lambda(\tilde{p},(0,a)) = \lambda(\tilde{p},(a,1)) = \min(\lambda(p_1),\lambda(p_2)) \geq \lambda_2(p) . \tag{3.2}$$

Let us now introduce the normalized eigenfunctions $y_1 = y(\tilde{p},(0,a))$ and $y_2 = y(\tilde{p},(a,1))$. We define z to be the combination of y_1 and y_2 in the interval $(0,1/2)$. The function z is a solution of the problem

$$-z'' + Pz = \lambda(\tilde{p})z , \quad t \in (0,1/2) , \quad z(0) = z(1/2) = 0 , \tag{3.3}$$

where

$$\int_0^{1/2} P = \int_0^1 \tilde{p}/2 \leq (B_1 + B_2)/2 = B/2 .$$

Using our formal notation, we see that $z = y(P,(0,1/2))$. The definition of P is clear from the construction in Section 2.

Let q be the extremal coefficient in $E(B/2,(0,1/2))$. We choose $\sigma \geq 1$ in such a way that $\int_0^{1/2} \sigma P = B/2$. Then we have

$$(3.4) \qquad \lambda(\tilde{p}) \leq \lambda(P,(0,1/2)) \leq \lambda(\sigma P,(0,1/2)) \leq \lambda(q,(0,1/2)) = \tilde{\lambda} .$$

The final step is to extend q to the whole interval $(0,1)$ via the definition

$$q(t) = q(t-1/2) , \quad t \in (1/2, 1) ,$$

and to let $\tilde{z}$ be the function $y(q,(0,1/2))$ in $(0,1/2)$, extended to the interval $(0,1)$ by reflecting the graph of $\tilde{z}$ in the point $(1/2,0)$. Then $\tilde{z}$ is a solution of the problem

$$(3.5) \qquad -(\tilde{z})'' + q\tilde{z} = \tilde{\lambda}\tilde{z}, \quad t \in (0,1) , \quad \tilde{z}(0) = \tilde{z}(1) = 0 ,$$

$q \in E(B)$ and $\tilde{z}$ has exactly one zero in $(0,1)$. From (3.2) and (3.4), we see that $\lambda_2(p) \leq \tilde{\lambda}$. We have found the maximal second eigenvalue when p varies in the class $E(B)$.

A computation gives $\tilde{\lambda}$ and $\tilde{z}$.

What can we say about uniqueness of the extremal coefficient It is clear from Theorem 1 that if q is extremal in $E(B,I)$, then q and the first eigenfunction of our problem are constant in a subinterval J of I . If $J_1 \subset (0,1/2)$ and $J_2 \subset (1/2,1)$ are these subintervals for the coefficient q in (3.5), it is clear that we can construct a new extremal coefficient $q_1 \in E(B)$ belonging to the eigenvalue $\tilde{\lambda}$ by taking a subinterval J_0 of J_1 and move it next to J_2: we obtain a new solution $\tilde{z}_1$ of the differential equation which is a second eigenfunction of q_1 and which has constant sign in $(0,1/2-|J_0|)$ and in $(1/2-|J_0|,1)$ the support intervals of q_1 are of course symmetric around the centres of these two intervals. Modulo this reshuffling of the support of the extremal coefficient for the second eigenvalue, the coefficient is unique. The situation is analogous for higher eigenvalues.

An informal discussion of these questions is given in [8a] (also cf. the erratum in [8b]).

4. The maximum of the nth eigenvalue, $n = 3, 4, \ldots$.
Proof of Theorem 2.

Without loss of generality we can confine ourselves to studying nonnegative coefficients $p \in F(B)$ (cf. Section 2).

Let $p \in E(B)$, let $\lambda = \lambda_n(p)$ be the nth eigenvalue and let y be an eigenfunction belonging to λ , i.e. y is a solution of (1.1) and y has exactly $n-1$ zeros in $(0,1)$. We assume that $n \geq 3$. This gives us n intervals $\{I_i\}_1^n$ where y has constant sign and vanishes at the endpoints. Let us put $\int_{I_i} p_+ = B_i$, $i = 1, 2, \ldots, n$. Clearly, we have $B = \Sigma_1^n B_i$.

Let p_i be extremal in $E(B_i, I_i)$, $i = 1, 2, \ldots, n$, and let μ be the minimum of the first eigenvalues $\lambda(p_i, I_i)$, $i = 1, 2, \ldots, n$. We note that $\lambda \leq \mu$ and find $s_i \in [0,1]$ such that $\lambda(s_i p_i, I_i) = \mu$, $i = 1, 2, \ldots, n$. As in the case $n = 2$, we consider the normalized eigenfunctions $y_i = y(s_i p_i, I_i)$, $i = 1, 2, \ldots, n$.

Let us assume that the numbering has been done in such a way that $|I_1| \leq |I_2| \leq \ldots \leq |I_n|$. If $|I_1| < |I_n|$, we combine

y_n and y_1 in an interval J_1 of length $(|I_n| + |I_1|)/2$,
y_1 and y_2 in an interval J_2 of length $(|I_1| + |I_2|)/2$,
...
y_{n-1} and y_n in an interval J_n of length $(|I_{n-1}| + |I_n|)/2$.

We place the intervals $\{J_i\}$ in this order on the interval $(0,1)$ and define

$$z(t) = (-1)^{i+1} z_i(t) , \quad i = 1, 2, \ldots, n ,$$

where $z_i(t) = y(P, J_i)$, $i = 1, 2, \ldots, n$ and P is defined in each interval J_i by the construction in Section 2. In particular we have

$$\int_0^1 P = \sum_{i=1}^n \int_{I_i} s_i p_i \leq \int_0^1 p_+ = B .$$

The function z is differentiable at those endpoints of the intervals $J_1, J_2, \ldots, J_n$ which are in $(0,1)$. We conclude that z solves the problem

(4.2) $-z'' + Pz = \mu z\ ,\quad t \in (0,1)\ ,\quad z(0) = z(1) = 0\ ,$

and that z has exactly $n-1$ zeros in $(0,1)$. Furthermore, we know that $\lambda_n(p) \le \mu$.

If $\min_i |J_i| < \max_i |J_i|$, we repeat the construction starting with z instead of y. In this way, we get a sequence $\{\{I_i^{(k)}\}_{i=1}^{n}\}_{k=1}^{\infty}$ such that $\min_i |I_i^{(k)}|$ is increasing in k and bounded from above. In the limit we must have

$$\lim_{k\to\infty} \min_i |I_i^{(k)}| = \lim_{k\to\infty} \max_i |I_i^{(k)}| = 1/n\ .$$

While the eigenvalues $\{\mu_k\}$ of the corresponding boundary value problems of type (4.2) form an increasing sequence, the sequence $\{\int_0^1 P_k\}$ is decreasing. Furthermore, we have $\|P_k\| \le \max_i \|P_i\|_\infty < \infty$ for all k. Using weak* compactness, we find P_0 which is the weak* limit of a subsequence of $\{P_k\}$ and $\mu_0 = \lim_{k\to\infty} \mu_k$. The limit Z of a subsequence of $\{z_k\}$ will be a solution of

$$-Z'' + P_0 Z = \mu_0 Z\ ,\quad t \in (0,1)\ ,\quad Z(0) = Z(1) = 0\ .$$

Furthermore, Z has exactly $n-1$ zeros in $(0,1)$ at i/n, $i = 1, \ldots, n-1$, and $P_0 \in E(\rho B)$ where $\rho \in [0,1]$.

Let $T_i = ((i-1)/n\ ,\ i/n)$ and $C_i = \int_{T_i} P_0$, $i = 1, 2, \ldots, n$. Let q_i be extremal in $E(C_i, T_i)$ and let γ be the minimum of the first eigenvalues $\lambda(q_i, T_i)$, $i = 1, 2, \ldots, n$. It is clear that $\mu_0 \le \gamma$.

From Theorem 1, we see that the maximum of the first eigenvalue is an increasing function of $\int_I p$, where p is the extremal coefficient in $E(\cdot, I)$ (where we assume that $|I|$ is fixed). Let $\Lambda_1^+(B,I)$ denote the maximum of the first eigenvalue over the class $E(B,I)$. Just as in the case $n = 2$, we find numbers $r_i \in [0,1]$ such that $\Lambda_1^+(r_i C_i, T_i) = \gamma$, $i = 1, 2, \ldots, n$. In each interval T_i, we define $\tilde{q}_i$ to be the extremal coefficient in the class $E(r_i C_i, T_i)$. We also define $\tilde{q} = \tilde{q}_i$ in T_i, $i = 1, 2, \ldots, n$, and $\tilde{Z}_i$ as the solution of

$$-(\tilde{Z}_i)'' + \tilde{q}\tilde{Z}_i = \gamma\, \tilde{Z}_i\ ,\quad t \in T_i\ ,\quad \tilde{Z}_i((i-1)/n) = \tilde{Z}(i/n) = 0.$$

In formal notation, we have $\tilde{z}_i = y(\tilde{q}_i, T_i)$.

We know that all intervals T_i have the same length, the form of the coefficients $\tilde{q}_i$ and that these n equations have the same first eigenvalue γ . It is clear from the computations at the end of the proof of Theorem 1 that the eigenvalue γ determines $\int_{T_i} \tilde{q}_i$ and the length of the support of $\tilde{q}_i$ uniquely provided that $|T_i| = 1/n$ for all i . Thus there exist functions Q and V on $(0,1/n)$ such that for $i = 1, 2, \ldots, n$,

$$\tilde{q}(t) = Q(t-(i-1)/n) \ , \quad \tilde{z}_i(t) = V(t-(i-1)/n), \ t \in T_i .$$

If $W(t) = (-1)^{i+1} \tilde{z}_i(t)$, $t \in T_i$, $i = 1, 2, \ldots, n$, we see that $W \in C^1(0,1)$ is a solution of

$$-W'' + \tilde{q}\, W = \gamma W \ , \quad t \in (0,1) \ , \quad W(0) = W(1) = 0 \ ,$$

which changes sign exactly at i/n , $i = 1, 2, \ldots, n-1$. Furthermore, we have $\lambda_n(p) \le \mu \le \mu_0 \le \gamma$ and

$$\int_0^1 \tilde{q} = \sum_i^n \int_{T_i} \tilde{q}_i \le \int_0^1 p_0 \le \int_0^1 p_+ = B \ .$$

Replacing $\tilde{q}$ in each interval T_i by the extremal coefficient in $E(B/n,T_i)$, we get an even larger n^{th} eigenvalue for the associated problem.

We have proved Theorem 2.

5. <u>The infimum of the nth eigenvalue</u>, $n = 2, 3, \ldots$. <u>Proof of Theorem 3</u>.

Without loss of generality, we confine ourselves to studying nonpositive coefficients $p \in F(B)$ (cf. Section 2).

Let $p \in F(B)$, let $\lambda = \lambda_n(p)$ be the nth eigenvalue and let y be an eigenfunction belonging to λ , i.e. y is a solution of (1.1) and y has exactly $n-1$ zeros in $(0,1)$. This gives us n intervals $\{I_i\}_1^n$ where y has constant sign and vanishes at the endpoints. We have $\lambda = \lambda(p,I_i)$, $i = 1, 2, \ldots, n$, and

(5.1) $\qquad \lambda \ge \max \lambda(-p_-, I_i) \ , \quad 1 \le i \le n \ .$

If q is extremal for $F(B,I)$, we know that $q = -B\delta_c$ where c is the midpoint of I.

If $\int_{I_i} p_- = B_i$, $i = 1, 2, \ldots, n$, we have $B = \Sigma_1^n B_i$. Let $p_i = -B_i \delta_{x_i}$ be extremal for $F(B_i, I_i)$ and let μ be the maximum of the first eigenvalues $\lambda(p_i, I_i)$. We note that $\lambda \geq \mu$. The next step is to find $s_i \in [0,1]$ such that $\lambda(s_i p_i, I_i) = \mu$, $i = 1, 2, \ldots, n$. Let $y_i = y(s_i p_i, I_i)$ be the normalized eigenfunctions, $i = 1, 2, \ldots, n$.

Let us assume that $|I_1| < |I_n|$ and that the numbering has been done in such a way that $|I_1| \leq |I_2| \leq \ldots \leq |I_n|$. Slightly extending our definition in Section 2, we combine

y_n and y_1 in an interval J_1 of length $(|I_n| + |I_1|)/2$
y_1 and y_2 in an interval J_2 of length $(|I_1| + |I_2|)/2$
. . .
y_{n-1} and y_n in an interval J_n of length $(|I_{n-1}| + |I_n|)/2$.

This means that the Dirac measure at a point in the interval J_k will have the coefficient $-(s_{k-1}B_{k-1} + s_k B_k)/2$, $k = 1, 2, \ldots, n$, (where $B_0 = B_n$).

REMARK. What we do is to paste together the left-hand half of the first function with the right-hand half of the second function. The new function will be continuous at the point where we change from the first to the second function. It will not be differentiable there since there is a Dirac measure in the associated differential equation.

We place the intervals $\{J_i\}_1^n$ in this order on $(0,1)$ and define

$$z(t) = (-1)^{i+1} z_i(t), \quad i = 1, 2, \ldots, n,$$

where $z_i(t) = y(P, J_i)$ for all i and P is defined above. In particular, we have

$$\int_0^1 P = \sum_1^n \int_{I_i} s_i p_i \geq -\sum_1^n B_i .$$

Arguing in the same way as in Section 4, we find $\mu_0 \leq \lambda$, P_0 and Z such that

$$-Z'' + P_0 Z = \mu_0 Z \;, \quad t \in (0,1) \;, \quad Z(0) = Z(1) = 0 \;,$$

where Z has exactly $n-1$ zeros in $(0,1)$ at i/n, $i = 1, 2, \ldots, n-1$, P_0 is nonpositive and $\int_0^1 P_0 \geq -B$. We can now continue as in Section 4 and conclude that the lower bound for the nth eigenvalue is assumed when the coefficient in the differential equation is p_n^- as defined in Theorem 3.

It is clear from our discussion that the extremal coefficient is unique. We have proved Theorem 3.

6. The extremum problem for L^r-norms, $r > 1$

This section has been written by H. Egnell.

The method given above works also for the following variant of our problem. Let

$$E_r(B) = \{p \in L^r(0,1) : \int_0^1 p_+^r = B^r\} \;,$$

$$F_r(B) = \{p \in L^r(0,1) : \int_0^1 p_-^r = B^r\} \;.$$

For $r > 1$, we wish to determine

(i) $$\max_{p \in E_r(B)} \lambda_n(p) = \Lambda_{n,r}^+(B) \;,$$

(ii) $$\min_{p \in F_r(B)} \lambda_n(p) = \Lambda_{n,r}^-(B) \;,$$

where $\{\lambda_n(p)\}_1^\infty$ are the eigenvalues of problem (1.1).

This problem has been studied in [1] and the special case $r = 2$ is given as an exercise (with hints) in [13]. The following result on the first eigenvalue is given in [4].

THEOREM D. _If_ $r > 1$ _and_ $r' = r/(r-1)$, _then_

$$\Lambda_{1,r}^+(B) = \inf_{u \in X} J_{B,r}^+(u) \;, \quad \Lambda_{1,r}^-(B) = \inf_{u \in X} J_{B,r}^-(u) \;,$$

where $X = H_0^1([0,1])$ _as before and_

$$J_{B,r}^+(u) = \Big(\int_0^1 (u')^2 + B\big(\int_0^1 |u|^{2r'}\big)^{1/r'}\Big) / \int_0^1 u^2 \;, \quad J_{B,r}^-(u) = J_{-B,r}^+(u) \;.$$

Furthermore, if u _is a nonnegative minimizer of_ $J_{B,r}^+$, _normalized so that_ $\|u\|_{2r'} = 1$, _then_ $p = \pm B u^{2(r'-1)}$ _gives the optimum_ *in (i) or (ii), respectively.*

Applying the method of proof in Theorems 2 and 3 and a simple change of coordinates, we find the solution of our problem:

THEOREM I. *If* $r > 1$, *then*

$$\Lambda^+_{n,r}(B) = n^2\,\Lambda^+_{1,r}(B/n^2)\ ,\quad \Lambda^-_{n,r}(B) = n^2\,\Lambda^-_{1,r}(B/n^2)\ ,\quad n = 2, 3, \ldots .$$

If u *is a minimizer of* $J^+_{B/n^2,r}$ *or* $J^-_{B/n^2,r'}$ *normalized so that* $\|u\|_{2r'} = 1$, *then the extremum in (i) or (ii) is assumed when* $p = p_n$ *has period* $1/n$ *and*

$$p_n(t) = \pm B|u(tn)|^{2(r'-1)}\ ,\quad t \in (0,1/n)\ ,$$

where the upper sign refers to the maximum problem and the lower sign refers to the minimum problem. Furthermore, these optimizers are unique.

COROLLARY I. $$\Lambda^+_{n,r} = (n\pi)^2 + 2B\left(\int_0^1 |\sin(\pi t)|^{2r'}dt\right)^{1/r'} + O(n^{-2})$$

$$\Lambda^-_{n,r} = (n\pi)^2 - 2B\left(\int_0^1 |\sin(\pi t)|^{2r'}dt\right)^{1/r'} + O(n^{-2})$$

Theorems D and I give the optimal n^{th} eigenvalue as the minimum of a functional. The relation between the extremal n^{th} eigenvalue and the extremal first eigenvalue enables us to deduce the asymptotic formulas in Corollary I.

In [1], Ashbaugh and Harrell give the nonlinear differential equation associated with the optimal n^{th} eigenvalue but there is no explicit formula for eigenvalues of the type given in Theorems D and I.

If $r = 1$, there is uniqueness only for the minimizer (cf. the proof of Theorem 2). To make the discussion at the end of Section 3 more complete, we give the following characterization of the maximizers: for functions $p \in E(B)$, we have $\lambda_n(p) = \Lambda^+_n(B)$ if and only if

$$p = \Lambda^+_n(B)\left\{\sum_{i=1}^{n} \chi(I_i)\right\}\ ,$$

where $\chi(I)$ denotes the characteristic function of the interval

I and we have

$$\text{dist}\ (I_i, I_{i+1}) = \pi/\sqrt{\Lambda_n^+(B)}\ , \quad i = 1, 2, \ldots, n-1\ ,$$

$$\text{dist}\ (0, I_1) = \text{dist}\ (I_n, 1) = \pi/(2\sqrt{\Lambda_n^+(B)})\ .$$

We assume that the intervals $\{I_i\}_1^n$ have been numbered in increasing order from left to right.

REFERENCES

1. M.S. Ashbaugh and E.M. Harrell II. Maximal and minimal eigenvalues and their associated nonlinear equations. Preprint 1985.

2. C. Bandle, Isoperimetric inequalities and applications. Pitman, London 1980.

3. G. Birkhoff and G.-C. Rota, Ordinary differential equations. 2nd ed. Xerox College Publ., Lexington, Mass. 1969.

4. H. Egnell, Extremal properties of the first eigenvalue of a class of elliptic eigenvalue problems. To appear, Ann. della Scuola Norm. Sup. di Pisa.

5. M. Essén, Optimization and rearrangements of the coefficient in the operator $d^2/dt^2 - p(t)^2$ on a finite interval. J. of Math. Anal. Appl. 115 (1986), 278-304.

6. M.Essén, Optimization and α-disfocality for ordinary differential equations. Can. J. of Math. XXXVII (1985), 310-323.

7. M. Essén, Optimization and rearrangements of the coefficient in the differential equation $y'' \pm qy = 0$. C.R. Acad. Sci. Canada VI (Febr. 84), 15-20.

8a. E.M. Harrell II, Hamiltonian operators with maximal eigenvalues. J. of Math. Phys. 25 (1984), 48-51.

8b. E.M. Harrell II, Erratum to [8a]. J. of Math. Phys. 27 (1986) 419.

9. A.G. Ramm, Question 5 (Part 2). Notices of the Amer. Math. Soc. 29 (1982), 328-329.

10. H.L. Royden, Real analysis, 2nd ed. MacMillan 1968.

11. B. Schwarz, On the extrema of the frequencies of nonhomogeneous strings with equi-measurable density. J. of Math. and Mech. 10 (1961), 401-422.

12. G. Talenti, Estimates for eigenvalues of Sturm-Liouville

problems. General Inequalities 4 (ed. by W. Walter), 341-350. Birkhaäuser 1984.

13. E. Trubowitz, Book in preparation.

Department of Mathematics
University of Uppsala
Thunbergsvägen 3
S-752 38 Uppsala, Sweden

International Series of
Numerical Mathematics, Vol. 80

LANDAU'S INEQUALITY FOR THE DIFFERENTIAL AND DIFFERENCE OPERATORS

Man Kam Kwong and A. Zettl

Abstract. We survey known results on the best constants in Landau's inequality relating norms of a function y and its first two derivatives y', y'' and between a sequence x and its first two differences Δx, $\Delta^2 x$. Some of these have been known for decades; others have just recently been found and are not yet published while still others have not been found yet. Extremals are also discussed as well as some related results.

1. INTRODUCTION.

We consider the inequalities

$$\|y'\|^2 < K\|y\|\ \|y''\| \tag{1.1}$$

and

$$\|\Delta x\|^2 < C\|x\|\ \|\Delta^2 x\|. \tag{1.2}$$

The norm in (1.1) may be any one of the classical $L^p(J)$ norms with $J = R^+ = (0,\infty)$ or $J = R = (-\infty,\infty)$:

$$\|y\|_p^p = \int_J |y(t)|^p dt, \qquad 1 < p < \infty,$$

$$\|y\|_\infty = \text{ess. sup } |y(t)|, \qquad t \in J.$$

The norm in (1.2) may be any one of the classical $\ell^p(M)$ norms with $M = Z = \{\ldots -2,-1,0,1,2,\ldots\}$ or $M = N = \{0,1,2,\ldots\}$:

$$\|x\|_p^p = \sum_{j\in M} |x_j|^p, \quad 1 < p < \infty,$$

This paper is in final form and no version of it will be submitted for publication elsewhere.

$$\|x\|_\infty = \sup|x_j|, \quad j \in M.$$

Here

$$x = (x_j) \quad j \in M,$$

$$\Delta x = (x_{j+1}-x_j) \quad j \in M,$$

$$\Delta^2 x = \Delta(\Delta x).$$

It is well known - see e.g. [Kwong & Zettl 1980] - that, given, $1 \leq p \leq \infty$, and $J = R$ or $J = R^+$ there exists a constant K such that (1.1) holds for all y in Y where

$$Y = Y(p,J) = \{y \in L^p(J): \ y' \text{ is absolutely continuous on all compact subintervals of } J, \ y'' \in L^p(J)\}.$$

Similarly it is known [Gindler & Goldstein 1981] that, given p, $1 \leq p \leq \infty$, and $M = Z$ or $M = N$ there exists a positive number C such that (1.2) holds for all x in $\ell^p(M)$. (Note that $x \in l^p(M)$ implies that $\Delta x \in \ell^p(M)$.)

Clearly then there is a smallest i.e. best such constant K and a best such constant C. These are denoted by $K = K(p,J)$ and $C = C(p,M)$ to emphasize their dependence on these quantities. The problem of determining the exact values of $K(p,J)$ and $C(p,M)$ has a long and distinguished history dating back at least as far as a paper of Landau in 1913 [Landau 1913]. It has received the attention of many authors. The reader is referred to the references of the papers mentioned below.

No values of $K(p,R)$, $K(p,R^+)$, $C(p,Z)$, $C(p,N)$ are known for any p other than $p = 1,2$, or ∞. Below we will summarize the known values of C and K (some of which have only recently been found), discuss some upper and lower bounds, and survey the state of knowledge regarding extremals. Also discussed are the possible values that the norms of y, y', y'' or x, Δx, $\Delta^2 x$ can assume. A number of open problems are mentioned. We believe that many of these not only are "accessible" but can be successfully attacked using only elementary methods, in the sense that they require only a knowledge of calculus and the basics of real variable theory including knowledge of the L^p and ℓ^p spaces.

2. KNOWN CONSTANTS.

In this section we summarize the known values of $K(p,J)$ and $C(p,M)$. No proofs are given here but we do give references where proofs can be found. These references are not necessarily the original ones.

$\underline{p = 1}$.

$C(1,Z) = 2$	Kwong and Zettl [1986]
$K(1,R) = 2$	Ditzian [1975]
$C(1,N) = 5/2$	Kwong and Zettl [1986]
$K(1,R^+) = 5/2$	Berdyshev [1971]

$\underline{p = 2}$.

$C(2,Z) = 1$	Copson [1979]
$K(2,R) = 1$	Hardy, Littlewood and Polya [1932]
$C(2,N) = 2$	Copson [1979]
$K(2,R^+) = 2$	Hardy and Littlewood [1932]

$\underline{p = \infty}$.

$C(\infty,Z) = 2$	Ditzian and Neumann [1986]
$K(\infty,R) = 2$	Hadamard [1914]
$C(\infty,N) = 4$	Certain and Kurtz [1977]
$K(\infty,R^+) = 4$	Landau [1913]

REMARK 2.1. *Note that in all above cases* $C = K$. *This leads to the questions*:

1. *Is* $K(p,R) = C(p,Z)$ *for all* p, $1 \le p \le \infty$?
2. *Is* $K(p,R^+) = C(p,N)$ *for all* p, $1 \le p \le \infty$?

While it may be tempting to conclude that the answer to both questions is yes and this may indeed be the case, it is interesting to note that for higher order versions of (1.1) and (1.2):

$$\|y^{(k)}\|^n \le K\|y\|^{n-k}\,\|y^{(n)}\|^k$$

$$\|\Delta^k x\|^n \le C\|x\|^{n-k}\,\|\Delta^n x\|^k$$

it is known [Ditzian 1983] that $C > K$ for certain values of n

and k when $p = \infty$.

3. EXTREMALS.

For the case $p = 2$ we have the following result:

THEOREM 3.1. (a) *There is no extremal for* $K(2,R)$.
(b) *The function* $y(t) = \exp(-t/2)\sin(\sqrt{3}t/2-\pi/3)$, $t > 0$ *is an extremal of* $K(2,R^+)$ *and all other extremals are given by*

$$ay(bt),\ a \in R,\ b \in R^+.$$

Proof. See [Hardy, Littlewood and Polya 1934] for part (a). For part (b) we give, following Kwong and Zettl [1980] an elementary proof of the statement on extremals and in addition show that $K(2,R^+) = 2$. It is sufficient to consider real valued functions.

Let $f, f' \in L^2(0,\infty)$. Then $\lim_{t\to\infty} f(t) = 0$ and

$$\int_0^\infty f(t)f'(t)dt = \lim_{t\to\infty}\int_0^t f(t)f'(t)dt$$
$$= \lim_{t\to\infty} \tfrac{1}{2}[f^2(t) - f^2(0)]$$
$$= -\tfrac{1}{2} f^2(0) \leq 0.$$

Let $y, y', y'' \in L^2(0,\infty)$. Then for any c in R^+ it is easy to verify the identity:

$$c^2|y'|^2 + (c^2y''+cy'+y)^2 = c^4|y''|^2 + |y|^2 + 2c(cy'+y)'(cy'+y).$$

Integrating over $(0,\infty)$ gives

$$(3.2)\qquad c^2\|y'\|_2^2 + \|c^2y'' + cy' + y\|_2^2 = c^4\|y''\|_2^2 + \|y\|_2^2 + 2c\int_0^\infty (cy'+y)'(cy'+y)dt$$

As seen above, the last integral is non-positive. Thus

$$c^2\|y'\|_2^2 \leq c^4\|y''\|_2^2 + \|y\|_2^2 .$$

Choosing $c^2 = \|y\|_2/\|y''\|_2$ gives

$$\|y'\|_2^2 \leq 2\|y\|_2 \; \|y''\|_2.$$

Thus $K(2,R^+) \leq 2$. Equality holds if and only if the terms left out in (3.2) are zero, namely

$$c^2 y'' + cy' + y = 0.$$

and

$$cy'(0) + y(0) = 0.$$

Solving this initial value problem gives the extremals as stated in (b). Therefore $K(2,R^+) = 2$ and (b) holds.

There are no extremals for $K(1,R)$ and $K(1,R^+)$ - see [Kwong and Zettl 1980]. However, there are extremals for $K(1,R)$ and $K(1,R^+)$ in a certain generalized sense. See Berdyshev [1971] or Kwong and Zettl [1980].

When $p = \infty$ the whole line case $K(\infty,R)$ has extremals, one example of which is an odd periodic function of period 4 given by

$$y(t) = \begin{cases} t(2-t), & 0 \leq t \leq 2 \\ (t-2)(t-4), & 2 \leq t \leq 4 \; . \end{cases}$$

This is an example of an Euler spline. Clearly $ay(bt)$ for any $a, b \in R$ is another extremal. There are more. Replace that part of y on $[1,\infty)$ with the constant function 1 and you have an essentially different extremal.

More generally, replace y on $[1,\infty)$ by any other function such that y on $(0,\infty)$ is absolutely continuous on compact subsets of $(0,\infty)$, bounded by 1 on $[1,\infty)$ and whose first and second derivatives on $[1,\infty)$ are not too large and you get another extremal. Thus it is clear that there are infinitely many linearly independent extremals in this case.

The half-line case $K(\infty,R^+)$ has the following extremal [Landau 1913]:

$$y(t) = \begin{cases} -1 + 6(t-c)^2, & 0 \le t \le c \\ -1 + 6(t-c)^2 - 8(t-c)^3 + 3(t-c)^4, & c \le t \le 1 + c \\ 0, & 1 + c \le t < \infty \end{cases}$$

where $c = 1/\sqrt{3}$.

There are other extremals besides $ay(bt)$, $a \in R$, $b \in R^+$. For instance, we can replace that part of y on $[1 + c,\infty)$ with another function which is bounded by 1, absolutely continuous on compact subsets of $(0,\infty)$ and whose first and second derivatives are not too large.

In the case of inequality (1.2) there is no extremal when $p = 2$ for either $C(2,Z)$ or $C(2,N)$. See Copson [1979] for an elementary proof.

When $p = \infty$ an extremal for $C(\infty,Z)$ is the 4-periodic sequence $\{\ldots,-1,-1,1,1,-1,-1,1,1,\ldots\}$. The question of the existence of an extremal for $C(\infty,N)$ is still open.

The existence of extremals when $p = 1$ for either $C(1,Z)$ or $C(1,N)$ seems to be an open problem.

4. BOUNDS.

Since, for $1 < p < \infty$, $p \neq 2$ the values of the constants $K(p,J)$ and $C(p,M)$ are not known the question of finding upper and lower bounds arises. It is well known [Ditzian 1983] that

$$C(p,Z) \ge K(p,R),\ C(p,N) \ge K(p,R^+),\ 1 \le p \le \infty.$$

Thus any lower bound of K is also a lower bound of C.

THEOREM 4.1. (Kwong and Zettl) *For any* $p \neq 2$, $1 \le p \le \infty$, *we have*

$$1 < K(p,R) \le C(p,Z),\ K(p,R) < K(p,R^+) \le C(p,R^+).$$

Proof. See Kwong and Zettl [1980].

THEOREM 4.2. (Franco, Kaper, Kwong and Zettl [1983].) *The best constant* $K(p,R)$ *is bounded above by* $U(p)$ *where*

$$U(p) = 2, \; 1 < p \le 3/2$$
$$U(p) = (p-1)^{-1}, \; 3/2 \le p \le 2$$
$$U(p) = (q-1)^{(2-q^n)q^{-n}} \left(\prod_{i=1}^{n} \left(\frac{q}{q^i - 1} - 1\right)^{q^{-i}} \right)^{2(q-1)},$$
$$2 < p < \infty, \; p^{-1} + q^{-1} = 1,$$

$n = [(\log_2 q)^{-1}]$ *where* $[\cdot]$ *denotes the greatest integer function. Note that* $n = 0$ *if* $1 < p < 2$, $n \ge 1$ *if* $2 \le p < \infty$ *and* $q^k \le 2$ *for all* $k = 1,2,\ldots,n$ *but* $q^{n+1} > 2$.

For purposes of illustration we list a few values of $U(p)$:

$$U(3) = 2^{1/3}$$
$$U(4) = \left(\frac{15}{7}\right)^{3/8} \doteq 1.22082962$$
$$U(5) = 4^{47/125}\left(\frac{11}{9}\right)^{8/25}\left(\frac{19}{61}\right)^{32/125} \doteq 1.33222966$$
$$U(6) = 5^{19/128}\left(\frac{19}{11}\right)^{5/18}\left(\frac{59}{91}\right)^{25/108} \doteq 1.39745611.$$

A table of numerically computed values of $U(p)$ and $L(p)$ is given in [Franco et al 1983].

5. VALUES OF THE NORMS OF y, y', y'' AND x, Δx, $\Delta^2 x$.

PROPOSITION 1. *Given* p *with* $1 \le p \le \infty$, $J = R$ *or* $J = R^+$ *and* $u_0 > 0$, $u_1 > 0$ *there exists a function* y *in* $Y(p,J)$ *such that*

$$\|y\|_p = u_0, \; \|y'\|_p = u_1. \tag{5.1}$$

Proof. This is elementary. Let y be a smooth function with compact support and consider $ay(bt)$. It can be shown by a direct and straightforward computation that a and b can be chosen so that (5.1) holds when $1 \le p < \infty$. The case $p = \infty$ is clear.

Can (5.1) be extended to

$$\|y\|_p = u_0, \; \|y'\|_p = u_1, \; \|y''\|_p = u_2? \tag{5.2}$$

In other words can the norms of y, y', y'' be assigned arbitrary positive values? That the answer to this question is no follows immediately from inequality (1.1). Can all values of $\|y\|$, $\|y'\|$ and $\|y''\|$ not ruled out by inequality (1.1) be achieved? The next result answers this question except, possibly, for the case of equality in (1.1).

THEOREM 5.1 (Ljubic 1964) *Let* $1 \leq p < \infty$, *let* $J = R$ *or* $J = R^+$. *If* u_i *are positive numbers satisfying*

$$u_1^2 < K(p,J)u_0u_2 \tag{5.3}$$

then there exists a function y *in* $Y(p,J)$ *such that*

$$\|y^{(i)}\|_p = u_i, \quad i = 0,1,2. \tag{5.4}$$

Before we give a proof we state

COROLLARY 1. *If* $u_1^2 = K(p,J)u_0u_2$ *then there is a function* y *satisfying* (5.2) *if and only if* $K(p,J)$ *has an extremal. Thus, in this case* (5.2) *holds when* $J = R^+$ *for* $1 < p \leq \infty$ *but not for* $p = 1$; *when* $J = R$ (5.2) *holds for* $p = \infty$ *but not for* $1 \leq p < \infty$.

Proof. We give a proof which is (i) different from that of Ljubic, (ii) elementary and (iii) new. The case $p = \infty$ is clear. Since p, $1 \leq p < \infty$, is fixed throughout this proof we omit the subscript p on the norm symbol.

Let

$$Q(y) = \|y'\|^2/(\|y\|\,\|y''\|) \quad \text{for } y \in Y(p,J), \quad y'' \not\equiv 0.$$

CLAIM. The range of Q contains the open interval $(0,K)$, $K = K(p,J)$. Let

$$Q(y_1) = a, \quad Q(y_2) = b, \quad a < b.$$

Then y_1 and y_2 are linearly independent. Note that Q restricted to the two dimensional set $S = \text{span}\,\{y_1,y_2\} - \{0\}$ is continuous. Since the continuous image of a connected set is connected it follows that $[a,b] \subset$ range of Q. From the definition of K it follows that b can be taken arbitrarily close to K. The claim is established if we can show that a can be taken arbitrarily small. Choose a smooth function y with compact support in R^+ whose graph is as follows:

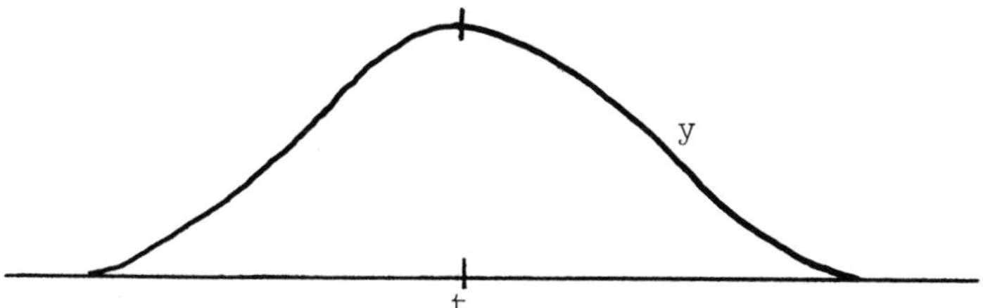

For any $h > 0$ let y_h be the function obtained from y by pulling apart its graph at $(t_0, y(t_0))$ a distance h:

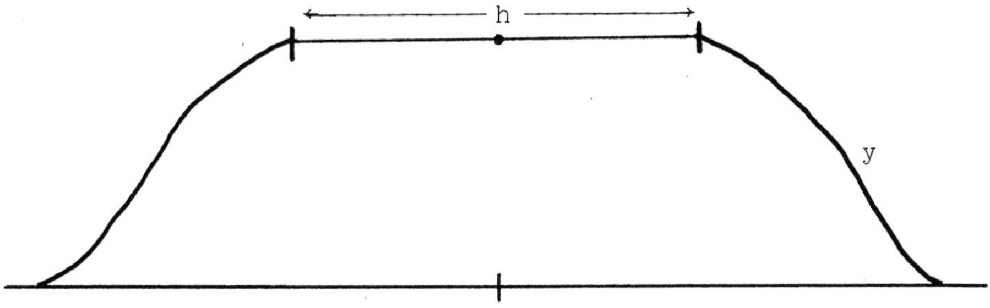

Then $\|y_h^{(i)}\| = \|y^{(i)}\|$, $i = 1,2$. Thus $Q(y_h)$ can be made arbitrarily small by choosing h and t_0 large enough.

Now choose z such that

$$Q(z) = u_1^2/u_0u_2$$

and consider $z_{ab}(t) = az(bt)$ for $a > 0$, $b > 0$. A simple computation shows that a and b can be chosen so that $\|z_{ab}\| = u_0$ and $\|z'_{ab}\| = u_1$. Since $Q(z_{ab}) = Q(z)$ it follows that $\|z''_{ab}\| = u_2$ and the proof is complete.

In contrast to (5.1) for functions y, for sequences x in ℓ^p the values of $\|x\|_p$ and $\|\Delta x\|_p$ cannot be assigned arbitrarily. This is easily seen. Let x be in $\ell^\infty(Z)$, $x = \{x_j\}$ with $\|x\|_\infty \le 1$. Then clearly $\|\Delta x\|_\infty \le 2$ since each x_j must satisfy: $-1 \le x_j \le 1$. Thus, for example, there is no sequence x in $1^\infty(Z)$ such that

$$\|x\|_\infty = 1 \quad \text{and} \quad \|\Delta x\|_\infty = 3.$$

For sequences x a more interesting question is the following:

QUESTION. Let $1 \le p \le \infty$, $M = Z = \{\ldots -2,-1,0,1,2,\ldots\}$ or $M = N = \{0,1,2,\ldots\}$. If x is $\ell^p(M)$, with $\|x\|_p \le 1$ what are all the possible values of $\|\Delta x\|_p$?

THEOREM 5.2 <u>If</u> $x \in \ell^p(M)$ <u>and</u> $\|x\|_p \le \ell$, <u>then</u>

(i) $\|\Delta x\|_p \le 2$,

(ii) $\|\Delta x\|_p$ <u>assumes all values in the interval</u> $(0,2)$ <u>as</u> x <u>ranges over</u> $\|x\|_p \le 1$;

(iii) <u>For both</u> $M = Z$ <u>and</u> $M = N$, <u>there exists</u> x <u>such that</u> $\|\Delta x\|_p = 2$ <u>if and only if</u> $p = \infty$.

Proof. Case 1. $p = \infty$. Let $x = \{\ldots 0,0,-r,r,0,0\ldots\}$ when $M = Z$ and when $M = N$ let $x = \{0,0,-r,r,0,0,\ldots\}$. In either case $\|x\|_\infty = r$ and $\|\Delta x\|_\infty = 2r$. Thus when r ranges from 0 to 1, $\|x_\infty\| \le 1$ and $\|\Delta x\|_\infty$ ranges from 0 to 2. In particular for $r = 1$ we have $\|x\|_\infty = 1$ and $\|\Delta x\| = 2$.

Case 2. $1 \le p < \infty$. Let $M = Z$. Choose $x = \{\ldots,0,0,-r \cdot a, ra, -ra, ra,\ldots,ra,0,0,\ldots\}$ where $r > 0$, $a = n^{-1/p}$ and there are n nonzero terms. Then

$$\|x\|_p = r$$

and

$$\|\Delta x\|_p^p = \frac{1}{n}\,[2+(n-1)2^p]r^p.$$

Thus $\|\Delta x\|_p \to 2r$ as $n \to \infty$. Therefore as r ranges from 0 to 1, $\|\Delta x\|_p$ assumes all values between 0 and 2.

To show that the norm of x cannot exceed 2, consider the shift operator L defined by

$$L\{x_j\} = \{x_{j+1}\}.$$

Then for any x in $\ell^p(M)$ with $\|x\|_p \leq 1$.

(5.5) $$\|\Delta x\| = \|x - Lx\| \leq \|x\| + \|Lx\| \leq 2, \quad 1 \leq p \leq \infty.$$

In case $p = \infty$ we saw from the simple example above that there exists an x in $\ell^\infty(M)$ for $M = Z$ or $M = N$ such that

$$\|x\|_\infty = 1 \quad \text{and} \quad \|\Delta x\|_\infty = 2.$$

CLAIM. For $1 \leq p < \infty$ if $\|x\|_p \leq 1$ then $\|\Delta x\|_p < 2$. In other words there is no "extremal" i.e. there is no $x \in \ell^p(M)$ with $\|x\|_p \leq 1$ such that $\|\Delta x\|_p = 2$. To prove this assume that for some x in $\ell^p(M)$ with $\|x\|_p \leq 1$ we have

$$\|\Delta x\|_p = 2.$$

Then from (5.5)

$$2 = \|\Delta x\|_p \leq \|Lx\|_p + \|x\|_p \leq 2$$

we may conclude that

$$\|\Delta x\|_p = \|Lx\|_p + \|x\|_p = 2$$

and

$$\|x\|_p = 1, \quad \|Lx\|_p = 1.$$

From equality in the triangle inequality $\|\Delta x\|_p = \|Lx\|_p + \|x\|_p$ it follows that x and Lx are linearly dependent and thus L must have an eigenvalue. But for $M = Z$, L has no eigenvalues. For $M = N$ let $x = \{x_j\}_{j=0}^\infty$. Then $\|x\|_p^p = |x_0|^p + \|Lx\|_p^p$. Hence $x_0 = 0$, since $\|x\|_p = \|2x\|_p$. This and $Lx = \lambda x$ implies $x_j = 0$ for all $j = 0, 1, 2, \ldots$ completing the proof when $N = Z$. For $M = N$ the proof is entirely similar.

Notes added in May 1986:

1. We have shown that the equality $C(p,Z) = K(p,R)$ does not hold for all p. This answers question 1 on p. 3.
2. We have shown that for each of the three constants $C(1,Z)$, $C(1,N)$ and $C(\infty,N)$ there is no extremal.

REFERENCES

1. M. K. Kwong and A. Zettl [1980], "Ramifications of Landau's inequality", Proc. Roy. Soc. Edinbrugh, 86A(1980), pp. 175-212.

2. Z. M. Franco, H. G. Kaper, M. K. Kwong and A. Zettl [1983], "Bounds for the best constant in Landau's ineqality on the line", Proc. Roy. Soc. Edinburgh, 95A(1983), pp. 257-262.

3. H. G. Kaper and B. E. Spellman [1986], "Best constants in norm inequalities for the difference operator", pre-print.

4. Z. Ditzian [1983], "Discrete and shift Kolmogorov type inequalities", Proc. Roy. Soc. Edinburg, 93A(1983), pp. 307-317.

5. H. A. Gindler and J. A. Goldstein [1981], "Dissipative operators and series inequalities", Bull. Austr. Math. Soc. 23(1981), pp. 429-442.

6. Z. Ditzian and D. J. Newman [1986], "Discrete Kolmogorov - type inequalities", preprint.

7. Z. M. Franco, H. G. Kaper, M. K. Kwong and A. Zettl [1985], "Best constants in norm inequalities for derivatives on a half-line", Proc. Roy. Soc. Edinburgh, 100A(1985), pp. 67-84.

8. H. G. Kaper and B. E. Spellman [1986], "Best constants in norm inequalities for the difference operator", pre-print.

9. E. T. Copson [1979], "Two series inequalities", Proc. Roy. Soc. Edinburgh 83A(1979), 109-114.

10. E. Landau [1913], "Einige Ungleichungen fur zweimal differenzierbare Funktionen", Proc. London Math. Soc.]3(1983), 43-49.

11. Z. Ditzian [1975], "Some remarks on inequalities of Landau and Kolmogorov", Eqn. Math. 12(1975), 145-151.

12. V. I. Berdyshev (1971], "The best approximation in $L(0,\infty)$ to the differentiation operator", Matematicheskie Zametki 5(1971), 477-481.

13. G. H. Hardy and J. E. Littlewood [1932]. "Some integral inequalities connected with the calculus of vairations", Quart. J. Math. Oxford Ser. 2, 3, (1932), 241-252.

14. M. W. Certain and T. G. Kurtz [1977], "Landau-Kolmogorov inequalities for semigroups and groups", Proc. Amer. Math. Soc. 63(1977), 226-230.

15. G. H. Hardy, J. E. Littlewood and G. Polya [1934], "Inequalities", Cambridge, 1934.

16. Ju. I. Ljubic [1964], "On inequalities between the powers of a linear operator", Transl. Amer. Math. Soc. Ser. (2) 40(1964), 39-84; translated from Akad. Nauk. SSSR Ser. Mat. 24(1960), 825-864.

Department of Mathematical Sciences
Northern Illinois University
DeKalb, IL 60115 USA

International Series of
Numerical Mathematics, Vol. 80

EIN EXISTENZSATZ FÜR GEWÖHNLICHE DIFFERENTIALGLEICHUNGEN IN GEORDNETEN BANACHRÄUMEN

Herrn Jan Błaż zum 60. Geburtstage gewidmet

Roland Lemmert, Raymond M. Redheffer* und Peter Volkmann

Abstract. An existence theorem for ordinary differential equations in Banach spaces will be given, where the right hand side is monotone increasing with respect to a cone.

1. EINLEITUNG

Nach A.N. Godunov [4] läßt sich der Existenzsatz von Peano (wenn nichts weiter als die Stetigkeit der rechten Seite der Differentialgleichung gefordert wird) auf keinen unendlich-dimensionalen Banachraum übertragen. Aus diesem Grunde verdienen Existenzsätze für gewöhnliche Differentialgleichungen in Banachräumen ein gewisses Interesse; eine grobe Übersicht über solche Sätze findet man in [10].

Nun sei E ein reeller Banachraum und $K \subseteq E$ ein Kegel (vgl. M.G. Kreĭn und M.A. Rutman [6]). Dann wird in E durch die Festsetzung

$$x \le y \iff y-x \in K \qquad (x,y \in E)$$

eine (reflexive, antisymmetrische, transitive) Ordnung $\le$ erzeugt. Zunächst sei K regulär im Sinne von M.A. Krasnosel'skiĭ [5], d.h. aus

---

*Anläßlich eines Gastaufenthaltes an der Universität Karlsruhe unter der Schirmherrschaft der Deutschen Forschungsgemeinschaft.

This paper is in final form and no version of it will be submitted for publication elsewhere.

$$x_1 \le x_2 \le x_3 \le \dots \le b$$

in E folgt die Konvergenz $x_n \to x_o$ (im Sinne der Norm) gegen ein Grenzelement $x_o \in E$. Unter diesen Bedingungen gilt nach V.Ja. Stecenko [7] ein Existenzsatz für das Anfangswertproblem

$$u(0) = a\,, \quad u' = f(t,u) \quad (0 \le t \le T)$$

in E, wobei f(t,x) bezüglich der Banachraum-Veränderlichen x als monoton wachsend vorausgesetzt wird. Der Beweis erfolgt mit Hilfe von sukzessiven Approximationen.

Mit einer beliebigen Menge A gilt nach [9] ein entsprechender Satz für den Banachraum

$$E = \ell_\infty(A) = \{x \mid x = (x_\alpha)_{\alpha \in A},\ x_\alpha \in \mathbb{R}\ (\alpha \in A),\ \|x\| = \sup_{\alpha \in A} |x_\alpha| < \infty\}$$

der auf A definierten, beschränkten, reellwertigen Funktionen, wenn dieser durch den (im Falle einer unendlichen Menge A nichtregulären) Kegel

$$K = \ell_\infty^+(A) = \{x \mid x = (x_\alpha)_{\alpha \in A} \in E,\ \ x_\alpha \ge 0\ (\alpha \in A)\}$$

geordnet ist. Der Beweis in [9] kann als Anwendung eines Fixpunktsatzes von A. Tarski ([8], Satz 1) aufgefaßt werden. (Abschnitt 3 auf S.203 von [9] ist übrigens falsch.)

Im Folgenden werden die genannten Ergebnisse zu einem einheitlichen Existenzsatze zusammengefaßt; dabei werden Banachräume E betrachtet, die der folgenden Voraussetzung entsprechen:

(V) <u>Es sei A eine Menge, und für jedes $\alpha \in A$ sei K_α ein regulärer Kegel in einem Banachraume E_α. Dann wird</u>

$$E = \{x \mid x = (x_\alpha)_{\alpha \in A},\ x_\alpha \in E_\alpha\ (\alpha \in A),\ \|x\| = \sup_{\alpha \in A} \|x_\alpha\| < \infty\}$$

<u>gebildet, und dieser Raum wird geordnet durch den Kegel</u>

$$K = \{x \mid x = (x_\alpha)_{\alpha \in A} \in E,\ x_\alpha \in K_\alpha\ (\alpha \in A)\}\ .$$

Ein Spezialfall ist z.B. $A = \{1\} \cup A'$ mit $E_\alpha = \mathbb{R}$, $K_\alpha = [0,\infty)$ für $\alpha \in A'$ (und $1 \notin A'$). Dann ergibt sich $E = E_1 \times \ell_\infty(A')$, $K = K_1 \times \ell_\infty^+(A')$, wobei E_1 ein Banachraum ist, welcher durch einen regulären Kegel K_1 geordnet wird.

Die oben erwähnten Beweise (aus [7] bzw. [9]) lassen sich nun nicht auf den Fall (V) übertragen. Daher wird unser Existenzsatz auf den nachstehenden Fixpunktsatz zurückgeführt, den man bei N. Bourbaki [2] sowie (mit Beweis) auch bei N. Dunford und J.T. Schwartz [3] findet, und dessen Beweis nach einer in [3] gemachten Anmerkung auf E. Zermelos zweiten Beweis [11] des Wohlordnungssatzes zurückgeht.

FIXPUNKTSATZ. *Es sei* $\leq$ *eine (reflexive, antisymmetrische, transitive) Ordnung in einer Menge* Ω, *und es gelte:*

(P1) $\Omega \neq \emptyset$.

(P2) *Ist* C *eine Kette in* Ω (*d.h. für* $\gamma, \delta \in C$ *ist* $\gamma \leq \delta$ *oder* $\delta \leq \gamma$), $C \neq \emptyset$, *so existiert* $\sup C \in \Omega$.

Ferner sei Φ *eine (mindestens) auf* Ω *definierte Funktion mit folgenden Eigenschaften:*

(P3) $\Phi(\Omega) \subseteq \Omega$.

(P4) $\omega \leq \Phi\omega \quad (\omega \in \Omega)$.

Dann besitzt Φ *in* Ω *einen Fixpunkt.*

Mit Hilfe dieses Satzes wird in [3] das Zornsche Lemma (nebst anderen Sätzen) aus dem Auswahlaxiom hergeleitet. Der Beweis des Fixpunktsatzes selbst kommt ohne Auswahlaxiom aus, und dementsprechend wird auch für unseren Existenzsatz das Auswahlaxiom nicht benötigt.

2. ZWEI LEMMATA

Das folgende Lemma 1 findet man bei V.A. Bondarenko [1]; wir geben hier einen ausführlicheren Beweis.

LEMMA 1. *Der Banachraum* E *sei durch einen regulären Kegel* K *geordnet, und es sei* C *eine Kette in* E, $C \neq \emptyset$. *Ferner sei* C *nach oben beschränkt, d.h. mit einem* $b \in E$ *sei*

$$x \leq b \qquad (x \in C) .$$

Dann existiert

(1) $$\bar{x} = \sup C ,$$

und zu jedem $\varepsilon > 0$ gibt es ein $x_\varepsilon \in C$ mit der Eigenschaft

(2) $$\|\bar{x}-x\| < \varepsilon \quad (x \in C,\ x \geq x_\varepsilon) .$$

Beweis. 1. Zunächst gibt es zu jedem $\varepsilon > 0$ ein $y_\varepsilon \in C$ mit der Eigenschaft

(3) $$\|x-y_\varepsilon\| < \varepsilon \quad (x \in C,\ x \geq y_\varepsilon) ,$$

denn anderenfalls müßte ein $\varepsilon > 0$ existieren, so daß für jedes $y \in C$ mit einem $x = x(y) \in C$, $x(y) \geq y$ die Ungleichung

$$\|x(y)-y\| \geq \varepsilon$$

richtig wäre. Wählt man $y_1 \in C$ beliebig und setzt man rekursiv

$$y_{n+1} = x(y_n) \quad (n = 1,2,3,\dots) ,$$

so erhält man sowohl

(4) $$y_1 \leq y_2 \leq y_3 \leq \dots \leq b$$

als auch

(5) $$\|y_{n+1}-y_n\| \geq \varepsilon \quad (n = 1,2,3,\dots) .$$

Da K regulär ist, folgt aus (4) die Konvergenz der Folge (y_n), aber das steht im Widerspruch zu (5).

2. Gemäß (3) gibt es ein $y_1 \in C$ mit

$$\|x-y_1\| < 1 \quad (x \in C,\ x \geq y_1) .$$

Da für $y \in C$ die Menge $C(y) = \{x \mid x \in C,\ x \geq y\}$ wieder eine nichtleere, nach oben beschränkte Kette ist, gilt (3) auch für $C(y)$ an Stelle von C, und dementsprechend können

$$y_1, y_2, y_3, \dots$$

rekursiv durch die Forderungen

(6) $$y_{n+1} \in C(y_n) , \quad \|x-y_{n+1}\| < \frac{1}{n+1} \quad (x \in C(y_n),\ x \geq y_{n+1})$$

bestimmt werden. Insbesondere gilt

$$y_1 \leq y_2 \leq y_3 \leq \dots \leq b ,$$

also existiert $\bar{x} = \lim_{n\to\infty} y_n$. Es läßt sich mit Hilfe von (6) leicht bestätigen, daß $\bar{x}$ die Eigenschaften (1), (2) besitzt.

LEMMA 2. Für den Banachraum E und seinen Ordnungskegel K sei (V) erfüllt. Es sei $T > 0$, $M \geq 0$ und C eine nichtleere Kette von Lip_M-Funktionen ω: $[0,T] \to E$; es gelte also:

1. Für $\omega, \tilde{\omega} \in C$ ist $\omega \leq \tilde{\omega}$ (d.h. $\omega(t) \leq \tilde{\omega}(t)$ für alle $t \in [0,T]$) oder $\tilde{\omega} \leq \omega$.

2. Jedes $\omega \in C$ genügt der Lipschitz-Bedingung

$$\|\omega(s)-\omega(t)\| \leq M|s-t| \quad (s,t \in [0,T]) .$$

Ferner existiere b: $[0,T] \to E$ und $\beta \in \mathbb{R}$ mit

$$(7) \qquad \omega(t) \leq b(t) , \quad \|\omega(t)\| \leq \beta \qquad (\omega \in C, \ 0 \leq t \leq T) .$$

Dann existiert

$$(8) \qquad \bar{\omega}(t) = \sup_{\omega \in C} \omega(t) \qquad (0 \leq t \leq T) ,$$

und $\bar{\omega}$: $[0,T] \to E$ ist eine Lip_M-Funktion.

Beweis. 1. Für $0 \leq t \leq T$, $\omega \in C$ ist zunächst

$$\omega(t) = (\omega_\alpha(t))_{\alpha \in A} \quad \text{mit} \quad \omega_\alpha(t) \in E_\alpha \quad (\alpha \in A) ,$$

und entsprechend gilt $b(t) = (b_\alpha(t))_{\alpha \in A}$ mit $b_\alpha(t) \in E_\alpha$. Nun ist für $0 \leq t \leq T$ und $\alpha \in A$ die Menge

$$C_\alpha(t) = \{\omega_\alpha(t) \mid \omega \in C\}$$

eine Kette in E_α, welche wegen (7) durch $b_\alpha(t)$ nach oben beschränkt ist. Nach Lemma 1 existiert also

$$(9) \qquad \bar{\omega}_\alpha(t) = \sup C_\alpha(t) .$$

Hieraus folgt mit Lemma 1 und (7)

$$\|\bar{\omega}_\alpha(t)\| \leq \beta ,$$

und daher ist

$$(10) \qquad \bar{\omega}(t) = (\bar{\omega}_\alpha(t))_{\alpha \in A}$$

(für jedes $t \in [0,T]$) ein Element des Raumes E. Aus (9), (10) folgt (8) ohne weiteres.

2. Es seien $s,t \in [0,T]$, und es sei $\alpha \in A$. Wegen (9) gibt es nach Lemma 1 zu $\varepsilon > 0$ ein $\omega^1 \in C$ mit

$$\|\bar{\omega}_\alpha(t) - \omega_\alpha(t)\| < \varepsilon \qquad (\omega \in C,\ \omega \geq \omega^1) \quad . \tag{11}$$

Entsprechend gibt es ein $\omega^2 \in C$ mit

$$\|\bar{\omega}_\alpha(s) - \omega_\alpha(s)\| < \varepsilon \qquad (\omega \in C,\ \omega \geq \omega^2) \quad . \tag{12}$$

Verwendet man (11), (12) für $\omega = \max\,\{\omega^1, \omega^2\}$: $[0,T] \to E$, so ergibt sich mit der Lipschitz-Bedingung für diese Funktion

$$\begin{aligned} \|\bar{\omega}_\alpha(s) - \bar{\omega}_\alpha(t)\| &\leq \|\bar{\omega}_\alpha(s) - \omega_\alpha(s)\| + \|\omega_\alpha(s) - \omega_\alpha(t)\| + \|\omega_\alpha(t) - \bar{\omega}_\alpha(t)\| \\ &\leq \varepsilon + M|s-t| + \varepsilon \ , \end{aligned}$$

und $\varepsilon \searrow 0$ liefert dann

$$\|\bar{\omega}_\alpha(s) - \bar{\omega}_\alpha(t)\| \leq M|s-t| \ .$$

Damit ist $\bar{\omega}_\alpha \in \mathrm{Lip}_M$, also wird durch (10) eine Lip_M-Funktion $\bar{\omega}$: $[0,T] \to E$ erklärt.

3. DER EXISTENZSATZ

Sind $\omega, \tilde{\omega}$: $[0,T] \to E$ Funktionen mit Werten in einem geordneten Banachraume E und gilt $\omega(t) \leq \tilde{\omega}(t)$ für alle $t \in [0,T]$, so schreiben wir (wie schon in Lemma 2) kurz $\omega \leq \tilde{\omega}$.

SATZ. Für den Banachraum E und seinen Ordnungskegel K sei die Voraussetzung (V) erfüllt. Es sei $T > 0$, $a \in E$, und es seien v,w: $[0,T] \to E$ stetige Funktionen mit $v \leq w$. Ferner sei

$$\Delta = \{(t,x) \mid 0 \leq t \leq T,\ v(t) \leq x \leq w(t)\}$$

und $f(t,x)$: $\Delta \to E$ eine stetige, beschränkte, bezüglich x wachsende Funktion; es gelte also

$$(t,x), (t,y) \in \Delta,\ x \leq y \implies f(t,x) \leq f(t,y) \ . \tag{13}$$

Für stetige ω: $[0,T] \to E$ mit $v \leq \omega \leq w$ sei noch

$$(\Phi\omega)(t) = a + \int_0^t f(s,\omega(s))\,ds \qquad (0 \leq t \leq T) \ , \tag{14}$$

und es werde

$$v \leq \Phi v, \Phi w \leq w \tag{15}$$

vorausgesetzt. Dann besitzt das Anfangswertproblem

(16) $u(0) = a\,, \quad u'(t) = f(t,u(t)) \quad (0 \le t \le T)$

eine kleinste Lösung $u = \underline{u}$ und eine größte Lösung $u = \bar{u}$.

Beweis. 1. Es wird also die Existenz einer kleinsten und einer größten Lösung der zu (16) äquivalenten Integralgleichung

(17) $u = \Phi u$

behauptet. Wir zeigen nur die Existenz von $\underline{u}$; die Existenz von $\bar{u}$ kann ebenso erledigt werden. Nach Voraussetzung ist

$$M = \sup_{(t,x)\in\Delta} \|f(t,x)\|$$

endlich. Setzt man noch

$$\beta = \|a\| + MT\,,$$

so liegen die Werte des Operators Φ in der Funktionenmenge

$$\Sigma = \{\omega \mid \omega\colon [0,T] \to E,\ \omega \in \mathrm{Lip}_M,\ \|\omega(t)\| \le \beta\ (0 \le t \le T)\}\,.$$

Außerdem gilt nach (13) für Elemente ω, $\tilde{\omega}$ des Definitionsbereiches von Φ die Implikation

(18) $\omega \le \tilde{\omega} \quad \Rightarrow \quad \Phi\omega \le \Phi\tilde{\omega}\,.$

2. Die Menge

$$H = \{h \mid h\colon [0,T] \to E \text{ stetig},\ v \le h \le w,\ \Phi h \le h\}$$

ist nicht leer (es ist $w \in H$), und die Menge

$$\Omega = \{\omega \mid \omega \in \Sigma,\ \omega \le \Phi\omega,\ v \le \omega \le h\ (h \in H)\}$$

erfüllt die Voraussetzungen (P1) - (P4) des in der Einleitung angegebenen Fixpunktsatzes ((P1), (P2) werden weiter unten bewiesen; der Nachweis von (P3) ist so einfach, daß er übergangen wird; (P4) gilt gemäß Definition von Ω). Demzufolge gilt für ein $\underline{u} \in \Omega$ die Beziehung

$$\underline{u} = \Phi\underline{u}\,,$$

also ist $\underline{u}$ eine Lösung von (17). Jede weitere Lösung u von (17) liegt in H, und mit Hilfe der Definition von Ω ergibt sich dann $\underline{u} \le u$, also ist $\underline{u}$ die kleinste Lösung von (17).

3. Es sind noch (P1), (P2) nachzuweisen. Dazu sei zunächst bemerkt, daß für eine stetige Funktion ω: $[0,T] \to E$ mit

(19) $$v \le \omega \le w$$

auch

(20) $$v \le \Phi\omega \le w$$

ausfällt. Aus (19) folgt nämlich wegen (18) $\Phi v \le \Phi\omega \le \Phi w$, und hieraus ergibt sich (20) mit Hilfe von (15).

Zu (P1): Aus $v \le \Phi v$ folgt

(21) $$\Phi v \le \Phi(\Phi v) .$$

Aus $v \le h$ $(h \in H)$ folgt $\Phi v \le \Phi h$, wegen $\Phi h \le h$ gilt also

(22) $$v \le \Phi v \le h \quad (h \in H) .$$

Da noch $\Phi v \in \Sigma$ ist, ergibt sich aus (21), (22) die Beziehung $\Phi v \in \Omega$.

Zu (P2): Es sei C eine Kette in Ω, $C \neq \emptyset$. Wegen $\Omega \subseteq \Sigma$ ist dann C eine in Σ gelegene Kette, die nach oben durch w beschränkt ist. Nach Lemma 2 wird also durch

$$\bar{\omega}(t) = \sup_{\omega \in C} \omega(t) \quad (0 \le t \le T)$$

eine Lip_M-Funktion $\bar{\omega}$: $[0,T] \to E$ erklärt. Man bestätigt nun ohne große Mühe, daß $\bar{\omega}$ in Ω liegt und daß $\bar{\omega} = \sup C$ gilt.

Damit ist der Satz bewiesen.

Nachstehende Folgerung ergibt sich auf übliche Weise.

FOLGERUNG. Für den Banachraum E und seinen Ordnungskegel K sei (V) erfüllt; überdies besitze K innere Punkte. Es sei $T > 0$, $U \subseteq E$ und a ein innerer Punkt von U. Die Funktion $f(t,x)$: $[0,T] \times U \to E$ sei stetig und bezüglich x wachsend. Dann existiert $T_1 \in (0,T]$, so daß das Anfangswertproblem

$$u(0) = a , \quad u' = f(t,u)$$

eine Lösung u: $[0,T_1] \to E$ besitzt.

Zum Beweise kann f als beschränkt vorausgesetzt werden (sonst werde U verkleinert). Man wähle p aus dem Inneren von K.

Für ein hinreichend kleines $\lambda > 0$ gilt

$$\{x \mid x \in E,\ a-\lambda p \le x \le a+\lambda p\} \subseteq U .$$

Man setze

$$v(t) = a-\lambda p , \quad w(t) = a+\lambda p \qquad (0 \le t \le T) .$$

Mit Φ gemäß (14) gilt dann für ein hinreichend kleines $T_1 \in (0,T]$

$$v(t) \le (\Phi v)(t), (\Phi w)(t) \le w(t) \qquad (0 \le t \le T_1) .$$

Daher ist der Satz (mit T_1 an Stelle von T) anwendbar.

LITERATUR

1. V.A. Bondarenko, Integral'nye neravenstva dlja uravnenija Vol'terra v banahovom prostranstve s konusom. Mat. Zametki 9 (1971), 151-160.

2. N. Bourbaki, Éléments de mathématique I: Théorie des ensembles, Fascicule de résultats. Hermann & Cie., Paris, 1939.

3. N. Dunford und J.T. Schwartz, Linear operators. Part I: General theory. Interscience Publishers, Inc., New York, 1957.

4. A.N. Godunov, O teoreme Peano v banahovyh prostranstvah. Funkcional'. Analiz Priložen. 9 (1975), Nr. 1, 59-60.

5. M.A. Krasnosel'skiǐ, Pravil'nye i vpolne pravil'nye konusy. Doklady Akad. Nauk SSSR 135 (1960), 255-257.

6. M.G. Kreǐn und M.A. Rutman, Lineǐnye operatory, ostavljajuščie invariantnym konus v prostranstve Banaha. Uspehi Mat. Nauk 3 (1948), Nr. 1(23), 3-95.

7. V.Ja. Stecenko, K-pravil'nye konusy. Doklady Akad. Nauk SSSR 136 (1961), 1038-1040.

8. A. Tarski, A lattice-theoretical fixpoint theorem and its applications. Pacific J. Math. 5 (1955), 285-309.

9. P. Volkmann, Équations différentielles ordinaires dans les espaces des fonctions bornées. Czechoslovak Math. J. 35(110) (1985), 201-211.

10. ——————, Existenzsätze für gewöhnliche Differentialgleichungen in Banachräumen. Mathematica ad diem natalem septuagesimum quintum data, Festschrift Ernst Mohr zum 75. Geburtstag, Fachbereich Math. Techn. Univ. Berlin (1985), 271-287.

11. E. Zermelo, Neuer Beweis für die Möglichkeit einer Wohlordnung. Math. Ann. 65 (1908), 107-128.

Roland Lemmert und Peter Volkmann, Mathematisches Institut I, Universität Karlsruhe, Postfach 6980, 7500 Karlsruhe 1, Westdeutschland.

Raymond M. Redheffer, Department of Mathematics, University of California, Los Angeles, CA 90024, U.S.A.

International Sereis of
Numerical Mathematics, Vol. 80

OPTIMAL BOUNDS FOR THE CRITICAL VALUE IN A SEMILINEAR BOUNDARY VALUE PROBLEM ON A SURFACE

René P. Sperb

Abstract. In this paper two optimal inequalities are proven for the critical value λ^* of the semilinear boundary value problem $\Delta u+\lambda f(u)=0$ in $\Omega \subset M$, $u=0$ on $\partial\Omega$. Here M is a two-dimensional Riemannian manifold with positive Gaussian curvature.

1. INTRODUCTION

The simplest model for a reaction-diffusion process under steady-state conditions leads to the problem

$$(1.1) \qquad \Delta u + \lambda f(u) = 0 \quad \text{in } \Omega, \quad u = 0 \quad \text{on } \partial\Omega.$$

In almost all cases of importance u is a nonnegative quantity (a concentration for an example). The kinetics of the reaction is described by the nonlinearity f(u) and the reaction vessel is some bounded domain Ω in R^n. Finally the positive parameter λ usually involves the diffusion coefficient and the rate constant of the reaction.

In this work we consider the case in which Ω is a bounded domain on a two-dimensional Riemannian manifold M (a surface). It should be noted that the "nonhomogeneous problem"

$$(1.2) \qquad \Delta u + \lambda p(x) f(u) = 0 \quad \text{in } \Omega, \ u = 0 \quad \text{on } \partial\Omega,$$

with $p(x) > 0$, $x =$ generic point in $\Omega \subset R^2$, is equivalent to (1.1). One just uses the fact that

---

This paper is in final form and no version of it will be submitted for publication elsewhere.

$$\frac{\Delta}{p(x)}$$

is the Laplace-Beltrami operator of a manifold whose metric tensor is

$$g_{ik} = p(x)\,\delta_{ik}.$$

We will confine ourselves to an important type of nonlinearity satisfying

(H) $\qquad f(0) > 0$ and f is increasing and convex.

For such nonlinearities it is known (see. e.g. the review article of Lions [2]) that the following situation occurs:

There is a critical value λ^* of λ such that there is at least one positive solution for $\lambda\in(0,\lambda^*)$ and no positive solution if $\lambda > \lambda^*$.

The main interest here is in obtaining optimal lower bounds for the critical value λ^*. Two such bounds will be derived: in Theorem 1 a lower bound for λ^* is derived which is best possible in the sense that equality holds if Ω is a geodesic strip on a sphere and in Theorem 2 the optimal domain is a geodesic disk on a sphere.

These bounds will be derived by the method of sub- and supersolutions. A supersolution $\bar{u}$ in problem (1.1) is any $C^2(\Omega)\wedge C^0(\bar{\Omega})$-function satisfying

(1.3) $\qquad \Delta\bar{u} + \lambda f(\bar{u}) \le 0 \quad$ in Ω, $\bar{u} \ge 0$ on $\partial\Omega$,

and for a subsolution $\underline{u}$ the inequality signs have to be reversed. It is known (see e.g. Amann [1]) that if $\underline{u}$ and $\bar{u}$ exist and $\underline{u} \le \bar{u}$ in Ω then there is also at least one solution u of (1.1) and one has

$$\underline{u} \le u \le \bar{u}.$$

For nonlinearities satisfying (H) we can always select $\underline{u} \equiv 0$ so that it remains to find a supersolution then.

In this work an idea of Payne [3] is extended to the case

under consideration. Payne used a supersolution of the form $\bar{u}(x) = v(t(x))$ where $t(x)$ is the solution of

(1.4) $$\Delta t + 1 = 0 \quad \text{in} \quad \Omega \subset R^2 \quad \text{on} \quad \partial\Omega,$$

and the function $v(t)$ is to be chosen appropriately. The main ingredient used was the fact that the function

$$P_1 = |\nabla t|^2 + t$$

assumes its maximum on $\partial\Omega$, while for convex domains the function

$$P_2 = |\nabla t|^2 + 2t$$

takes its maximum at a point of Ω where $\nabla t = 0$. Here the essential result to be used is that the quantities corresponding to P_1, P_2 are now

$$P_j = e^{jKt(x)}\left(|\nabla t|^2 + \frac{1}{K}\right), \quad j = 1, 2,$$

with K denoting a lower bound for the Gaussian curvature of M.

2. LOWER BOUNDS FOR λ^*

Let Ω be a bounded domain on a two-dimensional Riemannian manifold M and denote by K_G the Gaussian curvature of M and by k_g the geodesic curvature of $\partial\Omega$.

We first compare the solution of (1.1) for $\Omega \subset M$ with the solution for a geodesic strip S on a sphere of radius R, i.e. with the solution of

(2.1) $$\frac{1}{R^2}\,\frac{1}{\sin\vartheta}\,\frac{d}{d\vartheta}\left(\sin\vartheta\frac{dw}{d\vartheta}\right) + \lambda f(w) = 0 \quad \text{in} \quad \left(\vartheta_0, \frac{\pi}{2}\right),$$

with the boundary conditions

(2.2) $$\frac{dw}{d\vartheta}\left(\frac{\pi}{2}\right) = 0 = w(\vartheta_0).$$

The following result then holds

THEROEM 1. Assume that f satisfies (H) and furthermore (i) $K_G \geq K > 0$, $k_g \geq k$ and $K|\Omega| + |\partial\Omega| \geq 0'$ where $|\cdot|$ denotes the corresponding measure, (ii) $\vartheta_0 = \frac{\pi}{2} - \rho k^{1/2}$, with ρ = radius of

<u>largest geodesic circle contained in</u> Ω <u>and</u> $R = K^{-1/2}$.

$$\lambda^*(\Omega) \geq \lambda^*(S).$$

<u>Proof.</u> We first introduce a new variable s by setting

$$K^{1/2}s = \operatorname{ctg}\vartheta$$

and define

$$v(s) = w(\vartheta(s)),$$

$w(\vartheta)$ being a solution of (2.1). A straightforward calculation shows that $v(s)$ is a solution of

$$(2.3) \qquad (1+Ks^2)^{3/2}\frac{d}{ds}\left((1+Ks^2)^{1/2}\frac{dv}{ds}\right) + \lambda f(v) = 0 \quad \text{in } (0,s_0)$$

$$(2.4) \qquad \frac{dv}{ds}(0) = 0 = v(s_0),$$

with

$$K^{1/2}s_0 = \operatorname{ctg}\vartheta_0.$$

We now show that the function

$$\bar{u}(x) = v(s(x))$$

with

$$(2.5) \qquad s(x) = \left[\frac{1}{K}\left(e^{2K(t_m - t(x))} - 1\right)\right]^{1/2}$$

is a supersolution of problem (1.1). Here $t(x)$ is the solution of

$$(2.6) \qquad \Delta t + 1 = 0 \quad \text{in } \Omega \subset M, \quad t = 0 \quad \text{on } \partial\Omega,$$

and

$$t_m = \max t(x).$$

We calculate

$$\Delta\bar{u} = v'\Delta s + v''\,|\nabla s|^2,$$

with a prime for differentiation with respect to s and Δ and ∇ denoting Laplace-Beltrami operator and gradient on M, respectively. Further one finds

$$\nabla s = -\frac{1}{s}e^{2K(t_m-t)}\nabla t = -\left(\frac{1}{s} + Ks\right)\nabla t$$

$$\Delta s = -\left(\frac{1}{s} + Ks\right)\Delta t + \left(\frac{1}{s^2} - K\right)\nabla s\cdot\nabla t$$

$$= \left(\frac{1}{s} + Ks\right)\left(1 + \left(K - \frac{1}{s^2}\right)\right)|\nabla t|^2$$

At this point we make use of the differential equation for v(s) to find that

$$(2.7)\qquad \Delta\bar{u} + \lambda f(\bar{u}) = \left(v'\left(\frac{1}{s} + Ks\right) + \lambda f(v)\right)\left(1 - \frac{|\nabla t|^2}{s^2}\right).$$

It was proven in [4] that if assumption (i) of Theorem 1 holds then one has

$$(2.8)\qquad |\nabla t|^2 \le \frac{1}{K}\left(e^{2K(t_m-t)} - 1\right) = s^2.$$

It remains to show then that the other term in braces on the right of equation (2.7) is nonpositive. To this end we consider the function

$$(2.9)\qquad g(s) = s(1 + Ks^2)^{-1/2}\left(v'\left(\frac{1}{s} + Ks\right) + \lambda f(v(s))\right)$$

Obviously $g(0) = 0$ and using the differential equation (2.3) again it follows that

$$(2.10)\qquad g'(s) = -\lambda f(v)\,(1 + Ks^2)^{-3/2} + [\lambda s(1 + Ks^2)^{-1/2} f(v(s))]'$$

$$= \lambda s(1 + Ks^2)^{-1/2} v' \frac{df}{dv} \le 0$$

since $\frac{df}{dv} \ge 0$ was assumed in (H) and $v'(s) \le 0$ in $(0,s_0)$ as is easy to see from (2.3), (2.4). Hence $\bar{u}$ satisfies

$$\Delta\bar{u} + \lambda f(\bar{u}) \le 0 \quad \text{in } \Omega,$$

and in addition $\bar{u} = 0$ on $\partial\Omega$ if we choose

$$s_0 = [\tfrac{1}{K}(e^{2Kt_m} - 1)]^{1/2}. \tag{2.11}$$

For any value λ for which (2.1), (2.2), or equivalently (2.3), (2.4) have a positive solution there is at least one solution of problem (1.1) (with $\Omega \subset M$), and hence

$$\lambda^*(\Omega) \geq \lambda^*(S).$$

The endpoint s_0 given by (2.11) still involves the unknown quantity t_m.

Another simple supersolution argument shows that λ^* in (2.3) is a decreasing function of s_0. We may therefore use any upper bound for t_m to get a more explicit expression for s_0 or, equivalently, ϑ_0. It was shown in [4] that if (i) holds then

$$t_m \leq -\tfrac{1}{K}\log(\cos(\rho K^{1/2})), \tag{2.12}$$

$\rho =$ radius of the largest disk contained in Ω. Inserting this bound in (2.11) one gets

$$\vartheta_0 = \frac{\pi}{2} - \rho K^{1/2},$$

which completes the proof of Theorem 1.

REMARK. If Ω is a convex domain in the plane we can use similar arguments in the limiting case $K \to 0$. We may use the limiting cases of (2.3), (2.4) and (2.11) directly. One is then led to

COROLLARY. *Let Ω be a convex plane domain. Then one has*

$$\lambda^*(\Omega) \geq \lambda^*(S),$$

where S is a strip of width $2d$ with $d = \sqrt{2t_m}$.
As before an upper bound for t_m is needed. A number of such bounds are known, see e.g. [5].

In the special case $f(u) = e^u$ the inequality in the Corollary can be written as

$$(2.13)\qquad \lambda^*(\Omega) \geq \frac{0.44}{t_m},$$

with equality for a strip. If Ω is the unit disk it is known that $\lambda^* = 2$, while (2.13) gives $\lambda^* \geq 1.76$.

In Theorem 1 the optimal domain is a geodesic strip on a sphere. In the next result the optimal domain is a geodesic disk on a sphere.

THEOREM 2. *Assume that* f *satisfies* (H) *and* (i) $K_G \geq K > 0$, $k_g \geq k$ (ii) *Let* $\lambda^*(D)$ *be the critical value of a geodesic disk, i.e. of the problem*

$$(2.14)\qquad \frac{1}{R^2}\frac{1}{\sin\vartheta}\frac{d}{d\vartheta}(\sin\vartheta\frac{dw}{d\vartheta}) + \lambda f(w) = 0 \quad \text{in} \quad (0,\vartheta_1)$$

$$(2.15)\qquad \frac{dw}{d\vartheta}(0) = 0 = w(\vartheta_1)$$

with

$$\vartheta_1 = 2\arctan[(1+\frac{k^2}{K})^{1/2} - \frac{k}{K^{1/2}}],\quad R = K^{-1/2}.$$

Then

$$\lambda^*(\Omega) \geq \lambda^*(D)$$

holds.

Proof. We use a new variable z defined by

$$(2.16)\qquad K^{1/2}z = 2\tan\frac{\vartheta}{2}.$$

Then the function $v(z) = w(\vartheta(z))$ satisfies

$$(2.17)\qquad (1+\frac{Kz^2}{4})\frac{1}{z}\frac{d}{dz}(z\frac{dv}{dz}) + \lambda f(v) = 0$$

$$(2.18)\qquad \frac{dv}{dz}(0) = 0 = v(z_1)$$

with the obvious definition of z_1.

We now choose as a supersolution for (1.1) $\bar{u}(x) = v(z(x))$ with

$$(2.19)\qquad z(x) = [\frac{4}{K}(K\tau^2+1)e^{-Kt} - 1]^{1/2},\quad \tau = \max_{\partial\Omega}|\nabla t|.$$

One now finds that

$$\Delta z = \left(\frac{2}{z} + \frac{1}{2}Kz\right)\left[1 + \left(\frac{K}{2} - \frac{2}{z^2}\right)|\nabla t|^2\right] \tag{2.20}$$

Using (2.17) one is led to

$$\Delta\bar{u} + \lambda f(\bar{u}) = \left[v'\left(\frac{2}{z} + \frac{K}{2}z\right) + \lambda f(v)\right]\left(1 - \frac{4}{z^2}|\nabla t|^2\right). \tag{2.21}$$

It was shown in [4] that the function

$$P_1 = \left(|\nabla t|^2 + \frac{1}{K}\right)e^{Kt}$$

assumes its maximum on $\partial\Omega$ which is equivalent to the inequality

$$1 - \frac{4}{z^2}|\nabla t|^2 \geq 0 \quad \text{in } \Omega. \tag{2.22}$$

It remains to show that the first quantity in braces on the right of (2.21) is nonpositive. To this end consider

$$g(z) = \left[v'\left(\frac{2}{z} + \frac{K}{2}z\right) + \lambda f(v(z))\right]\frac{Kz^2}{4 + Kz^2}. \tag{2.23}$$

Clearly $g(0) = 0$ and using (2.17) one finds

$$g'(z) = v'\left(\frac{\lambda z^2}{4 + Kz^2}\frac{df}{dv} + \frac{1}{2}\right) \leq 0,$$

implying that $g(z) \leq 0$, that is

$$\Delta\bar{u} + \lambda f(\bar{u}) \leq 0 \quad \text{in } \Omega.$$

We choose the endpoint in (2.18) as $z_1 = 2\tau$, and the same reasoning as in the proof of Theorem 1 shows that we may take any upper bound for τ. We then use the bound given in [4] stating that under assumption (i) one has

$$\frac{z_1}{2} = \tau \leq \left[\left(\frac{k}{K}\right)^2 + \frac{1}{K}\right]^{1/2} - \frac{k}{K}, \tag{2.24}$$

with equality for a geodesic circle on a sphere. Solving for ϑ_1 one finds the value given in Theorem 2 which completes the proof.

3. CONCLUDING REMARKS

(a) Theorems 1 and 2 can be applied to problem (1.2) with the following interpretations of the quantities appearing there:

$K_G = \frac{\Delta(\log p)}{-2p}$, Δ = ordinary Laplacian

$k_g = \varkappa + \frac{1}{2}\frac{\partial}{\partial n}(\log p)$, $\varkappa$ = ordinary curvature of $\partial\Omega$

n = outward normal

$$|\Omega| = \int_\Omega p\,dx , \quad |\partial\Omega| = \oint_{\partial\Omega} \sqrt{p}\,ds$$

$\rho = \max_{x_1\in\Omega} \min_{x_0\in\partial\Omega} \int_\gamma \sqrt{p}\,ds$, γ = simple arc in Ω connecting x_0 and x_1, with line element ds.

(b) The supersolutions chosen here satisfy the boundary conditions. Hence one has also

(3.1) $$|\nabla u| \le |\nabla \bar{u}| \text{ on } \partial\Omega.$$

This inequality can be integrated in various ways. Inserting the supersolution used to prove Theorem 1 this yields

(3.2) $$\oint_{\partial\Omega} |\nabla u|ds \le K\frac{dw}{d\vartheta}\Big|_{\vartheta_0} \tan\vartheta_0 \oint_{\partial\Omega} |\nabla t|ds = K\frac{dw}{d\vartheta}\Big|_{\vartheta_0} \cdot \tan\vartheta_0 |\Omega| .$$

In some applications the quantity

(3.3) $$\eta(\Omega) = \frac{1}{|\Omega|}\int_\Omega f(u)dx$$

is of interest. From (3.2) it follows that for any $\lambda\in(0,\lambda^*(S))$

(3.4) $$\eta(\Omega) \le \frac{K}{\lambda}\frac{dw}{d\vartheta}\Big|_{\vartheta_0} \cdot \tan\vartheta_0 = \eta(S) ,$$

with S denoting the geodesic strip defined in Theorem 1.

(c) Theorems 1 and 2 can be generalized to higher dimensions. However, the results are no longer optimal in a similar sense.

(d) The method can be modified somewhat if one has Robin

boundary conditions instead of Dirichlet boundary conditions.

(e) Theorems 1 and 2 also hold for the first eigenvalue λ_1 of the fixed membrane problem on M, i.e.

(3.5) $$\Delta u + \lambda u = 0 \quad \text{in } \Omega \subset M, \quad u = 0$$

One just has to replace λ^* by λ_1 then.

REFERENCES

1. H. Amann, Fixed points equations and nonlinear eigenvalue problems in ordered Banach spaces, SIAM Rev. 18 (1976), 620-709.

2. P.L. Lions, On the Existence of Positive Solutions of Semilinear Elliptic Equations, SIAM Rev. 24 (1982), 441-467.

3. L.E. Payne, Bounds for solutions of a class of quasilinear elliptic boundary value problems in terms of the torsion function, Proc. Roy. Soc. Edinburgh 88a (1981), 251-256.

4. R.P. Sperb, Isoperimetric Ineqalities in a Boundary Value Problem on a Riemannian Manifold, J. of Appl. Math. & Phys. (ZAMP) 31 (1981), 740-753.

5. R.P. Sperb, Maximum Principles and their Applications, Math. in Science & Eng. 157, Academic Press, New York 1981.

René P. Sperb, Seminar für Angewandte Mathematik, ETH-Zentrum, CG-8092 Zürich, Switzerland

International Series of
Numerical Mathematics, Vol. 80

SOME INEQUALITIES OF SOBOLEV TYPE ON TWO-DIMENSIONAL SPHERES

Giorgio Talenti

Abstract. We offer the sharp form of two Sobolev type inequalities on two-dimensional spheres. An approach to more general inequalities is outlined.

1. INTRODUCTION AND STATEMENT OF RESULTS

Sobolev inequalities basically amount to statements of the following type. Let

$$\int_{M^n} |Du(x)|^p H^n(dx) = 1 \tag{1}$$

and suppose either sprt u - the support of u - is compact and M^n is not compact, or M^n is compact and

$$\text{mv } u = \frac{1}{H^n(M^n)} \int_{M^n} u(x) H^n(dx) \tag{2}$$

- the mean value of u over M^n - vanishes. Then either

$$\int_{M^n} |u(x)|^q H^n(dx) \tag{3a}$$

or

$$\sup \{|u(x)| : x \in M^n\} \tag{3b}$$

is bounded by

(4) a constant independent of u

according to whether $1 \le p < n$ or $p > n$. Here M^n stands for a n-dimensional Riemannian manifold (subject to suitable conditions: a smooth n-dimensional submanifold of some Euclidean space, say), H^n denotes an appropriate (i.e. Hausdorff) n-dimensional measure, D stands for gradient (i.e. covariant differentiation along M^n), vertical bars $|\ |$ denote absolute value of scalars or length of vectors, exponent q is related to exponent p and dimension n by

(5) $$\frac{1}{q} = \frac{1}{p} - \frac{1}{n} .$$

u is a smooth map from M^n into $\mathbb{R}$, the set of real numbers.

Sobolev inequalities are a basic tool in the theory of partial differential equations and in calculus of variations. Exhaustive references on this subject are Adams [1], T. Aubin [3], Maz'ja [6]. Sobolev inequalities have been mostly investigated in case M^n is an open subset G of n-dimensional Euclidean space $\mathbb{R}^n$ and the competing functions have compact support in G. In this case, the <u>Sobolev constant</u> - i.e. the least value of (4) - is known [2][9] to be

(6a) $$\frac{n^{-1/p}}{\sqrt{\pi}} \left(\frac{p-1}{n-p}\right)^{1-1/p} \left[\frac{\Gamma(n)\Gamma(1+n/2)}{\Gamma(n/p)\Gamma(1+n-n/p)}\right]^{1/n}$$

if $1 \le p < n$, and is estimated by

(6b) $$\frac{n^{-1/p}}{\sqrt{\pi}} \left[\Gamma\left(1 + \frac{n}{2}\right)\right]^{1/n} \left(\frac{p-1}{p-n}\right)^{1-1/p} \left[H^n(G)\right]^{1/n-1/p}$$

if $p > n$ - such an estimate actually <u>equals</u> the Sobolev constant if ($p > n$ and) G is a ball.

In the present paper we look for Sobolev constants in case

(7) $$n = 2$$

and M^n is

(8) $$S^2 = \{x \in \mathbb{R}^3 : x_1^2 + x_2^2 + x_3^2 = 1\} ,$$

the unit two-dimensional sphere in Euclidean three-dimensional space $\mathbb{R}^3$.

Note that

(9) $$H^2(dx) = \sin\theta \, d\theta \, d\phi$$

as x runs over S^2, and

(10) $$|Du| = \sqrt{\left(\frac{\partial u}{\partial \theta}\right)^2 + \left(\frac{1}{\sin\theta}\frac{\partial u}{\partial \phi}\right)^2}$$

as u maps S^2 into $\mathbb{R}$. Here ϕ and θ - longitude and colatitude - are geographical coordinates on S^2 (i.e., $x_1 = \cos\phi \sin\theta$, $x_2 =$

$\sin\phi \sin\theta$, $x_3 = \cos\theta$, $0 \le \phi < 2\pi$, $0 \le \theta \le \pi$).

Our main results sound as follows.

THEOREM 1. *The following inequality*

(11) $$\int_{S^2} [u(x) - \text{mv}\, u]^2 H^2(dx) \le \frac{1}{4\pi} \left[\int_{S^2} |Du(x)| H^2(dx)\right]^2$$

holds for any u *that maps* S^2 *smoothly into* $\mathbb{R}$. *Inequality* (11) *is sharp*.

THEOREM 2. *If* $p > 2$ *and*

(12a) $$C = (2\pi)^{-1/p} \left\{\sqrt{\pi}\, \frac{\Gamma\left(1 - \frac{p}{2p-2}\right)}{\Gamma\left(1 - \frac{1}{2p-2}\right)}\right\}^{1-\frac{1}{p}}$$

then

(12b) $$\max u - \min u \le C\left[\int_{S^2} |Du(x)|^p H^2(dx)\right]^{\frac{1}{p}}$$

for any u *that maps* S^2 *smoothly into* $\mathbb{R}$. *Inequality* (12) *is sharp*.

2. PROOF OF THEOREM 1

Without loss of generality, we assume u is positive. Let U(t) be the level sets of u,

$$U(t) := \{x \in S^2 : u(x) > t\} \subset S^2 .$$

The layer-cake formula

(13) $$u = \int_0^\infty \mathbf{1}_{U(t)}\, dt$$

- where $\int$ is taken in Bochner sense and $\mathbf{1}_E$ stands for characteristic function of the set E - holds for any real-valued nonnegative function u that is integrable over S^2. It shows that any such function u is a superimposition of the characteristic functions of U(t).

Formula (13) implies

(14) $$\text{mv}\, u = \frac{1}{4\pi} \int_0^\infty \mu(t)\, dt ,$$

where

(15) $$\mu(t) = H^2(U(t))$$

is the distribution function of u.

Equations (13) and (14) and Minkowski inequality give

$$||u - \text{mv}\, u||_{L^2(S^2)} \leq \int_0^\infty ||1_{U(t)} - \frac{\mu(t)}{4\pi}||_{L^2(S^2)}\, dt \ ,$$

where $||\cdot\cdot||_{L^2(S^2)}$ is a shorthand for $[\int_{S^2} (\ldots)^2 H^2(dx)]^{1/2}$. Then

(16) $$||u - \text{mv}\, u||_{L^2(S^2)} \leq \int_0^\infty \sqrt{\mu(t)[4\pi - \mu(t)]}\ \frac{dt}{\sqrt{4\pi}} \ ,$$

since

$$||1_E - \frac{1}{4\pi} H^2(E)||_{L^2(S^2)} = \sqrt{H^2(E)[1 - \frac{1}{4\pi} H^2(E)]}$$

for every measurable subset E of S^2.

Recall that

(17) $$H^2(E)\cdot H^2(S^2 \setminus E) \leq [H^1(\partial E)]^2$$

for every measurable subset E of S^2 (whose boundary is rectifiable). (17) is the isoperimetric inequality on S^2, see [7] for example.

Recall also that

(18) $$\int_{S^2} |Du(x)| H^2(dx) = \int_0^\infty H^1(\partial U(t))\, dt \ ,$$

if u is nonnegative and Lipschitz continuous. (18) is Federer's coarea formula, see [4].

Putting together (16), (17) and (18) gives

(19) $$||u - \text{mv}\, u||_{L^2(S^2)} \leq \frac{1}{\sqrt{4\pi}} \int_{S^2} |Du(x)| H^2(dx) \ .$$

the desired inequality.

Sobolev inequality (11) has been derived from isoperimetric inequality (17), via Federer's coarea formula. Vice versa, Sobolev inequality (11) implies isoperimetric inequality (17). In fact,

$$\int_{S^2} |Du(x)| H^2(dx)$$

approaches

$$H^1(\partial E)$$

as u (is smooth and) approaches 1_E - the characteristic function of a measurable subset E of S - in such a way that $\|u-1_E\|_{L^2(S^2)}$ approaches zero (see [9, § 1] for instance). Under the same circumstances,

$$\int_{S^2} [u(x) - \text{mv } u]^2 H^2(dx)$$

approaches

$$H^2(E)[1 - \frac{1}{4\pi} H^2(E)] ,$$

clearly.

As is well-known and easy to see, equality holds in isoperimetric inequality (17) if E is a <u>spherical cap</u> - i.e., the intersection of S^2 and a half-space. Thus, the above argument tells us that the ratio between the two sides of (11) approaches 1 if u approaches the characteristic function of a spherical cap. In other words, inequality (11) is sharp.

Theorem 1 is fully proved.

3. A LEMMA

Let

$$u^* = \int_0^\infty 1_{[0,\mu(t)]} \, dt , \tag{20}$$

the <u>decreasing rearrangement</u> of u in the sense of Hardy and Littlewood [5, Chapter 10] - i.e. the decreasing right-continuous map from $[0,4\pi]$ into $[0,\infty[$ which is equidistributed with u. Here u is a smooth map of S^2 into $[0,\infty]$ and μ is the distribution function of u, see formula (15). It can be shown that u^* is locally Lipschitz continuous; moreover, we have the following

LEMMA. *If p is any exponent* ≥ 1, *then*

$$\int_{S^2} |Du(x)|^p H^2(dx) \geq \int_0^{4\pi} [s(4\pi-s)]^{p/2}\left[-\frac{du^*}{ds}(s)\right]^p ds . \tag{21}$$

This lemma is a special case of symmetrization theorem [8].

4. PROOF OF THEOREM 2

For convenience, let u be positive. The very definition of u^* tells us that

(22a) $$\max u = u^*(0) ,$$

(22b) $$\min u = u^*(4\pi-) .$$

Thus,

(23) $$\max u - \min u = \int_0^{4\pi} \left[-\frac{du^*}{ds}(s)\right] ds ,$$

since u^* is absolutely continuous. Hence Hölder inequality gives

(24a) $$\max u - \min u \le C\left[\int_0^{4\pi} [s(4\pi-s)]^{p/2} \left[\frac{du^*}{ds}(s)\right]^p ds\right]^{1/p} ,$$

where

(24b) $$C = \left[\int_0^{4\pi} [s(4\pi-s)]^{-p/2(p-1)} ds\right]^{1-1/p} ,$$

the right-hand side of (12a). Inequality (12) follows from (24) and the previous lemma.

An inspection shows that equality holds in (12) if (and only if) u satisfies

(25) $$u^*(s) = (\text{const.})\int_{s/4\pi}^{1} [t(1-t)]^{-p/2(p-1)} dt + (\text{another const.}).$$

Equation (25) holds if

(26a) $$u(x) = \int_\theta^\pi (\sin t)^{-1/(p-1)} dt ,$$

a function which is invariant under a group of rotations over S^2 - here θ is the colatitude of x - and obeys

(26b) $$u(x) = u^*\left(4\pi \sin^2\frac{\theta}{2}\right) .$$

Theorem 2 is demonstrated.

5. GENERALIZATIONS

THEOREM 3. *If* $p \ge 1$, $q \ge 1$, $\frac{1}{q} + \frac{1}{2} > \frac{1}{p}$ *and* C *is defined by*

$$\left[(4\pi)^{\frac{1}{p}-\frac{1}{q}}C\right]^{\frac{p}{p-1}} = \int_0^1 [t(1-t)]^{\frac{p}{p-1}\left(\frac{1}{q}-\frac{1}{2}\right)}[t^{q-1}+(1-t)^{q-1}]^{\frac{p}{q(p-1)}}dt,$$

<u>then</u>

$$\left[\int_{S^2}|u(x)-\text{mv } u|^q H^2(dx)\right]^{\frac{1}{q}} \le C\left[\int_{S^2}|Du(x)|^p H^2(dx)\right]^{\frac{1}{p}} .$$

A proof of theorem 3 starts as in § 2 and leads at once to the following inequality:

$$||u-\text{mv}\,u||_{L^q(S^2)} \le$$
$$(4\pi)^{-\frac{1}{q}}\int_0^\infty [4\pi-\mu(t)]^{\frac{1}{q}}\mu(t)^{\frac{1}{q}}\left[\left(1-\frac{\mu(t)}{4\pi}\right)^{q-1}+\left(\frac{\mu(t)}{4\pi}\right)^{q-1}\right]^{\frac{1}{q}} dt .$$

By (20), the distributional derivative of u^* is

$$C_0^\infty(]0,4\pi[) \ni \phi \to -\int_0^\infty \phi(\mu(t))\,dt .$$

Consequently we have:

$$\int_0^{4\pi}\phi(s)\left[-\frac{du^*}{ds}(s)\right]ds = \int_0^\infty \phi(\mu(t))\,dt ,$$

provided ϕ is sufficiently well-behaved in $]0,4\pi[$ and near the end points.

Therefore,

$$||u-\text{mv}\,u||_{L^q(S^2)} \le$$
$$(4\pi)^{-\frac{1}{q}}\int_0^\infty [(4\pi-s)s]^{\frac{1}{q}}\left[\left(1-\frac{s}{4\pi}\right)^{q-1}+\left(\frac{s}{4\pi}\right)^{q-1}\right]\left[-\frac{du^*}{ds}(s)\right] ds .$$

The conclusion follows, via Hölder inequality and the lemma from § 3.

REFERENCES

1. R.A. Adams, Sobolev spaces. Academic Press 1975.

2. T. Aubin, Problèmes isopérimétriques et espaces de Sobolev. J. Diff. Geometry <u>11</u> (1976), 573-598.

3. T. Aubin, Nonlinear analysis on manifolds. Springer-Verlag, 1982.

4. H. Federer, Curvature measures. Trans. Amer. Math. Soc. 93 (1959).

5. G.H. Hardy, J.E. Littlewood and G. Pólya, Inequalities. Cambridge Univ. Press, 1964.

6. V.G. Maz'ja, Sobolev spaces. Springer-Verlag, 1985.

7. R. Osserman, The isoperimetric inequality. Bull. Amer. Math. Soc. 84 (1978).

8. E. Sperner jr., Zur Symmetrisierung von Funktionen auf Sphären. Math. Z. 134 (1973), 317-327.

9. G. Talenti, Best constant in Sobolev inequality. Ann. Mat. Pura Appl. 110 (1976), 353-372.

Giorgio Talenti, Istituto Matematico dell'Università, Viale Morgagni 67/A, I-50134 Firenze, Italy

Inequalities in Economics, Optimization and Applications

Schwarzwald house with upper cow-byre

International Series of
Numerical Mathematics, Vol. 80

ENTROPIES, GENERALIZED ENTROPIES, INEQUALITIES, AND THE MAXIMUM ENTROPY PRINCIPLE

János Aczél and Bruno Forte

Abstract. Different entropy concepts (increasingly general or particular depending on one's point of view), the underlying knowledge (conditions, information) and inequalities associated to them are reviewed. Among the latter is the maximum entropy principle. It is proposed that entropies (in particular newly introduced ones) should be justified by showing that their maximization yields useful probability distributions, appropriate for the practical situation.

Apart from Physics, the first entropy measure was introduced by Hartley [8]. Here, as in what follows, we consider the entropy as measure of uncertainty about which of n events occurs from a partition $\{E_1, E_2, \ldots, E_N\}$ of the sure event. *Hartley's entropy* is just

(1) $$\log_2 N$$

which looks pretty primitive, but is the only reasonable measure *if we know only the number of events*. Hartley has indeed consciously refused to work with probabilities. *If we know* at least *how many* (say, n) events *have nonzero probabilities*, we get the modification

(2) $$\log_2 n$$

of Hartley's entropy used by Aczél, Forte and Ng [4].

This paper is in final form and no version of it will be submitted for publication elsewhere.

If the probabilities $p_1, p_2, \ldots, p_N$ *of* $E_1, E_2, \ldots, E_N$ *are known,* we arrive at the *Shannon entropy* [11]

$$(3) \qquad -\sum_{k=1}^{N} p_k \log_2 p_k \quad \text{(extended by } 0.\log 0 = 0$$

by def. if some $p_k = 0$ is permitted)

or at its generalizations, for instance the *Rényi entropies* [10]

$$(4) \qquad \frac{1}{1-\alpha} \log_2 \sum_{k=1}^{N} p_k^{\alpha} \quad \text{(with } 0^{\alpha} = 0 \text{ by def. if}$$

$p_k = 0$ permitted), when $\alpha \neq 1$

(they have the Shannon entropy as limit when $\alpha \to 1$ and have other desirable properties). The Hartley entropy $\log_2 N$ is the special case $p_1 = p_2 = \ldots = p_N = \frac{1}{N}$ of both (3) and (4) without 0 probabilities, as is $\log_2 n$ if 0 probabilities are permitted. Each is also the maximum of such Shannon and Rényi ($1 \neq \alpha > 0$) entropies.

The Shannon and Rényi entropies for $\alpha > 0$ ($\alpha \neq 1$) are (in a sense exact) lower bounds for average codeword lengths

$$(5) \qquad \sum_{k=1}^{N} p_k n_k$$

or for exponential mean codeword lengths

$$\frac{\alpha}{1-\alpha} \log_2 \sum_{k=1}^{N} p_k 2^{(1-\alpha)n_k/\alpha} ,$$

respectively, (where the individual codeword lengths n_k are positive integers and $\sum_{k=1}^{N} 2^{-n_k} \leq 1$).

These facts are related to other *inequalities* associated with these and other entropies, for instance the "how to keep the expert honest" inequality (see e.g. [3]): An expert gives $q_1, q_2, \ldots, q_N$ as probabilities of the events (weather, market situations, etc.) $E_1, E_2, \ldots, E_N$, which in reality (or to the best of his knowledge) are $p_1, p_2, \ldots, p_N$. It is agreed that he gets paid the amount $f(q_k)$ if E_k happens. So his expected gain is $\sum_{k=1}^{N} p_k f(q_k)$.

How should the 'payoff function' f be chosen so that the expert's expected gain be maximal if he told the truth? Clearly f should then satisfy the inequality

$$(6) \qquad \sum_{k=1}^{N} p_k f(q_k) \le \sum_{k=1}^{N} p_k f(p_k).$$

It turns out that, without any further supposition on f, this inequality is satisfied for variable N or for fixed $N > 2$ if, and only if, $f(p) = c \log_2 p + b$ $(c \ge 0)$ so that the right hand side of (6) will be

$$c \sum_{k=1}^{N} p_k \log_2 p_k + b,$$

linking the subject to the Shannon entropy. Indeed, the "if" part of the above statement is equivalent to what is sometimes called the Shannon inequality:

$$-\sum_{k=1}^{N} p_k \log_2 q_k \ge -\sum_{k=1}^{N} p_k \log_2 p_k \quad \left(\sum_{k=1}^{N} p_k = \sum_{k=1}^{N} q_k = 1\right),$$

which in turn implies the above result that the Shannon entropy is a lower bound of the average codeword length $\sum_{k=1}^{N} p_k n_k$. There exist similar results with regard to the Rényi entropies.

If now *we know about the events* $E_1, E_2, \ldots, E_N$ *more than just their probabilities,* we arrive at *inset entropies*

$$(7) \qquad c \sum_{k=1}^{N} p_k \log p_k + \sum_{k=1}^{N} p_k g(E_k)$$

where g is an arbitrary real valued function of the events, while c is an arbitrary constant (also the Hartley, Shannon and Rényi entropies could be multiplied by constants for most purposes). We can get to (7) among others again by the "how to keep the expert honest" method [1]: If we allow the payoff function f to depend also upon the events E_k (not just their probabilities), then the previous 'keeping the expert honest' inequality is replaced by

$$(8) \qquad \sum_{k=1}^{N} p_k f(q_k, E_k) \le \sum_{k=1}^{N} p_k f(p_k, E_k)$$

and the general solution (again also for fixed $N > 2$ and without further suppositions on f) will be $f(p,E) = c \log p + g(E)$ so that the right hand side of (8) becomes exactly (7).

Examples of other applications of inset entropies can be found, for instance, in the theory of gambling where Meginnis [9] interprets the second sum in

$$(9) \qquad c \sum_{k=1}^{N} p_k \log p_k + \sum_{k=1}^{N} p_k g(E_k)$$

as the expected gain and the first as the "joy in gambling". Also, for the so-called continuous (partial) analogue of the Shannon entropy $-\sum_{k=1}^{N} p_k \log p_k$,

$$-\int_{\alpha}^{\beta} \rho(t) \log \rho(t) dt,$$

(where ρ is the probability density function), the approximating sums of the integral are not Shannon entropies but inset entropies (9);

$$(10) \qquad -\sum_{k=1}^{N} \rho(\tau_k) \log \rho(\tau_k)(t_k - t_{k-1})$$

$$= -\sum_{k=1}^{N} \frac{F(t_k)-F(t_{k-1})}{t_k - t_{k-1}} \log \frac{F(t_k)-F(t_{k-1})}{t_k - t_{k-1}} (t_k - t_{k-1})$$

$$= -\sum_{k=1}^{N} p_k \log p_k - \sum_{k=1}^{N} p_k \log t(E_k),$$

where F is the probability distribution function, so $p_k = F(t_k) - F(t_{k-1})$ $(k = 1,2,\ldots,n)$ are the probabilities,

$E_k = (t_{k-1}, t_k)$ and $\ell(E_k) = t_k - t_{k-1}$ [2]. For applications of (10) to geographical and economical analysis, see for instance [5,6].

Further generalizations to entropies associated with random variables have been made by Forte (for inst. [7]).

Here we draw two consequences from the above:

(i) *All* (above) *entropies are conditional* on what we know about the events, the entropies being the measures of the remaining uncertainty about which of the events will happen. It is remarkable that, while each of the Hartley, modified Hartley, Shannon and inset entropies contains the previous ones as *special cases*, also each corresponds to more knowledge, that is, *more conditions* on the events.

(ii) All these entropies are connected to *inequalities* (and equations, see for instance [3]).

In another sense, the *maximum entropy principle*, of course, also relies on inequalities: we are looking among probability distributions $(p_1, p_2, \ldots, p_N)$, satisfying certain conditions (equations) for the one which *maximizes* a "suitable" entropy (makes it the largest, hence satisfying an inequality).

Perhaps the best known example is that the normal distribution maximizes the Shannon entropy $-\sum_{k=1}^{N} p_k \log p_k$ under the conditions

$$(11) \qquad p_1+p_2+\ldots+p_N = 1, \quad \alpha_1 p_1+\alpha_2 p_2+\ldots+\alpha_N p_N = 0,$$
$$\alpha_1^2 p_1+\alpha_2^2 p_2+\ldots+\alpha_N^2 p_N = \sigma^2,$$

where $\alpha_1, \alpha_2, \ldots, \alpha_N$ are the possible (real) values of a random variable, while σ^2 is its variance (also given). So the Shannon entropy is a "suitable" entropy. Equations like (11) are again *conditions representing our* (partial) *knowledge*, this time about the otherwise unknown probabilities $p_1, p_2, \ldots, p_N$. There are two interpretations of the above argument: (I) It is usually considered to 'justify' the normal distribution, because the normal distribution maximizes the Shannon entropy (under appropriate conditions). (II) We say that it can be interpreted also as *'justification' of the Shannon entropy,* because the Shannon entropy is maximized by the normal distribution which is what we should get (under the same conditions), based on experience and usefulness.

We propose that "suitable" *entropies should be introduced preferably as expressions, the maximization of which gives "useful" probability distributions.* Entropies as measures of conditional uncertainty must take into account *all kinds of information* provided by the problem, be they mathematical, scientific or "real life". The *maximum entropy principle can be used to define* some of those *entropies.* There remains much to do in this respect, even with regard to the above and other more or less generally used entropies.

Similar remarks may be made concerning other information measures, for instance about the minimization of directed divergences.

This research has been supported in part by Natural Sciences and Engineering Research Council of Canada grants. A previous version of this paper has been presented at the Calgary workshop on maximum entropy and Bayesian methods in applied statistics.

REFERENCES

1. J. Aczél, A mixed theory of information, V. How to keep the (inset) expert honest. J. Math. Anal. Appl. 75 (1980), 447-453.

2. J. Aczél, A mixed theory of information, VI. An example at last: A proper discrete analogue of the continuous Shannon measures of information (and its characterization). Univ. Beograd. Publ. Elektrotehn. Fak. Ser. Mat. Fiz. Nr. 602-633 (1978-80), 65-72.

3. J. Aczél and Z. Daróczy, On measures of information and their characterizations. Academic Press, New York - San Francisco - London, 1975.

4. J. Aczél, B. Forte and C.T. Ng, Why Shannon and Hartley entropies are natural. Adv. in Appl. Probab. 6 (1974), 131-146.

5. D.F. Batten, Spatial analysis of interacting economics. Kluwer-Nijhoft, Boston-Hague-London, 1983.

6. M. Batty, Speculations on an information theoretical approach to spatial representation. In Spatial representation and spatial interaction. Nijhoft, Leiden-Boston, 1978, pp.115-147.

7. B. Forte, Subadditive entropies for a random variable. Bol. Un. Mat. Ital. (5) 14B (1977), 118-133.

8. R.V. Hartley, Transmission of information. Bell Systems Tech. J. 7 (1928), 535-563.

9. J.R. Meginnis, A new class of symmetric utility rules for gambles, subjective marginal probability functions, and a generalized Bayes rule. Bus. Econ. Stat. Sec. Proc. Amer. Stat. Assoc. 1976, 471-476.

10. A. Rényi, On measures of entropy and information. Proc. 4th Berkeley Symp. Math. Stat. and Prob. 1960, Vol. 1, Univ. of Calif. Press, Berkeley, CA, 1961, pp.547-561.

11. C.E. Shannon, A mathematical theory of communication. Bell Systems Technical J. 27 (1948), 379-423, 623-656.

János Aczél, Centre for Information Theory, University of Waterloo, Waterloo, Ontario, Canada N2L 3G1.

Bruno Forte, Centre for Information Theory, University of Waterloo, Waterloo, Ontario, Canada N2L 3G1.

International Series of
Numerical Mathematics, Vol. 80

INEQUALITIES AND MATHEMATICAL PROGRAMMING, III

S. Iwamoto, R.J. Tomkins and Chung-lie Wang

Abstract. Three equivalent mathematical programming problems concerning monotonic infinite sequences with suitable constraints are solved by means of pertinent inequalities. The continuous version of the inequalities as well as some variants of discrete and continuous inequalities are also studied.

1. INTRODUCTION

The close relationship between inequalities and mathematical programming problems is well known (see, e.g. [4,5,13,19-22]). A mathematical programming approach is commonly used to establish important inequalities. Moreover, if a certain inequality with a mathematical programming problem could somehow be established first, then it could be used to solve the mathematical programming problem. The goal of this paper is to establish some variants of Hölder's inequality and then to indicate how they may be used to solve certain optimization problems.

Very recently, Mudholkar, Freimer and Subbaiah [14] established the following generalized Hölder inequality by means of a mathematical programming approach. Let $a_1 \geq a_2 \geq \dots \geq a_n \geq 0$, $b_1 \geq b_2 \geq \dots \geq b_n \geq 0$, m with $1 \leq m \leq n$ fixed, $p > 1$, $q = p/(p-1)$. Then

$$\sum_{j=1}^{n} a_j b_j \leq \left(\sum_{j=1}^{m} a_j^p\right)^{1/p} \left(\sum_{j=1}^{k} b_j^q + (m-k)C_k^q\right)^{1/p}, \tag{1}$$

where $B_i = b_i + b_{i+1} + \dots + b_n$, $C_k = B_{k+1}/(m-k)$, and k with $0 \leq k < m$ is the integer such that

$$b_k \geq C_k \quad \text{and} \quad b_{k+1} < C_k . \tag{2}$$

In this form the inequalities appear in [15], while the inequality in [14] is obtained from (1)(2) by replacing k+1 by m-k.

Moreover, in [15] Freimer and Mudholkar gave a simple example to illustrate the inequality (1): For any $n \geq 2$, taking $m = p = 2$, (1) implies that

$$\sum_{j=1}^{n} a_j b_j \leq A_o , \quad \text{for } b_2 < B_2 \quad \text{with} \quad k = 0 \tag{3}$$

and

$$\sum_{j=1}^{n} a_j b_j \leq A_1 , \quad \text{for } b_1 < B_1/2 \quad \text{with} \quad k = 1 , \tag{4}$$

where

$$A_o = (a_1^2 + a_2^2)^{1/2} B_1/\sqrt{2}$$

and

$$A_1 = (a_1^2 + a_2^2)^{1/2} (b_1^2 + B_2^2)^{1/2} .$$

It should be noticed that the inequality (3) can also be established by the use of a majorization theorem (see, e.g. [12] or later). Also note that $A_o \leq A_1$.

In the development of the theory of inequalities, Hölder's inequality and its generalizations (or variants) have captured the attention of many investigators (see, e.g. [1-5, 12, 13, 16-18]). If $a_j, b_j \geq 0$, $j = 1, \ldots, n$, and $q = p/(p-1)$ with $p > 1$, then the basic form of Hölder's inequality states that (see, e.g. [4, 5, 13])

$$\sum_{j=1}^{n} a_j b_j \leq \left(\sum_{j=1}^{n} a_j^p \right)^{1/p} \left(\sum_{j=1}^{n} b_j^q \right)^{1/q} . \tag{5}$$

The inequality sign in (5) is reversed for $p < 1$. (In case $p < 0$, we assume $a_j, b_j > 0$.) In either case, equality holds if and only if $a_j^p = c b_j^q$, $j = 1, \ldots, n$, for some constant c or $b_1 = \ldots = b_n = 0$.

In [14, 15], refined mathematical problems were solved by means of Lagrange multipliers (with or without introducing the Kuhn-Tucker conditions) in order to establish inequality (1). Furthermore, Freimer and Mudholkar [15] gave, among others, a brief but comprehensive account of the continuous version of the inequality (1).

In view of the findings of the examples (3)-(4), we shall

introduce a majorization theorem so that we can use it in conjunction with the usual Hölder inequality (5) to establish the inequality (1). And then the associated mathematical programming problem will be easily solved. This is the motivation of this paper.

In Section 2, we shall state problems and summarize notation and lemmas that will be used. In Section 3, we shall present discrete inequalities for infinite sequences. In the following sections we shall successively establish the continuous versions of the discrete inequalities and conclude with some remarks.

2. NOTATIONS AND LEMMAS

Let us begin by displaying some notations and symbols that we shall need:

$$\ell = \{t = \{t_j\} \mid t_n \geq t_{n+1} \geq 0 \text{ for } n = 1,2,3,\ldots, \sum_{j=1}^{\infty} t_j < \infty\} ,$$

$$\ell_n = \{t \in \ell \mid t_k = 0 \text{ for } k \geq n+1\} ,$$

$$B_i = \sum_{j=i}^{\infty} b_j , \quad \text{for } b \in \ell ,$$

$$q = p/(p-1) , \text{ for } p \neq 1 .$$

In [14, 15], the finiteness of the integer n played no essential role. So we consider here only the infinite case and leave the finite case as a natural consequence.

We cite Lemma 2.1 of Freimer and Mudholkar [15] with a slight modification (together with their short and elegant proof) as follows.

LEMMA 1. <u>Let</u> $b \in \ell$ <u>and</u> m <u>be</u> <u>a</u> <u>positive</u> <u>integer</u>. <u>Then</u> <u>there</u> <u>exists</u> <u>a</u> <u>unique</u> <u>integer</u> k, $0 \leq k < m$, (<u>choose</u> $b_o > B_1/m$ <u>for</u> <u>convenience</u>) <u>such</u> <u>that</u>

$$b_k > B_{k+1}/(m-k) \geq b_{k+1} . \tag{6}$$

<u>Proof.</u> Let

$$\beta_r = (m-r)b_r - B_{r+1}$$

be defined for $r = 1, \ldots, m$. Then

$$\beta_r - \beta_{r+1} = (m-r)(b_r - b_{r+1}) \geq 0 \quad \text{for} \quad r = 1, \ldots, m-1$$

and $\beta_m = -B_{m+1} < 0$ ensure the existence of a unique k such that $\beta_k > 0 \geq \beta_{k+1}$ which is equivalent to (6).

We cite the inequality A.2.a from Marshall and Olkin [12, p.445] and add its reverse inequality for our purpose as follows (cf. [12] for a proof).

LEMMA 2. *The inequality*

(7) $$\sum_{j=1}^{n} u_j x_j \leq \sum_{j=1}^{n} v_j x_j$$

holds for every $x \in \ell_n$ *if and only if*

(8) $$\sum_{j=1}^{t} u_j \leq \sum_{j=1}^{t} v_j , \qquad t = 1, \ldots, n-1 ,$$

and

(9) $$\sum_{j=1}^{n} u_j = \sum_{j=1}^{n} v_j .$$

The inequality (7) *is reversed if the order of the sequence* x *is reversed*.

We now conclude this section with the following three equivalent mathematical programming problems (cf. [14, 15]). For a positive integer m, a real p, and $b \in \ell$:

(I) $$\operatorname*{opt}_{a \in \ell} \sum_{j=1}^{\infty} a_j b_j$$

subject to

(10) $$\sum_{j=1}^{\infty} a_j = 1 ;$$

(II) $$\operatorname*{opt}_{a \in \ell_n} \sum_{j=1}^{m-1} a_j b_j + a_m B_m$$

subject to (10);

(III) $$\operatorname*{opt}_{a \in \ell_{k+1}} \sum_{j=1}^{k} a_j b_j + a_{k+1} B_{k+1}$$

subject to

$$\sum_{j=1}^{k} a_j^p + (m-k)a_{k+1}^p = 1 .$$

In (I) - (III), opt = max if $p > 1$, while opt = min if $p < 1$ and the order of the sequence a is reversed. In (III), k is the integer specified in Lemma 1.

3. DISCRETE INEQUALITIES

We extend the inequality (1) to the case of infinite sequences as follows.

THEOREM 1. Let b be an element of ℓ and m be any positive integer. Then the inequality

$$(1) \qquad \sum_{j=1}^{\infty} a_j b_j \le \left(\sum_{j=1}^{m} a_j^p\right)^{1/p} \left(\sum_{j=1}^{k} b_j^q + (m-k)C_k^q\right)^{1/q}$$

holds for $p > 1$ and $a \in \ell$, where k is given by Lemma 1 and $C_k = B_{k+1}/(m-k)$. The sign of inequality in (1) is reversed if $p < 1$ and the order of the sequence a is reversed. In either case, equality holds in (1) if

$$(11) \qquad \frac{a_1^p}{b_1^q} = \cdots = \frac{a_k^p}{b_k^q} = \frac{a_{k+1}^p}{C_k^q} \quad \text{and} \quad a_{k+1} = a_{k+2} = \cdots$$

Proof. From (6) with $b \in \ell$, it follows

$$b_{m-1} \le \cdots \le b_{k+1} \le B_{k+1}/(m-k)$$

or

$$(12) \qquad b_{m-1}/B_{k+1} \le \cdots \le b_{k+1}/B_{k+1} \le 1/(m-k)$$

Noting (12), we set

$$(13) \qquad \begin{aligned} u_j &= b_{k+j}/B_{k+1}, \quad j = 1,\ldots,m-k-1, \quad u_{m-k} = B_m/B_{k+1}, \\ v_j &= 1/(m-k), \quad j = 1,\ldots,m-k, \\ x_j &= a_{k+j}, \quad j = 1,\ldots,m-k . \end{aligned}$$

Suppose $a \in \ell_{m-k}$. Since the corresponding values of u_j and v_j given in (13) satisfy the conditions (8) and (9) of Lemma 2, we obtain

$$(14) \qquad \sum_{j=k+1}^{m-1} a_j b_j/B_{k+1} + a_m B_m/B_{k+1} \le \sum_{j=k+1}^{m} a_j/(m-k) .$$

For $a, b \in \ell$, we have

$$\sum_{j=1}^{\infty} a_j b_j \le \sum_{j=1}^{m-1} a_j b_j + a_m B_m \tag{15}$$

with equality, if $a_m = a_{m+1} = \dots$.

From (14) and (15) we have for a

$$\sum_{j=1}^{\infty} a_j b_j \le \sum_{j=1}^{k} a_j b_j + \sum_{j=k+1}^{m} a_j B_{k+1}/(m-k) \tag{16}$$

with equality, if $a_{k+1} = a_{k+2} = \dots$.

Finally, a direct application of the usual Hölder inequality (5) on the right-hand side of (16) yields (1) for $p > 1$.

Similarly, the reverse inequality of the inequality (1) is readily established for $p < 1$ by means of the second part of Lemma 2 and the reversed order of the sequence a. In either case, it is clear that equality holds under condition (11).

REMARK. In view of the establishment of the inequality (1) in Theorem 1, the common optimal value of the three equivalent mathematical programming problems stated in Section 2 is, in turn, found to be

$$E_k = E_k(b) = \left(\sum_{j=1}^{k} b_j^q + (m-k) C_k^q \right)^{1/q} .$$

Moreover, it is interesting to notice that

$$E_k \le E_{k+1} \quad \text{for} \quad q > 1 \tag{17}$$

and

$$E_k \ge E_{k+1} \quad \text{for} \quad q < 1 . \tag{18}$$

Since the function t^q is convex for $q > 1$ and concave for $q < 1$, we have

$$(m-k) C_k^q \le b_{k+1}^q + (m-k-1) C_{k+1}^q , \quad q > 1 , \tag{19}$$

and

$$(m-k) C_k^q \ge b_{k+1}^q + (m-k-1) C_{k+1}^q , \quad q < 1 . \tag{20}$$

It is apparent that the inequalities (17) and (18) are equivalent to the inequalities (19) and (20), respectively.

4. CONTINUOUS INEQUALITIES

In order to establish continuous analogues of the results in Section 3, we first summarize a key notion from Freimer and Mudholkar [15, p.64].

LEMMA 3 ([15]). *Let b be a continuous, positive and nonincreasing integrable function on the half line* $[0,\infty)$ *and M be a real number. Then there exists a number K,* $0 \le K < M$, *such that*

$$b(K) \le \int_K^\infty b(t)dt/(M-K) . \tag{21}$$

Proof. Define

$$B(t) = (M-t)b(t) - \int_t^\infty b(s)ds$$

for $0 \le t \le M$. If $B(0) \le 0$, then (21) holds with $K = 0$. So we may assume $B(0) > 0$. But

$$B(M) = -\int_M^\infty b(s)ds < 0$$

and the intermediate value theorem ensures the existence of a K such that $B(K) = 0$. The conclusion of the lemma is clear.

For our purposes, we need a lemma similar to the sufficient portion of Lemma 2 as follows.

LEMMA 4. *If u is a positive integrable function on* $[K,M]$ *with*

$$\int_K^t u(s)ds \le \frac{t-K}{M-K} Q, \quad K \le t \le M , \tag{22}$$

where $Q = \int_K^M u(s)ds + c$ *for some constant c, then*

$$\int_K^M u(t)f(t)dt + f(M)c \le \frac{Q}{M-K} \int_K^M f(t)dt \tag{23}$$

for every nonincreasing differentiable function f on $[K,M]$. *The sign of inequality in* (23) *is reversed if f is a nondecreasing function. In either case, equality holds in* (23) *if f is a constant function.*

Proof. Using integration by parts, we have

$$(24)\qquad \int_K^M f(t)dt\left(\int_K^t u(s)ds\right) + f(M)c$$

$$= f(t)\int_K^t u(s)ds\Big|_K^M + f(M)c - \int_K^M \left(\int_K^t u(s)ds\right)df(t)$$

$$\le f(M)Q - \int_K^M \frac{t-K}{M-K}\,Q\,df(t)$$

$$= \frac{Q}{M-K}\int_K^M f(t)dt\ .$$

The inequality sign in (24) is reversed if f is a nondecreasing function. The lemma is thus proved.

We now state and prove a continuous analogue of Theorem 1 as follows.

THEOREM 2. *Let* b *be a continuous, positive and nonincreasing integrable function on* $[0,\infty)$ *and* M *be a positive real number. Then*

$$(25)\qquad \int_0^\infty a(t)b(t)dt \le \left(\int_0^M a(t)^p dt\right)^{1/p}\left(\int_0^M \hat{b}(t)^q dt\right)^{1/q}$$

for every nonincreasing differentiable function a *on* $[0,\infty)$ *and* $p > 1$, *where*

$$\hat{b}(t) = \begin{cases} b(t)\ , & 0 \le t \le K \\ \int_K^\infty b(t)dt/(M-K)\ , & K \le t \le M \end{cases}$$

and K *is given in Lemma* 3. *The inequality sign in* (25) *is reversed if* $p < 1$ *and* a *is a nondecreasing differentiable function. In either case, equality holds if* $a(t)^p = c_1\hat{b}(t)^q$, $0 \le t \le M$ (*where* c_1 *is a constant*), *and* $a(t) = a(K)$, $t \ge K$.

Proof. Since a and b are nonincreasing functions on $[0,\infty)$, we have

$$(26)\qquad \int_0^\infty a(t)b(t)dt \le \int_0^K a(t)b(t)dt + \int_K^M a(t)b(t)dt + a(M)\int_M^\infty b(t)dt\ .$$

Noting the real K defined in Lemma 3, the nonincreasingness of the function b implies

(27) $$\int_K^t b(s)ds \le (t-K)b(K) \le \frac{t-K}{M-K}\int_K^\infty b(s)ds \quad .$$

Now setting $u = b$, $f = a$ and $c = \int_M^\infty b(t)dt$ in Lemma 4, from (26) and (27) it follows that

(28) $$\int_0^\infty a(t)b(t)dt \le \int_0^M a(t)\hat{b}(t)dt \quad .$$

Finally, a direct application of the usual Hölder inequality for integral (see, e.g. [4, 5, 12]) on the right-hand side of (28) yields (25) for $p > 1$. The conclusion of the theorem is thus clear.

REMARK. For the continuous case, we can likewise set up three equivalent mathematical programming problems (similar to those discrete ones in Section 2) as follows: For a positive real M, a real p, a differentiable function a and an integrable, positive and nonincreasing function b on $[0,\infty)$:

($\hat{\text{I}}$) $$\underset{a}{\text{opt}} \int_0^\infty a(t)b(t)dt$$

subject to

(29) $$\int_0^M a(t)\, dt = c \qquad (c > 0) \; ;$$

($\text{I}\hat{\text{I}}$) $$\text{opt} \int_0^M a(t)b(t)dt + a(M)\int_M^\infty b(t)dt$$

subject to (29);

($\text{I}\hat{\text{I}}\text{I}$) $$\text{opt} \int_0^K a(t)b(t)dt + a(K)\int_K^\infty b(t)dt$$

subject to

$$\int_0^K a(t)^p dt + (M-K)a(K)^p = c \; .$$

In ($\hat{\text{I}}$) - ($\text{I}\hat{\text{I}}\text{I}$), opt = max if a is a nondecreasing function and $p > 1$, while opt = min if a is a nondecreasing function and $p < 1$, where K is given in Lemma 3.

By Theorem 2, the common optimal value of ($\hat{\text{I}}$) - ($\text{I}\hat{\text{I}}\text{I}$) is found to be

$$(30)\qquad E(K) = \left(\int_0^M \hat{b}(t)^q dt\right)^{1/q} = \left(\int_0^K b(t)^q dt + (M-K)\left(\frac{\int_K^\infty b(t)dt}{M-K}\right)^q\right)^{1/q}.$$

It is useful to notice that $E(K)$ is an increasing function of K for $q>1$, while $E(K)$ is a decreasing function of K for $q<1$. In fact, from (30) follows

$$E'(K) = \frac{1}{q}\left(\int_0^M \hat{b}(t)^q dt\right)^{(\frac{1}{q})-1} [\ldots],$$

where

$$[\ldots] = b(K)^q - F^q + qF^{q-1}(F-b(K))$$

with $F = \int_K^\infty b(t)dt/(M-K)$. From (21), $b(K) \le F$. We now apply the mean value theorem for derivatives to the function t^q and obtain that $[\ldots]>0$ for $q>1$, while $[\ldots]<0$ for $q<1$. The conclusion of the argument is now clear.

REMARK. In view of the monotonicity of $E(K)$ shown above, $E(K)$ provides the least upper bound (on greatest lower bound) of $\int_0^\infty a(t)b(t)dt$ with K given in Lemma 3.

5. SOME VARIANTS

There are many variants (or generalizations) of the Hölder inequality (5) (and its continuous form) in the literature (see, e.g. [1-5, 12, 13, 16-18, 22]). In this connection, we first cite a variant of (5) from [17, p.554] (see [1, 16, 18] also) as follows. If $a_j, b_j > 0$, $j = 1,\ldots,n$, and $q = pr/(p-r)$, $p,r>0$, then the inequality

$$(31)\qquad \left(\sum_{j=1}^n a_j^r b_j^r\right)^{1/r} \le \left(\sum_{j=1}^n a_j^p\right)^{1/p} \left(\sum_{j=1}^n b_j^q\right)^{1/q}$$

holds for $p>r$. The sign of inequality in (31) is reversed for $p<r$. In either case, equality holds if and only if $a_j^p = cb_j^q$, $j = 1,\ldots,n$, for some constant c.

For the continuous version of (31), let positive continuous

functions a and b be defined on [0,N] with p, q, r given above. Then the inequality

$$(32)\qquad \left(\int_0^N a(t)^r b(t)^r dt\right)^{1/r} \le \left(\int_0^N a(t)^p dt\right)^{1/p}\left(\int_0^N b(t)^q dt\right)^{1/q}$$

holds for $p > r$. The sign of inequality in (32) is reversed for $p < r$. In either case, equality holds if and only if $a^p = cb^q$ on [0,N] for some constant c.

From (31) and (32) in association with the groundwork laid out above, we readily transliterate Theorem 1 and 2 into some variants as follows.

THEOREM 3. *Let b be an element of ℓ and m be a positive integer. Then the inequality*

$$(33)\qquad \left(\sum_{j=1}^{\infty} a_j^r b_j^r\right)^{1/r} \le \left(\sum_{j=1}^{m} a_j^p\right)^{1/p}\left(\sum_{j=1}^{k} b_j^q + (m-k)\tilde{B}_k^q\right)^{1/q}$$

holds for $p > r$ and $a \in \ell$, where k is given in a lemma (similar to Lemma 1) and $\tilde{B}_k^r = \sum_{j=k+1}^{\infty} b_j^r/(m-K)$. The sign of inequality in (33) is reversed if $p < r$ and the order of the sequence a is reversed.

THEOREM 4. *Let b be a continuous nonincreasing L_r-integrable function on $[0,\infty)$ and M be a positive real number. Then the inequality*

$$(34)\qquad \left(\int_0^{\infty} a(t)^r b(t)^r dt\right)^{1/r} \le \left(\int_0^M a(t)^p dt\right)^{1/p}\left(\int_0^M \tilde{b}(t)^q dt\right)^{1/q}$$

holds for every nonincreasing differentiable function a on $[0,\infty)$ and $p > r$, where

$$\tilde{b}(t) = \begin{cases} b(t)^r, & 0 \le t \le K \\ \left(\int_0^{\infty} b(t)^r dt\right)^{1/r}/(M-K), & K \le t \le M \end{cases}$$

and K is given in a lemma (similar to Lemma 3). The sign of inequality in (34) is reversed if $p < r$ and a is a nondecreasing differentiable function.

NOTE: Since the monotonicity of a^r and b^r (for $r > 0$) is hereditary for a and b, Lemma 1 and 3 can be readily adopted for

a^r and b^r correspondingly. So, we omit straight-forward details. The equality conditions for Theorems 3 and 4 respectively are evident and thus omitted. (Indeed, ℓ is now considered to be the space of all sequences p-th power summable.)

6. CONCLUDING REMARKS

Here we have established the discrete and continous cases (1) and (25) of the refined Hölder inequality introduced in [14, 15] and some variants. With these inequalities, we have solved the corresponding mathematical programming problems (I) - (III) and ($\hat{\text{I}}$) - (I$\hat{\text{I}}$I). It has been known that by means of inequalities, many mathematical programming problems can be easily solved (see, e.g. [19-22]). It has also been recognized that the Bellman dynamic programm approach [6] can be used to solve mathematical programming problems (see, e.g. [7-9, 21]) in order to avoid the complexities of multidimensional analysis [6, p.7]. In the connections, some extensions of the present work are further considered (see 10, 11). It is conceivable that continuous mathematical programming problems ($\hat{\text{I}}$) - (I$\hat{\text{I}}$I) can be established by a continuous dynamic programming approach as introduced in [8, 9]. However, this will not be explored here.

In view of the results given here and in [19-22], a continous development of the idea for solving mathematical programming problems by pertinent inequalities appears to be promising.

ACKNOWLEDGEMENT. The work of the last two authors was supported by grants from the Natural Sciences and Engineering Research Council of Canada.

REFERENCES

1. J. Aczél and E.F. Beckenbach, On Hölder's inequality. In: E.F. Beckenbach (ed.), General Inequalities 2 (pp.145-150), Birkhäuser Verlag, Basel and Stuttgart, 1980.

2. E.F. Beckenbach, On Hölder's inequality. J. Math. Anal. Appl. 15 (1966), 21-29.

3. E.F. Beckenbach, A "workshop" on Minkowski's inequality. In: O. Shisha (ed.), Inequalities I (pp.37-55), Academic Press, New York, 1967.

4. E.F. Beckenbach and R. Bellman, Inequalities. Springer-Verlag, New York, 1971.

5. G.H. Hardy, J.E. Littlewood and G. Pólya, Inequalities. Cambridge Univ. Press, London, 2nd Edition, 1952.

6. R. Bellman, Dynamic Programming. Princeton Univ. Press, Princeton, N.J., 1957.

7. S. Iwamoto, Dynamic programming to inequalities. J. Math. Anal. Appl. 58 (1977), 687-704.

8. S. Iwamoto and Chung-lie Wang, Continuous dynamic programming approach to inequalities. J. Math. Anal. Appl. 96 (1983), 119-129.

9. S. Iwamoto and Chung-lie Wang, Continuous dynamic programming approach to inequalities II. J. Math. Anal. Appl. 118 (1986), 279-286.

10. S. Iwamoto, R.J. Tomkins and Chung-lie Wang, On sensitivity analysis on an ordered allocation problem with a parametric low bound. In preparation.

11. S. Iwamoto, R.J. Tomkins and Chung-lie Wang, In an ordered allocation process two-phase dynamic programming approach. In preparation.

12. A.W. Marshall and I. Olkin, Inequalities: Theory of Majorization and its Applications. Academic Press, New York, 1979.

13. D.S. Mitrinović, Analytic Inequalities. Springer-Verlag, New York, 1970.

14. G.S. Mudholkar, M. Freimer and P. Subbaiah, An extension of Hölder's inequality. J. Math. Anal. Appl. 102 (1984), 435-441.

15. M. Freimer and G.S. Mudholkar, A Class of Generalizations of Hölder's Inequality. Inequalities in Statistics and Probability, IMS Lecture Notes - Monograph Series 5 (1984), 59-67.

16. Chung-lie Wang, Variants of the Hölder inequality and its inverses. Canad. Math. Bull. Vol. 20 (3) (1977), 377-384.

17. Chung-lie Wang, On development of inverses of the Cauchy and Hölder inequalities. SIAM Review, Vol. 21, No. 4 (1979), 550-557.

18. Chung-lie Wang, A survey on basic inequalities. Canadian Math. Soc. Notes 12 (1980), 8-12.

19. Chung-lie Wang, Inequality and Mathematical Programming. In: E.F. Beckenbach and W. Walter (ed.), General Inequalities 3 (pp.149-164), Birkhäuser Verlag, Basel and Stuttgart, 1983.

20. Chung-lie Wang, Inequalities and Mathematical Programming, II. In: W. Walter, General Inequalities 4 (pp.381-393), Birkhäuser Verlag, Basel and Stuttgart, 1984.

21. Chung-lie Wang, The Principle and Models of Dynamic Programming. J. Math. Anal. Appl. 118 (1986), 287-308.

22. Chung-lie Wang, Beckenbach inequalities and its variants. J. Math. Anal. Appl., to appear.

S. Iwamoto, Department of Economic Engineering, Faculty of Economic Kyushu University, Fukuoka 812, Japan

R.J. Tomkins and Chung-lie Wang, Department of Mathematics and Statistics, University of Regina, Regina, Saskatchewan, S4S 0A2, Canada

International Series of
Numerical Mathematics, Vol. 80

FUNCTIONS GENERATING SCHUR-CONVEX SUMS

C.T. Ng

Abstract. This paper provides a representation theorem for functions f with Schur-convex sums $\Sigma_1^n f(x_i)$.

1. INTRODUCTION

Let $X = (x_1, x_2, \ldots, x_n)$ and $Y = (y_1, y_2, \ldots, y_n)$ be n-tuples of real numbers. Then X is said to be majorized by Y, written $X < Y$, if $[x_1, x_2, \ldots, x_n] = [y_1, y_2, \ldots, y_n]P$ for some doubly stochastic $n \times n$ (real) matrix P. Contained in the works of Schur [9], Hardy, Littlewood and Pólya [1], and Karamata [2], is the fact that if a function f is convex on an interval $I \subseteq \mathbb{R}$, then $\Sigma_1^n f(x_i) \le \Sigma_1^n f(y_i)$ whenever $X < Y$ on I^n.

The above majorization between n-tuples of real numbers has been extended to multivariate majorization; and Schur's majorization theorem has a natural extension.

DEFINITION 1 ([7], 3A1 and 15A2). Let $X = (x_1, x_2, \ldots, x_n)$ and $Y = (y_1, y_2, \ldots, y_n)$ be n-tuples of vectors $x_i, y_i \in \mathbb{R}^m$. Then X is said to be majorized by Y, written $X < Y$, if $[x_1, x_2, \ldots, x_n] = [y_1, y_2, \ldots, y_n]P$ for some doubly stochastic $n \times n$ matrix P. Here $[x_1, x_2, \ldots, x_n]$ denotes the $m \times n$ matrix whose i-th column vector is x_i. A real-valued function φ defined on a subset $S \subseteq (\mathbb{R}^m)^n$ is said to be Schur-convex on S if $\varphi(X) \le \varphi(Y)$ whenever $X < Y$ on S.

THEOREM 2 ([7], 15B1). *If* $X < Y$, *then* $\Sigma_1^n f(x_i) \le \Sigma_1^n f(y_i)$ *for all (continuous) convex functions* $f: \mathbb{R}^m \to \mathbb{R}$.

This paper is in final form and no version of it will be submitted for publication elsewhere.

In the next section we give a necessary and sufficient condition on f under which $\Sigma_1^n f(x_i)$ is Schur-convex.

2. FUNCTIONS GENERATING SCHUR-CONVEX SUMS

Let f be a (finite) real-valued function defined on a convex set $D \subseteq \mathbb{R}^m$. Then f is said to be midconvex on D if $f(\lambda x+(1-\lambda)y) \le \lambda f(x)+(1-\lambda)f(y)$ for all $x,y \in D$ and $\lambda = 1/2$; and convex if the inequality holds for all $\lambda \in [0,1]$. A function $A: \mathbb{R}^m \to \mathbb{R}$ is additive if $A(x+y) = A(x)+A(y)$ for all $x,y \in \mathbb{R}^m$. For real-valued functions f and g with possibly different domains, the sum $f+g$ refers to the pointwise sum; and equality $f = g$ refers to the pointwise equality, over the intersection of their domains.

THEOREM 3. *Let $D \subseteq \mathbb{R}^m$ be nonvoid, convex and open, and let $f: D \to \mathbb{R}$ be a function. Then the following statements are equivalent:*

(3.1) *For some fixed $n \ge 2$, the sum $\Sigma_1^n f(x_i)$ is Schur-convex on D^n;*

(3.2) *f satisfies the functional inequality*

$$(*) \qquad f(\lambda x+(1-\lambda)y)+f((1-\lambda)x+\lambda y) \le f(x)+f(y)$$

for all $x,y \in D$, $\lambda \in [0,1]$;

(3.3) *f has the representation $f = C+A$, where $C: D \to \mathbb{R}$ is convex and $A: \mathbb{R}^m \to \mathbb{R}$ is additive;*

(3.4) *For all fixed $n \ge 2$, the sum $\Sigma_1^n f(x_i)$ is Schur-convex on D^n.*

Proof. Suppose for some fixed $n \ge 2$, the sum $\Sigma_1^n f(x_i)$ is Schur-convex on D^n. By holding $x_3,x_4,\ldots,x_n$ fixed in D, the Schur-convexity of $\Sigma_1^2 f(x_i)$ on D^2 follows. The inequality $f(x_1)+f(x_2) \le f(y_1)+f(y_2)$ for $(x_1,x_2) < (y_1,y_2)$ coincides with $(*)$ term for term. Thus (3.1) implies (3.2).

Suppose f satisfies $(*)$. Let $a \in \mathbb{R}^m$ be arbitrarily given and consider the function

(3.5) $$f_a(x) = f(x)+f(a-x)$$

defined for all $x \in D \cap (a-D) =: D_a$. We shall establish the continuity of f_a; and for this purpose we may assume that its domain D_a is nonvoid. Because D is open and convex, so is D_a. Furthermore, D_a is symmetric about the point $2^{-1}a$. Hence $2^{-1}a$ is an interior point of D_a. Since f satisfies (*), so does f_a satisfy (*) on D_a. By specifying $\lambda = 1/2$ in (*) we obtain the midconvexity of f and f_a. Let $b \in D_a$ be any point other than $2^{-1}a$; and consider the closed line segment L joining b and $a-b$. Notice that L is a line segment contained in D_a, having $2^{-1}a$ as its midpoint. If $\ell \in L$, then $\ell = \lambda b+(1-\lambda)(a-b)$ for some $\lambda \in [0,1]$, and $a-\ell = (1-\lambda)b+\lambda(a-b)$. Since f_a satisfies (*) on L, we get $f_a(\ell)+f_a(a-\ell) \le f_a(b)+f_a(a-b)$. But, by the definition of f_a, $f_a(\ell) = f_a(a-\ell)$ and $f_a(b) = f_a(a-b)$, so $f_a(\ell) \le f_a(b)$ follows. Since $\ell \in L$ is arbitrary, this proves that f_a is bounded from above on L. We now fix $b_1,b_2,\dots,b_m \in D_a$ such that $b_1-2^{-1}a, b_2-2^{-1}a,\dots,b_m-2^{-1}a$ are linearly independent. Each b_j defines a line segment L_j on which f_a is bounded from above. Thus f_a is bounded from above on their union $\cup L_j$. Being midconvex, f_a is then bounded from above on the midpoint closure of $\cup L_j$, which, due to the affine independence of the balanced segments L_j at $2^{-1}a$, is a neighbourhood of $2^{-1}a$. By the theorem of Bernstein-Doetsch ([6],p.145), the midconvex f_a, being locally bounded from above at $2^{-1}a$, is continuous on D_a.

Let $c \in \mathbb{R}^m$ be arbitrarily given and consider the difference $\Delta_c f$ defined by

(3.6) $$\Delta_c f(x) = f(x+c)-f(x)$$

on $D \cap (D-c)$, which is open and convex but possibly void. We shall obtain the continuity of $\Delta_c f$ on $D \cap (D-c)$. For this purpose, let $x_0 \in D \cap (D-c)$ be given. Let us arbitrarily select a point $d \in D+x_0$. Then x_0, x_0+c and $d-x_0$ are all in D. Since D is open, there exists a neighbourhood $V(x_0)$ of x_0 such that x, $x+c$ and $d-x$ are all in D whenever $x \in V(x_0)$. Thus every term in the identity

$$f(x+c)-f(x) = [f(x+c)+f(d-x)]-[f(x)+f(d-x)]$$

is defined for all $x \in V(x_0)$. The function $[f(x+c)+f(d-x)]$ in x, being the composition of a translation $x \to x+c$ and the continuous f_{d+c}, is continuous at x_0. The function $[f(x)+f(d-x)]$ in x is but f_d and is also continuous at x_0. Thus their difference is continuous at x_0. This proves the continuity of $\Delta_c f$ at x_0.

By a theorem of De Bruijn and Kemperman ([3], Theorem 5.1), the continuity of the differences $\Delta_c f$ implies the existence of an additive $A: \mathbb{R}^m \to \mathbb{R}$ such that $f-A$ is continuous on D. Since $C := f-A$ is again midconvex, its continuity then yields its convexity. Now $f = C+A$, proving that (3.2) implies (3.3).

Suppose f has the representation $f = C+A$ with convex C and additive A. By THEOREM 2, $\Sigma_1^n C(x_i)$ is Schur-convex on D^n. Since $X < Y$ on D^n implies $\Sigma_1^n x_i = \Sigma_1^n y_i$, and since $\Sigma_1^n A(x_i) = A(\Sigma_1^n x_i)$, $\Sigma_1^n A(y_i) = A(\Sigma_1^n y_i)$, we immediately have the Schur-convexity of $\Sigma_1^n A(x_i)$ on D^n. Now $\Sigma_1^n f(x_i) = \Sigma_1^n C(x_i) + \Sigma_1^n A(x_i)$ yields the Schur-convexity of $\Sigma_1^n f(x_i)$. Hence (3.3) implies (3.4).

As (3.4) trivially implies (3.1), this completes the proof.

REMARK 4. The decomposition of an f into the sum C+A is not unique. If $C'+A'$ is another such decomposition, then $C-C' = A'-A$ (on D). Since $C-C'$ is continuous and $A'-A$ is additive, we obtain the linearity of $A'-A =: T$ on $\mathbb{R}^m$. Thus $C' = C-T$ and $A' = A+T$ with linear T. In short, the decomposition of f is unique up to a linear function. In this sense, the decomposition $f = C+A$ is interesting only when A is discontinuous.

3. WRIGHT CONVEXITY AND FUNCTIONS WITH INCREASING DIFFERENCES

Weinberger [10] proved an inequality which was conjectured by Payne and Weinstein. Wright [11] observed that the inequality

$$(\dagger) \qquad f(x+\delta)-f(x) \le f(y+\delta)-f(y)$$

for $x < y$, $\delta > 0$ (in $\mathbb{R}$) is fundamental in obtaining such type of inequalities; and asked if midconvexity of f implies (†).

Kenyon [4] and Klee [5] furnished examples of midconvex functions failing (†). Roberts and Varberg [8] referred to (†) as Wright-convexity of f. Needless to say, (†) corresponds to that the difference $\Delta_\delta f(x) = f(x+\delta)-f(x)$ is increasing in x for each $\delta > 0$. Kuczma ([6], Chapter VII, §3) has pointed out that convexity of f is sufficient, but not necessary; and that midconvexity of f is necessary, but not sufficient, to imply that $\Delta_\delta f$ is increasing for $\delta > 0$.

Since (†) actually coincides with (*) on real intervals, THEOREM 3 offers a characterization of such functions:

COROLLARY 5. Let $I \subseteq \mathbb{R}$ be an open interval, and let $f: I \to \mathbb{R}$ be a function. Then the difference $\Delta_\delta f$ is increasing for each fixed $\delta > 0$ if, and only if, $f = C+A$ where $C: I \to \mathbb{R}$ is convex and $A: \mathbb{R} \to \mathbb{R}$ is additive.

4. A COMPARISON WITH Λ-CONVEXITY

For $\Lambda \subseteq [0,1]$, nonvoid in $]0,1[$, let $C(\Lambda)$ denote the class of all (Λ-convex) functions $f: \mathbb{R} \to \mathbb{R}$ satisfying $f(\lambda x+(1-\lambda)y) \le \lambda f(x)+(1-\lambda)f(y)$ for all $x,y \in \mathbb{R}$ and $\lambda \in \Lambda$. Under set inclusion, $C(\Lambda)$ is decreasing with respect to Λ. Let $\mathbb{Q}$ be the rationals. Then the class $C(\mathbb{Q} \cap [0,1])$ of all midconvex functions and the class $C([0,1])$ of all convex functions are the extreme situations. Since the class $C = \{f \mid f$ satisfies (*) on $\mathbb{R}\}$ is strictly between $C(\mathbb{Q} \cap [0,1])$ and $C([0,1])$, one may like to know if it is possible to have a subset Λ such that $C(\Lambda)$ coincides with C. In fact, there is no such correspondence. Suppose Λ is such that $C(\Lambda)$ contains C. Then $C(\Lambda)$ contains all additive functions $A: \mathbb{R} \to \mathbb{R}$, and hence the inequality $A(\lambda x+(1-\lambda)y) \le \lambda A(x)+(1-\lambda)A(y)$ holds for all $\lambda \in \Lambda$, $x,y \in \mathbb{R}$ and all additive A. In particular, with $x = 1$ and $y = 0$, $A(\lambda) \le \lambda A(1)$ must hold for all $\lambda \in \Lambda$ and all additive A. Since for each irrational λ, we can construct an additive A with arbitrarily assigned values for $A(\lambda)$ and $A(1)$, in particular $A(\lambda) \le \lambda A(1)$ can be violated, therefore $\Lambda \subseteq \mathbb{Q} \cap [0,1]$. But then $C(\Lambda) \supseteq C(\mathbb{Q} \cap [0,1]) \supsetneq C$. Thus once $C(\Lambda) \supseteq C$, the inclusion has to be proper. In fact, we have shown that once $C(\wedge)$ contains all additive functions, then it must be the class of all midconvex functions.

ACKNOWLEDGEMENT. This research has been supported by the Natural Sciences and Engineering Research Council of Canada under Grant A-8212.

REFERENCES

1. G.H. Hardy, J.E. Littlewood and G. Pólya, Inequalities. Cambridge University Press, Cambridge, 1934.

2. J. Karamata, Sur une inégalité relative aux fonctions convexes. Publ. Math. Univ. Belgrade 1 (1932), 145-148.

3. J.H.B. Kemperman, A general functional equation. Trans. Amer. Math. Soc. 86 (1957), 28-56.

4. H. Kenyon, Note on convex functions. Amer. Math. Monthly 63 (1956), 107.

5. V.L. Klee Jr., Solution of a problem of E.M. Wright on convex functions. Amer. Math. Monthly 63 (1956), 106.

6. M. Kuczma, An Introduction to the Theory of Functional Equations and Inequalities. PWN Warszawa-Kraków-Katowice, 1985.

7. A.W. Marshall and I. Olkin, Inequalities: Theory of Majorization and Its Applications. Academic Press, New York, 1979.

8. A. Wayne Roberts and Dale E. Varberg, Convex Functions. Academic Press, New York, 1973.

9. I. Schur, Über eine klasse von mittelbildungen mit anwendungen die determinanten. Theorie Sitzungsber. Berlin Math. Gesellschaft 22 (1922), 9-20. [Issai Schur Collected Works (A. Brauer and H. Rohrbach, eds.) Vol.II, 416-427. Springer -Verlag, Berlin, 1973.]

10. H.F. Weinberger, An inequality with alternating signs. Proc. Nat. Acad. Sci. USA 38 (1952), 611-613.

11. E.M. Wright, An inequality for convex functions. Amer. Math. Monthly 61 (1954), 620-622.

C.T. Ng, Centre for Information Theory, University of Waterloo, Waterloo, Ontario, Canada, N2L 3G1

International Series of
Numerical Mathematics, Vol. 80

HOW TO MAKE FAIR DECISIONS?

Zsolt Páles

Abstract. In this note we investigate fair decision functions, that is symmetric decision functions which satisfy the compromise principle and neglect odd ball opinions. The main purpose of our theory is to prove that a nonconstant decision function is fair if and only if it is a result of a generalized Gauß-type least square method.

1. INTRODUCTION

Consider a family P of n persons $p_1,\dots,p_n$ (for instance, let P be a parliament) who are make real valued decisions about several political and financial questions. In our model the process of decision making is the following: First any member p of P forms his *opinion* $\omega = \omega(p)$ about the question to be decided. This opinion ω belongs to an abstract set Ω that is called the set of all possible opinions. In practice, this set Ω is usually a subset of a real vector space or it is the set {YES,NO}. In the second step, summarizing all the opinions, a *decision function* d calculates the desired decision, that is, if we have opinions $\omega_1,\dots,\omega_n$, then $d(\omega_1,\dots,\omega_n) \in [-\infty,\infty]$ is the corresponding decision.

The properties of the function $\omega\colon P \to \Omega$ essentially depend on the family of persons belonging to P. Thus ω cannot be well described from a mathematical point of view. Therefore, in the present paper, we restrict ourselves to the investigation of decision functions.

There is a lot of contributions to the theory of decision making (e.g., [2], [3], [4], [5], [8], [9]), but our approach and

---

This paper is in final form and no version of it will be submitted for publication elsewhere.

our results are quite different from those in [2], [3], [4], [5], [8], [9].

Let Ω be an abstract set (without any structure). A function d is called a *decision function* on Ω if it is defined on the set $\bigcup_{n=1}^{\infty}\Omega^n =: \Omega^*$ and if it takes extended real values only.

A decision function $d\colon \Omega^* \to [-\infty,\infty]$ is called *symmetric* if

$$d(\omega_1,\dots,\omega_n) = d(\omega_{i_1},\dots,\omega_{i_n})$$

holds for all $n \in \mathbb{N}$, $\omega_1,\dots,\omega_n \in \Omega$ and for all permutations $i_1,\dots,i_n$ of $1,\dots,n$.

We say that the decision function d satisfies the *compromise principle* (or in other words, it is *internal*) if for all $n,m \in \mathbb{N}$, $\omega_1,\dots,\omega_{n+m} \in \Omega$ the decision corresponding to $\omega_1,\dots,\omega_{n+m}$ is between the decisions corresponding to $\omega_1,\dots,\omega_n$ and $\omega_{n+1},\dots,\omega_{n+m}$, i.e.,

$$\min\,(d(\omega_1,\dots,\omega_n),d(\omega_{n+1},\dots,\omega_{n+m}))$$

$$\le d(\omega_1,\dots,\omega_{n+m})$$

$$\le \max\,(d(\omega_1,\dots,\omega_n),d(\omega_{n+1},\dots,\omega_{n+m}))\ .$$

A decision function d is said to *neglect odd ball opinions* (or in other words, it is *regular*) if for all $n \in \mathbb{N}$, $\omega_0,\dots,\omega_n \in \Omega$ the decision corresponding to $\omega_0,\underbrace{\omega_1,\dots,\omega_1}_{k\text{-times}},\dots,\underbrace{\omega_n,\dots,\omega_n}_{k\text{-times}}$ is close to the decision corresponding to $\omega_1,\dots,\omega_n$ provided that k is large enough, i.e.,

$$\lim_{k\to\infty} d(\omega_0,\underbrace{\omega_1,\dots,\omega_1}_{k\text{-times}},\dots,\underbrace{\omega_n,\dots,\omega_n}_{k\text{-times}}) = d(\omega_1,\dots,\omega_n)\ .$$

A decision function is called *fair* if it is symmetric, internal and regular.

Now we present some simple examples for various decision functions.

EXAMPLE 1. Let Ω be the set of real numbers and let d be the arithmetic mean, i.e.,

$$d(\omega_1,\dots,\omega_n) = \frac{\omega_1+\dots+\omega_n}{n}\ .$$

Then an easy calculation shows that d is a fair decision function. Similarly, one can check that the power means and also quasiarithmetic means are fair decision functions, too.

EXAMPLE 2. Let $\Omega = \mathbb{R}$ and define d by

$$d(\omega_1,\ldots,\omega_n) = \omega_n \ .$$

Then d is not symmetric, but it is internal and regular.

EXAMPLE 3. Let $\Omega = \mathbb{R}_+$ and let d be the average of the arithmetic and geometric mean, i.e.,

$$d(\omega_1,\ldots,\omega_n) = \frac{1}{2}\left(\frac{\omega_1+\ldots+\omega_n}{n} + \sqrt[n]{\omega_1\cdots\omega_n}\right) .$$

Then d is symmetric and regular, however, it is not internal as the following example shows:

$$d(1,81) = d(25,25) = 25 \ , \quad \text{but} \quad d(1,81,25,25) = 24 \ .$$

EXAMPLE 4. Let $\Omega = \mathbb{R}$ and

$$d(\omega_1,\ldots,\omega_n) = \min\,(\omega_1,\ldots,\omega_n) \ .$$

Then it is easy to see that d is symmetric and internal, but it is not regular, since

$$1 = \lim_{k\to\infty} d(1,\underbrace{2,\ldots,2}_{k\text{-times}}) \neq d(2) = 2 \ .$$

EXAMPLE 5. Let $\Omega = \{Y,N\}$ and let $a \neq b$ be fixed real values. If $\omega_1,\ldots,\omega_n \in \Omega$, then let $d(\omega_1,\ldots,\omega_n)$ be equal to a if the number of Y's is greater than the number of N's among $\omega_1,\ldots,\omega_n$, and let it be b in the other cases. For instance,

$$d(Y,N,Y,Y) = a \quad \text{and} \quad d(Y,N) = b \ .$$

That is, the value of the decision is a if the opinion Y(es) has majority, and the decision is b in the opposite case.

Now one can check that this decision function d is symmetric and internal, but it is not regular since

$$a = \lim_{k\to\infty} d(Y,\underbrace{Y,\ldots,Y}_{k\text{-times}},\underbrace{N,\ldots,N}_{k\text{-times}}) \neq d(Y,N) = b \ .$$

2. A GENERALIZED LEAST SQUARE METHOD

It is well known that the arithmetic mean can be obtained as a result of the least square method due to Gauß. Namely, if $\omega_1,\dots,\omega_n$ are fixed real values, then the unique solution $t = t_o$ of the minimum problem

$$(\omega_1-t)^2 + \dots + (\omega_n-t)^2 \to \min$$

is exactly the arithmetic mean of $\omega_1,\dots,\omega_n$.

Let $I =]\alpha,\beta[\subset \mathbb{R}$ be an open interval ($-\infty \le \alpha < \beta \le \infty$). A convex function $f\colon I \to \mathbb{R}$ is called *strongly* [resp. *strictly*] convex if it is nonconstant [resp. nonlinear] on each proper subinterval of I.

Let Ω be an abstract set. A function $D\colon \Omega \times I \to \mathbb{R}$ is said to be an *admissible generating function* if

$$(1) \qquad t \mapsto f(t) := D(\omega_1,t) + \dots + D(\omega_n,t), \quad t \in I,$$

is strongly convex for all fixed $n \in \mathbb{N}$ and $\omega_1,\dots,\omega_n \in \Omega$. (We remark that if $t \to D(\omega,t)$ is strictly convex for all fixed $\omega \in \Omega$, then (1) is strongly convex; however, the strict convexity is not necessary.)

Employing an admissible generating function D, we can define a decision function d_D in the following way: Let $n \in \mathbb{N}$ and $\omega_1,\dots,\omega_n \in \Omega$ be fixed and consider the function f defined in (1). By the strong convexity of f, there exists a unique element $t_o \in [\alpha,\beta]$ such that f is strictly decreasing on $]\alpha,t_o[$ and is strictly increasing on $]t_o,\beta[$. This element t_o is denoted by $d_D(\omega_1,\dots,\omega_n)$, and the function $d_D\colon \Omega^* \to [-\infty,\infty]$ is called the D-*decision function*.

If $\Omega = \mathbb{R}$ and $D(\omega,t) = (\omega-t)^2$, then D is clearly an admissible generating function and d_D is the arithmetic mean.

Our fist result summarizes the properties of D-decision functions.

THEOREM 1. <u>Let</u> D <u>be an admissible generating function</u> (<u>on</u> $\Omega \times I$). <u>Then</u> d_D <u>is a fair decision function</u>, <u>i.e.</u>, <u>it is symmetric</u>, <u>internal and regular</u>.

<u>Proof.</u> The symmetry of d_D is obvious. To prove the internality, let $\omega_1,\dots,\omega_{n+m} \in \Omega$ be arbitrarily fixed. Denote

$$t_o = d_D(\omega_1,\ldots,\omega_n),\ t_1 = d_D(\omega_{n+1},\ldots,\omega_{n+m}),\ t_2 = d_D(\omega_1,\ldots,\omega_{n+m}) .$$

By the definition of d_D, (1) is strictly decreasing on $]\alpha,t_o[$, and

$$t \mapsto D(\omega_{n+1},t) + \ldots + D(\omega_{n+m},t)$$

is strictly decreasing on $]\alpha,t_1[$. Hence

$$t \mapsto D(\omega_1,t) + \ldots + D(\omega_{n+m},t)$$

is necessarily strictly decreasing on $]\alpha, \min(t_o,t_1)[$. This yields $\min(t_o,t_1) \leq t_2$. A similar argument shows that $t_2 \leq \max(t_o,t_1)$. Thus we have proved the internality of d_D.

To check regularity, let $\omega_o,\omega_1,\ldots,\omega_n \in \Omega$ be arbitrary and let

$$t_o = d_D(\omega_1,\ldots,\omega_n), \quad t_k = d_D(\omega_o,\underbrace{\omega_1,\ldots,\omega_1}_{k\text{-times}},\ldots,\underbrace{\omega_n,\ldots,\omega_n}_{k\text{-times}}) .$$

If $\alpha < t_o$, then choose t' with $\alpha < t' < t_o$ arbitrarily. Since (1) is strictly decreasing on $]\alpha,t_o[$, hence, for sufficiently large values of k,

$$t \mapsto D(\omega_o,t) + kD(\omega_1,t) + \ldots + kD(\omega_n,t)$$

is strictly decreasing on $]\alpha,t'[$. This proves that $t' \leq t_k$ if k is large enough. Since t' was arbitrary, hence $t_o \leq \liminf t_k$, and this is also valid in the case $t_o = \alpha$. Similarly, we can see that $\limsup t_k \leq t_o$ holds. Thus $\lim t_k = t_o$, which was to be proved.

3. A HAHN-BANACH THEOREM

Let $S = (S,+)$ be an arbitrary Abelian group. If A is a subsemigroup of S then the *algebraic interior* (or in other words, the *core*) of A is the set (see [6], [7])

$$\operatorname{cor} A := \{a \in A \mid \forall s \in S\ \exists n \in \mathbb{N}:\ na + s \in A\} .$$

It is easy to check that $\operatorname{cor} A$ is also a subsemigroup of S. If $A = \operatorname{cor} A$, then we say that A is *algebraically open*.

The following observation will be very useful in the determination of the algebraic interior of subsemigroups: If S is additively generated by a subset $T \subset S$, then

$$\operatorname{cor} A = \{a \in A \mid \forall t \in T\ \exists n \in \mathbb{N}:\ na + t \in A\} . \tag{2}$$

Proof. Denote by B the right hand side of (2). The inclusion cor $A \subseteq B$ is obvious. To prove the reversed inclusion, let $a \in B$ and let $s \in S$ be arbitrary. Since S is generated by T, there exist $t_1, \dots, t_k \in T$ such that $s = t_1 + \dots + t_k$. Since $a \in B$, we find $n_1, \dots, n_k \in \mathbb{N}$ such that $n_1 a + t_1, \dots, n_k a + t_k \in A$. Adding these elements we obtain

$$(n_1 + \dots + n_k)a + s \in A,$$

thus $a \in \operatorname{cor} A$.

Now we state a Hahn-Banach theorem for Abelian semigroups. This result is a special case of a more general statement in [6].

LEMMA. *Let S be an Abelian semigroup and let A and B be disjoint subsemigroups of S. Assume that one of the following conditions is satisfied:*

(i) $\operatorname{cor} A \neq \emptyset$ *and* $\operatorname{cor} B \neq \emptyset$;

(ii) *S is a group,* $\operatorname{cor} A \neq \emptyset$ *and* $B \neq \emptyset$.

Then there exists an additive function $f: S \to \mathbb{R}$ *such that*

(3) $$f|_A \le 0 \quad \text{and} \quad 0 \le f|_B$$

and furthermore

(4) $$f|_{\operatorname{cor} A} < 0 \qquad \text{and} \qquad 0 < f|_{\operatorname{cor} B} .$$

We remark that f cannot be identically zero, since in both cases $\operatorname{cor} A \neq \emptyset$, whence the first inequality in (4) implies that $f \neq 0$.

4. CHARACTERIZATION OF FAIR DECISION FUNCTIONS

In this section we prove a converse of Theorem 1, using the above Hahn-Banach separation theorem.

THEOREM 2. *Let* $d: \Omega^* \to [-\infty, \infty]$ *be a fair decision function. Then either d is constant or there exists an admissible generating function* $D: \Omega \times I \to \mathbb{R}$, *where* $I =]\inf d, \sup d[$, *such that* $d = d_D$.

Proof. Let d be a nonconstant fair decision function. We are going to construct D in three steps.

Step 1. Denote by $S = (S,+)$ the free Abelian semigroup generated by the elements of Ω. (This semigroup S is isomorphic to the semigroup of nonnegative integer valued functions that are defined on Ω and have finite support.)

Using d, we define a function $m: S \setminus \{0\} \to [-\infty,\infty]$ in the following way: If $s = \omega_1 + \dots + \omega_n$ then let

$$m(s) = d(\omega_1, \dots, \omega_n) .$$

Since d is a symmetric function, the value of m(s) does not depend on the representation of s. The internality of d implies that

$$\min (m(s),m(t)) \le m(s+t) \le \max (m(s),m(t)) \tag{5}$$

holds for all $s,t \in S$. The regularity of d means that

$$\lim_{k\to\infty} m(\omega_0+ks) = m(s) \tag{6}$$

for all $s \in S$ and $\omega_0 \in \Omega$.

Let $y \in I$ be an arbitrarily fixed element and define

$$A := A_y := \{s \in S \setminus \{0\} \mid m(s) < y\} ,$$

$$B := B_y := \{s \in S \setminus \{0\} \mid m(s) > y\} .$$

Then clearly A and B are disjoint subsets of S, and furthermore $A \neq \emptyset$ and $B \neq \emptyset$. We show that A and B are algebraically open subsemigroups of S.

If $s,t \in A$ then $m(s),m(t) < y$. Thus it follows from the second inequality in (5) that $m(s+t) < y$, i.e. $s+t \in A$. To prove the relation $A = \mathrm{cor}\, A$, observe that the set Ω additively generates S. If $s \in A$, then $m(s) < y$. Thus, if $\omega_0 \in \Omega$ is fixed, then (6) implies that $m(\omega_0+ks) < y$, i.e. $\omega_0+ks \in A$ if k is large enough. This shows that $s \in \mathrm{cor}\, A$.

Analogously, one can see that B is also an algebraically open subsemigroup of S.

Thus we have proved that the conditions (i) of the lemma are satisfied. Hence there exists an additive function $f = f_y: S \to \mathbb{R}$ such that (4) holds, i.e.

$$(7) \qquad f_y(s) \begin{matrix} < 0\,, & \text{if} \quad m(s) < y\,, \\ > 0\,, & \text{if} \quad m(s) > y\,. \end{matrix}$$

Step 2. Now we are going to show the existence of a negative valued function $g: I \to \mathbb{R}$ such that

$$(8) \qquad y \mapsto g(y)f_y(s)\,, \qquad y \in I\,,$$

is an increasing function for all fixed $s \in S$.

For $y \in I$ define $\delta_y: I \to \mathbb{R}$ by

$$\delta_y(x) = \begin{cases} 1\,, & \text{if} \quad x = y\,, \\ 0\,, & \text{if} \quad x \neq y\,. \end{cases}$$

Let G be the set of all real-valued functions on I with finite support:

$$G = \{ \sum_{i=1}^{n} r_i \delta_{y_i} \mid n \in \mathbb{N},\ r_1,\dots,r_n \in \mathbb{R},\ y_1,\dots,y_n \in I\}\,.$$

Then G is a group, moreover, it is a real linear space.

Let Δ be the set of all pairs of functions $p: I \to \mathbb{R}$ and $s: I \times I \to S$ satisfying the following conditions: p is a nonegative valued real function with nonempty finite support, furthermore, if either $y \geq z$ or $p(y)p(z) = 0$, then $s(y,z) = 0$.

Denote by C the set of elements

$$c = c_{p,s} := \sum_{y \in I} \left(p(y) - f_y\left(\sum_{z \in I} s(z,y)\right) + f_y\left(\sum_{z \in I} s(y,z)\right)\right)\delta_y\,,$$

where $(p,s) \in \Delta$ is arbitrary.

If (p,s) and (p',s') are in Δ, then $(p+p',s+s')$ is also in Δ. Therefore $c_{p,s}+c_{p',s'} = c_{p+p',s+s'}$ belongs to C. This shows that C is a subsemigroup of G.

If $q > 0$ and $x \in I$ then, letting $p = q\delta_x$ and $s \equiv 0$, we see that $c_{p,s} = q\delta_x$ is in C. However we can prove more, namely that

$$(9) \qquad \{q\delta_x \mid q \in \mathbb{R}_+,\ x \in I\} \subset \operatorname{cor} C\,.$$

Since the set of elements $r\delta_y$ $(r \in \mathbb{R},\ y \in I)$ additively generates G, it is enough to show that, for all fixed $q \in \mathbb{R}_+$, $r \in \mathbb{R}$, $x,y \in I$,

$$(10) \qquad nq\delta_x + r\delta_y \in C$$

if n is large enough.

If $x = y$ then choose n such that $nq+r$ be positive, thus (10) is satisfied.

Now consider the case $y < x$. Then the left hand side of (10) can be rewritten in the following way:

$$nq\delta_x + r\delta_y = (nq+f_x(s_o)-f_x(s_o))\delta_x + (r-f_y(s_o)+f_y(s_o))\delta_y \,. \tag{11}$$

Choose $s_o \in S$ such that $r-f_y(s_o)$ be positive. (For instance, if $s_1 \in A_y$ then by (7), $f_y(s_1) < 0$. Thus, for sufficiently large k, $r-f_y(ks_1) > 0$.) Now choose $n \in \mathbb{N}$ such that $nq+f_x(s_o)$ be also positive. Define $p\colon I \to \mathbb{R}$ and $s\colon I \times I \to S$ by

$$p = (nq+f_x(s_o))\delta_x + (r-f_y(s_o))\delta_y \,,$$

$$s(u,v) = \begin{cases} s_o \,, & \text{if } u = y,\ v = x \,, \\ 0 & \text{in the other cases .} \end{cases}$$

Then clearly $(p,s) \in \Delta$ and $c_{p,s}$ is equal to the right hand side of (11). This validates (10).

In the case $x < y$ a similar argument shows (10) for large values of n. Thus (9) is proved.

Our last task is to prove that $c_{p,s} \le 0$ cannot be satisfied for any $(p,s) \in \Delta$. We prove this statement using induction for the cardinality of the support of p.

If the support of p is a single element $y \in I$, then $c = c_{p,s} = p(y)\delta_y$ which shows that c is positive at y.

Assume now that if the support of p has at most n elements, then the statement is valid. Let $p\colon I \to \mathbb{R}$ be a nonnegative function with support $\{y_1,\dots,y_{n+1}\}$, $y_1 < \dots < y_{n+1}$. Assume on the contrary that there exists a function s such that $(p,s) \in \Delta$ and $c_{p,s} \le 0$. Then

$$p(y_i) - f_{y_i}\left(\sum_{j=1}^{i-1} s(y_j,y_i)\right) + f_{y_i}\left(\sum_{j=i+1}^{n+1} s(y_i,y_j)\right) \le 0 \tag{12}$$

for $i \in \{1,\dots,n+1\}$. Particularly, for $i = n+1$, we have

$$p(y_{n+1}) - f_{y_{n+1}}\left(\sum_{j=1}^{n} s(y_j,y_{n+1})\right) \le 0 \,,$$

i.e.

$$f_{y_{n+1}}\left(\sum_{j=1}^{n} s(y_j,y_{n+1})\right) > 0 \,.$$

Then, applying (7), we obtain

$$m\left(\sum_{j=1}^{n} s(y_j,y_{n+1})\right) \geq y_{n+1} > y_n ,$$

whence, by (7) again, we get

$$f_{y_n}\left(\sum_{j=1}^{n} s(y_j,y_{n+1})\right) > 0 .$$

Substract this inequality from (12) for $i = n$. Then

$$(13) \qquad p(y_n) - f_{y_n}\left(\sum_{j=1}^{n-1} (s(y_j,y_n)+s(y_j,y_{n+1}))\right) < 0 .$$

Define now the functions $p'\colon I \to \mathbb{R}$ and $s'\colon I \times I \to S$ by

$$p'(u) = \begin{cases} p(u) , & \text{if } I \ni u \neq y_{n+1} , \\ 0 , & \text{if } \quad u = y_{n+1} \end{cases}$$

and

$$s'(u,v) = \begin{cases} s(u,v) , & \text{if } u \in I,\ v \in I \setminus \{y_n,y_{n+1}\} \\ s(u,y_n)+s(u,y_{n+1}), & \text{if } u \in I \setminus \{y_n\},\quad v = y_n, \\ 0 & \text{in the other cases.} \end{cases}$$

Then $(p',s') \in \Delta$, and the inequalities (12) for $i \in \{1,\dots,n-1\}$ and (13) can be rewritten in the following common form:

$$p'(y_i) - f_{y_i}\left(\sum_{j=1}^{i-1} s'(y_j,y_i)\right) + f_{y_i}\left(\sum_{j=i+1}^{n} s'(y_i,y_j)\right) \leq 0$$

for $i \in \{1,\dots,n\}$. These inequalities yield $c_{p',s'} \leq 0$ which is a contradiction, since the support of p' has at most n elements.

It follows from the statement verified above that C does not contain the zero element of G. That is, C and $D = \{0\}$ are disjoint subsemigroups of G and the core of C is nonempty by (9). Thus the lemma can be applied in the case (ii). Hence there exists an additive function $F\colon G \to \mathbb{R}$ such that $F|_C \leq 0$ and $F|_{\mathrm{cor}\, C} < 0$. By the latter inequality and (9) we obtain

$$(14) \qquad F(q\delta_x) < 0$$

if $q > 0$ and $x \in I$. Fix $x \in I$ arbitrarily. Then the function

$$h(q) := F(q\delta_x) , \qquad q \in \mathbb{R} ,$$

is additive and by (14) it is bounded above on an interval. Therefore h is necessarily a linear function (see [1]), that is,

$$F(q\delta_x) = h(q) = qh(1) = qF(\delta_x) .$$

This shows that F is also a linear function.

Let $p_1, p_2 > 0$ any $y_1, y_2 \in I$ with $y_1 < y_2$, further let $s \in S$. Then it is easily seen that

$$(p_1 + f_{y_1}(s))\delta_{y_1} + (p_2 - f_{y_2}(s))\delta_{y_2} \in C .$$

Thus $F|_C \le 0$, and the linearity of F yields

$$(p_1 + f_{y_1}(s))F(\delta_{y_1}) + (p_2 - f_{y_2}(s))F(\delta_{y_2}) \le 0 .$$

Taking the limits $p_1, p_2 \to 0$ we get

(15) $$F(\delta_{y_1})f_{y_1}(s) \le F(\delta_{y_2})f_{y_2}(s)$$

for all $s \in S$ and $y_1 < y_2$.

Define now the function $g: I \to \mathbb{R}$ by

$$g(y) := F(\delta_y) , \qquad y \in I ,$$

then (14) shows that g is negative valued, and the inequality (15) implies that (8) is an increasing function for all $s \in S$.

Step 3. Now we can construct the desired admissible generating function D.

Let $t^* \in I$ be a fixed element and let

$$D(\omega, t) = \int_{t^*}^{t} g(y) f_y(\omega) dy , \qquad \omega \in \Omega , \qquad t \in I .$$

By the construction of g we can see that $t \to D(\omega, t)$ is convex for all fixed ω, since the integrand is increasing.

Let $\omega_1, \ldots, \omega_n$ be arbitrarily fixed values. We show that (1) is strongly convex and $t_0 := d(\omega_1, \ldots, \omega_n) = d_D(\omega_1, \ldots, \omega_n)$.

Using (7) and the negativity of g we obtain that

$$g(y)f_y(\omega_1) + \ldots + g(y)f_y(\omega_n) \begin{cases} > 0 , & \text{if } t_0 < y , \\ < 0 , & \text{if } t_0 > y . \end{cases}$$

This means that the function

$$y \mapsto g(y)f_y(\omega_1) + \ldots + g(y)f_y(\omega_n)$$

is strictly negative on $]\alpha, t_o[$ and is strictly positive on $]t_o, \beta[$, where $\alpha = \inf I$, $\beta = \sup I$. Therefore, integrating it, we find that (1) is strictly decreasing on $]\alpha, t_o[$ and is strictly increasing on $]t_o, \beta[$. This shows that (1) is strongly convex and $t_o = d_D(\omega_1, \dots, \omega_n)$.

The proof of the theorem is complete.

Summarizing our results we can state the following

COROLLARY. A decision function is fair if and only if either it is constant or it is determined by an admissible generating function.

REFERENCES

1. J. Aczél, Lectures on functional equations and their applications. Academic Press, New York - London, 1966.

2. J. Aczél, On weighted synthesis of judgements. Aequationes Math. 27 (1984), 288-307.

3. J. Aczél and C. Alsina, On synthesis of judgements. Manuscript (23 pages).

4. J. Aczél, Pl. Kannappan, C.T. Ng and C. Wagner, Functional equations and inequalities in 'rational group decision making'. In: E.F. Beckenbach and W. Walter (ed.), General Inequalities 3, Birkhäuser Verlag, Basel - Boston - Stuttgart, 1983, 239-246.

5. J. Aczél and T.L. Saaty, Procedures for synthesizing ratio judgements. J. Math. Psych. 27 (1983), 93-102.

6. Zs. Páles, A generalization of the Dubovitskii-Milyutin separation theorem for commutative semigroups. Submitted to Arch. Math.

7. Zs. Páles, Hahn-Banach theorem for separation of semigroups and its applications. Submitted ot Aequationes Math.

8. T.L. Saaty, Decision making for leaders. Lifetime Learning Publications, Belmont, Calif., 1982.

9. C. Wagner, Allocations, Lehrer models and the consensus of probabilities. Theory and Decisions 3 (1981),

Zsolt Páles, Department of Mathematics, Kossuth Lajos University, H-4010 Debrecen, Pf. 12, Hungary

International Series of
Numerical Mathematics, Vol. 80

ON THE τ_T-PRODUCT OF SYMMETRIC AND SUBSYMMETRIC DISTRIBUTION FUNCTIONS

B. Schweizer

Abstract. Let F be a non-defective distribution function (d.f.) and let $\bar{F}$ be the d.f. defined by $\bar{F}(x) = 1 - \ell^+F(-x)$. Then F is symmetric if $F = \bar{F}$; and we say that F is subsymmetric if $F \leq \bar{F}$. If T is a continuous t-norm and τ_T is the induced triangle function, then for any non-defective d.f.'s F and G we have $\tau_T(\bar{F},\bar{G}) \leq \overline{\tau_T(F,G)}$. It follows that if F and G are subsymmetric then $\tau_T(F,G)$ is subsymmetric. However, $\tau_T(F,G)$ is symmetric for all symmetric F and G only if $T = \text{Min}$.

For any non-decreasing function f defined on the real line $\mathbf{R}$, let ℓ^-f and ℓ^+f be the functions defined by

$$\ell^-f(x) = f(x-) \quad \text{and} \quad \ell^+f(x) = f(x+).$$

Clearly ℓ^-f is left-continuous and ℓ^+f is right-continuous.

A non-defective, one-dimensional distribution function (d.f.) is a function $F:\mathbf{R} \to [0,1]$ that is non-decreasing, left-continuous and such that $\lim_{x\to -\infty} F(x) = 0$ and $\lim_{x\to\infty} F(x) = 1$. We let $\mathcal{D}$ denote the set of all d.f.'s. A sequence $\{F_n\}$ in $\mathcal{D}$ is said to converge weakly to F in $\mathcal{D}$ (and we write $F_n \xrightarrow{w} F$) if $F_n(x) \to F(x)$ at every continuity point x of the limit function F.

Let j denote the identity function on $\mathbb{R}$, and for any F in $\mathcal{D}$, let $\bar{F}$ be the function defined on $\mathbb{R}$ by

$$(1) \qquad \bar{F} = 1 - (\ell^+F) \circ (-j) = 1 - \ell^-(F \circ (-j)).$$

This paper is in final form and no version of it will be submitted for publication elsewhere.

Clearly $\bar{F}$ is in $\mathcal{D}$ and $\bar{\bar{F}} = F$. We also have the following results which are easy consequences of the symmetry between F and $\bar{F}$ and which we state without proof.

LEMMA 1. *Let* d *denote either the usual Lévy metric* [3] *or the modified Lévy metric defined in* [4]. *Then, for any* F *and* G *in* $\mathcal{D}$, $d(F,G) = d(\bar{F},\bar{G})$.

LEMMA 2. *Let* F *be in* $\mathcal{D}$ *and let* $\{F_n\}$ *be a sequence in* $\mathcal{D}$ *such that* $F_n \xrightarrow{w} F$. *Then* $\bar{F}_n \xrightarrow{w} \bar{F}$.

LEMMA 3. *For any* F, G *in* $\mathcal{D}$, $G \leqq F$ *if and only if* $\bar{F} \leqq \bar{G}$.

The distribution function F is *symmetric* if $F = \bar{F}$; and we will say that F is *subsymmetric* if $F \leqq \bar{F}$ or, equivalently, if

$$F(x) + \ell^{+}F(-x) \leqq 1 , \tag{2}$$

for all x in $\mathbb{R}$. It follows at once that if F is subsymmetric and $G \leqq F$, then G is subsymmetric. Moreover, if a sequence $\{F_n\}$ of subsymmetric (resp., symmetric) distribution functions in $\mathcal{D}$ converges weakly to F in $\mathcal{D}$, then F is subsymmetric (resp., symmetric).

A *t-norm* T is a mapping from $[0,1] \times [0,1]$ to $[0,1]$ which is non-decreasing in each place, commutative, associative, and such that $T(1,x) = x$ for all x in $[0,1]$. It follows that $T(x,y) \leqq \text{Min}(x,y)$ for all x,y in $[0,1]$, whence Min is the unique maximal t-norm.

For any t-norm T, the *t-conorm* of T is the function T^* from $[0,1] \times [0,1]$ to $[0,1]$ defined by

$$T^*(x,y) = 1 - T(1 - x, 1 - y). \tag{3}$$

Clearly, $\text{Min}^* = \text{Max}$ and, for any t-norm T and all x,y in $[0,1]$,

$$T(x,y) \leqq \text{Min}(x,y) \leqq \text{Max}(x,y) \leqq T^*(x,y). \tag{4}$$

For any continuous t-norm T, the *t-function* induced by T is the mapping τ_T from $\mathcal{D} \times \mathcal{D}$ into $\mathcal{D}$ which is defined via

$$\tau_T(F,G)(x) = \sup_{u+v=x} T(F(u),G(v)), \tag{5}$$

for any F, G in $\mathcal{D}$ and any x in $\mathbb{R}$. It is known (see [4, Section 7.2]) that τ_T is non-decreasing (with respect to the usual pointwise ordering of functions in $\mathcal{D}$), continuous (with respect to weak convergence), commutative, associative, and such that $\tau_T(\varepsilon_0,F) = F$ for all F in $\mathcal{D}$, where ε_0 is given by

$$\varepsilon_0(x) = \begin{cases} 0, & x \leqq 0, \\ 1, & x > 0. \end{cases}$$

The principal aim of this note is to show that if F and G are subsymmetric and if T is a continuous t-norm, then $\tau_T(F,G)$ is subsymmetric (i.e., that the set of all subsymmetric d.f.'s is closed under τ_T) and, in addition, that $\tau_T(F,G)$ is symmetric for all symmetric F and G only if T = Min. We begin with an inequality which is of interest in its own right.

THEOREM 1. *Let T be a continuous t-norm and let τ_T be the induced t-function. Then, for any F, G in $\mathcal{D}$,*

$$\tau_T(\bar{F},\bar{G}) \leqq \overline{\tau_T(F,G)}; \tag{6}$$

and equality holds for all F, G in $\mathcal{D}$ if and only if T = Min.

Proof. Suppose first that F and G are continuous, so that $\bar{F} = 1 - F \circ (-j)$ and $\bar{G} = 1 - G \circ (-j)$. Then, for any x in $\mathbb{R}$, we have

$$\begin{aligned} \tau_T(\bar{F},\bar{G})(x) &= \sup_{u+v=x} T(\bar{F}(u),\bar{G}(v)) \\ &= \sup_{u+v=x} T(1 - F(-u), 1 - G(-v)) \\ &= \sup_{u+v=x} [1 - T^*(F(-u),G(-v))] \\ &= 1 - \inf_{u+v=x} T^*(F(-u),G(-v)) \\ &= 1 - \inf_{u+v=-x} T^*(F(u),G(v)). \end{aligned}$$

By (4), $T(x,y) \leqq T^*(x,y)$ for all x,y in $[0,1]$. Thus, for any x in $\mathbf{R}$,

(7) $$\sup_{u+v=x} T(F(u),G(v)) \leqq \inf_{u+v=x} T^*(F(u),G(v)).$$

Now using (5), (7) and the fact that $\tau_F(F,G)$ is a continuous d.f., yields

$$\tau_T(\bar{F},\bar{G})(x) \leqq 1 - \tau_T(F,G)(-x) = \overline{\tau_T(F,G)}(x)$$

and establishes (6) for continuous distribution functions.

Next, let F and G be any, not necessarily continuous, d.f.'s in $\mathcal{D}$. Then there exist sequences $\{F_n\}$ and $\{G_n\}$ of continuous d.f.'s in $\mathcal{D}$ such that $F_n \overset{w}{\to} F$ and $G_n \overset{w}{\to} G$ [1, p. 270]. By the above argument,

$$\tau_T(\bar{F}_n,\bar{G}_n) \leqq \overline{\tau_T(F_n,G_n)} ,$$

for $n = 1,2,\ldots;$ and by Lemma 2, $\bar{F}_n \overset{w}{\to} \bar{F}$ and $\bar{G}_n \overset{w}{\to} \bar{G}$. Thus, since τ_T is continuous on $\mathcal{D}$, using Lemma 2 again, we have

$$\tau_T(\bar{F},\bar{G}) = \lim_{n\to\infty} \tau_T(\bar{F}_n,\bar{G}_n) \leqq \lim_{n\to\infty} \overline{\tau_T(F_n,G_n)} = \overline{\tau_T(F,G)},$$

and this establishes the inequality (6) for all F, G in $\mathcal{D}$.

Turning to the case of equality, we first recall that a special case of the basic duality theorem established in [2] (see also [4, Section 7.7]) states that if $T = \mathrm{Min}$, then equality holds in (7). Thus if F and G are continuous, then

(8) $$\tau_{\mathrm{Min}}(\bar{F},\bar{G}) = \overline{\tau_{\mathrm{Min}}(F,G)},$$

whence, by continuity, (8) is valid for all F, G in $\mathcal{D}$.

Lastly, if $T \neq \mathrm{Min}$, then there exists a number c, $0 < c < 1$, such that $T(c,c) < c$. Suppose first that $c \leqq 1/2$ and let F_c be the d.f. defined by

$$(9)\qquad F_c(x) = \begin{cases} 0, & x \leq -1, \\ c, & -1 < x \leq 0, \\ 1-c, & 0 < x \leq 1, \\ 1, & 1 < x . \end{cases}$$

Note that F_c is symmetric, so that $F_c = \bar{F}_c$. A simple calculation yields that $\tau_T(F_c,F_c)(3/2) = T(1-c,1) = 1-c$ and that $\tau_T(F_c,F_c)(-3/2) = T(c,c)$; and it is readily seen that $\tau_T(F_c,F_c)$ is continuous at $x = -3/2$. Thus we have

$$\begin{aligned}\tau_T(\bar{F}_c,\bar{F}_c)(3/2) &= 1 - c < 1 - T(c,c) \\ &= 1 - \tau_T(F_c,F_c)(-3/2) = \overline{\tau_T(F_c,F_c)}(3/2)\end{aligned}$$

and equality in (6) fails to hold in this case.

If $c > 1/2$, then defining F_{1-c} via (9), we find that $\tau_T(F_{1-c},F_{1-c})(1/2) = \mathrm{Max}[T(c,c),1-c]$ and that $\tau_T(F_{1-c},F_{1-c})$ is continuous at $x = -1/2$ and equal to $T(1-c,c)$. If $\mathrm{Max}[T(c,c),1-c] = T(c,c)$, then we have

$$\tau_T(\bar{F}_{1-c},\bar{F}_{1-c})(1/2) = T(c,c) < c$$

and

$$\overline{\tau_T(F_{1-c},F_{1-c})}(1/2) = 1 - T(1-c,c) \geq 1 - (1-c) = c.$$

If $\mathrm{Max}[T(c,c),1-c] = 1-c$, then we have $\tau_T(\bar{F}_{1-c},\bar{F}_{1-c})(1/2) = 1-c$; and since $1-c < c$, it follows that $T(1-c,c) \leq T(c,c) < c$, whence $\overline{\tau_T(F_{1-c},F_{1-c})}(1/2) = 1 - T(1-c,c) > 1-c$. Thus equality in (6) fails again; and this completes the proof of Theorem 1.

THEOREM 2. *Let T be a continuous t-norm, let τ_T be the induced t-function, and suppose that F and G are subsymmetric. Then $\tau_T(F,G)$ is subsymmetric.*

Proof. Since $F \leq \bar{F}$, $G \leq \bar{G}$ and τ_T is non-decreasing, using Theorem 1 we have,

$$\tau_T(F,G) \leq \tau_T(\bar{F},\bar{G}) \leq \overline{\tau_T(F,G)} .$$

COROLLARY 1. *If* F *and* G *are symmetric, then* $\tau_{Min}(F,G)$ *is symmetric*.

Furthermore, since the distribution functions F_c defined in (9) are symmetric, we have

COROLLARY 2. *The distribution function* $\tau_T(F,G)$ *is symmetric for all symmetric distribution functions* F *and* G *only if* $T = \mathrm{Min}$.

In conclusion we note that if F is the d.f. of a random variable X, then $\bar{F}$ is the d.f. of $-X$, and that if F and G are given d.f.'s, then there exist independent random variables X and Y such that F is the d.f. of X and G is the d.f. of Y. The convolution, $F * G$, of F and G is then the d.f. of $X + Y$. Moreover, $-X$ and $-Y$ are also independent; and since $(-X) + (-Y) = -(X + Y)$, it follows that for any d.f.'s F and G in $\mathcal{D}$, $\bar{F} * \bar{G} = \overline{F * G}$. The simple argument in proof of Theorem 2 now yields that if F and G are subsymmetric, then $F * G$ is also subsymmetric. (The fact that the convolution of symmetric d.f.'s is symmetric is well-known.)

REFERENCES

1. P. Billingsley, Probability and Measure. John Wiley & Sons, New York, [1979].

2. M. J. Frank and B. Schweizer, On the duality of generalized infimal and supremal convolutions, Rend. Mat. (6) 12 [1979], 1-23.

3. B. V. Gnedenko and A. N. Kolmogorov, Limit Distributions for Sums of Independent Random Variables, Addison-Wesley, Reading MA, [1954].

4. B. Schweizer and A. Sklar, Probabilistic Metric Spaces, Elsevier Science Publishing Co., New York, [1983].

Berthold Schweizer, Department of Mathematics and Statistics, University of Massachusetts, Amherst, MA 01003 U.S.A.

Problems and Remarks

Schwarzwald house with lower cow-byre

International Series of
Numerical Mathematics, Vol. 80

THE BEHAVIOUR OF COMPREHENSIVE CLASSES OF MEANS UNDER EQUAL INCREMENTS OF THEIR VARIABLES

J. Aczél, L. Losonczi and Zs. Páles

Abstract. In [HN] intricate separate proofs are given for certain properties of the arithmetic, the geometric, the harmonic and the quadratic mean values. We give here simpler elementary proofs of these properties for wide classes of means containing, for instance, all root-mean-powers.

We denote vectors with ν positive components by Latin letters, scalars (positive real numbers) by Greek letters. If a scalar τ stands in place of a vector then the vector $(\tau,\dots,\tau)$ is meant. For instance

$$\sigma x + \tau = \sigma(\xi_1,\dots,\xi_\nu) + \tau = (\sigma\xi_1+\tau,\dots,\sigma\xi_\nu+\tau) .$$

PROPOSITION 1. *If* $\mu\colon \mathbb{R}_+^\nu \to \mathbb{R}_+$ *is a mean value, that is,*

$$\mu(\tau) = \mu(\tau,\dots,\tau) = \tau \tag{1}$$

and satisfies the Minkowski inequality

$$\mu(x+y) \le \mu(x) + \mu(y) , \tag{2}$$

then

$$\mu(x+\tau) \le \mu(x) + \tau \tag{3}$$

and

$$\tau \mapsto \mu(x+\tau) - \tau \quad \textit{decreases (in the wider sense)} . \tag{4}$$

If the inequality sign is reversed in (2) *then it reverses also in* (3) *and decrease is replaced by increase in* (4).

Proof. The inequality (3) follows from (1) and (2) by substituting $y = \tau = (\tau,\dots,\tau)$ into (2). - Moreover, if $\rho < \sigma$, then $\tau = \sigma - \rho > 0$, and from (3) we get

$$\mu(x+\sigma)-\sigma = \mu(x+\rho+(\sigma-\rho))-(\sigma-\rho)-\rho \le \mu(x+\rho)+(\sigma-\rho)-(\sigma-\rho)-\rho ,$$

that is (4). - The proof goes in the same way if the inequalities are reversed.

Of course (see e.g. [HLP]) root-mean-powers of degree ≥ 1 satisfy (2) and thus (3) and (4), while those of degree ≤ 1 satisfy the reverse inequalities. All satisfy (1).

PROPOSITION 2. *If* $\mu: \mathbb{R}_+^\nu \to \mathbb{R}_+$ *satisfies* (1) *and*

$$\mu(\tau x) = \tau\mu(x) , \tag{5}$$

furthermore μ *is differentiable at* $1 = (1,\dots,1)$ *and*

$$\frac{\partial\mu}{\partial\xi_\kappa}(1) = \pi_\kappa \quad (\kappa = 1,\dots,\nu) , \tag{6}$$

that is, $p = (\pi_1,\dots,\pi_\nu)$ *is the gradient of* μ *at* 1, *then*

$$\lim_{\tau\to\infty} [\mu(x+\tau)-\tau] = (p,x) = \sum_{\kappa=1}^{\nu} \pi_\kappa\xi_\kappa . \tag{7}$$

Proof. From (5) with $\sigma = 1/\tau$ and from (1) and (6) we get, since μ is differentiable at 1,

$$\lim_{\tau\to\infty} [\mu(x+\tau)-\tau] = \lim_{\tau\to\infty} \tau[\mu\left(\frac{1}{\tau} x + 1\right)-1] = \lim_{\sigma\to 0} \frac{\mu(1+\sigma x)-\mu(1)}{\sigma}$$

$$= \sum_{\kappa=1}^{\nu} \frac{\partial\mu}{\partial\xi_\kappa}(1)\xi_\kappa = \sum_{\kappa=1}^{\nu} \pi_\kappa\xi_\kappa .$$

Of course, root-mean-powers satisfy (1), (5), and are differentiable at 1. - In particular, for *weighted* root-mean-powers $\mu(x) = \mu(\xi_1,\dots,\xi_\kappa) = \left(\sum_{\kappa=1}^{\nu} \pi_\kappa\xi_\kappa^\rho\right)^{1/\rho}$ we have $\frac{\partial\mu}{\partial\xi_\kappa}(1) = \pi_\kappa$. - In any case, from (1) and (6), $\sum_{\kappa=1}^{\nu} \pi_\kappa = 1$ and, since μ maps $\mathbb{R}_+^\nu$ into $\mathbb{R}_+$, $\pi_\kappa > 0$ $(\kappa = 1,\dots,\nu)$. So *the right hand side of* (7) *is a weighted arithmetic mean*. *If* μ, *in addition to satisfying* (1) *and* (5), *is* symmetric and differentiable at 1 (for instance the symmetric root-mean-power), then $\pi_\kappa = \frac{1}{\nu}$ $(\kappa = 1,\dots,\nu)$ and the right hand side of (7) is the symmetric arithmetic mean $\frac{1}{\nu}\sum_{\kappa=1}^{\nu} a_\kappa$.

REFERENCES

[HN] L. Hoehn and I. Niven, Averages on the move. Math. Mag. 58 (1985), 151-156.

[HLP] G.H. Hardy, J.E. Littlewood and G. Pólya, Inequalities. Cambridge University Press, 1952.

J. Aczél, Department of Pure Mathematics, University of Waterloo, Waterloo, Ontario, Canada, N2L 3G1

L. Losonczi and Zs. Páles, Institute of Mathematics, L. Kossuth University, H-4010 Debrecen, Hungary

International Series of
Numerical Mathematics, Vol. 80

A PROBLEM CONCERNING SOME METRICS

Claudi Alsina

Let T be a continuous t-norm, i.e., T is a continuous two-place function from $[0,1]\times[0,1]$ into $[0,1]$ such that for all x, y, z in $[0,1]$ we have:

(i) $T(x,1)=x$;

(ii) $T(x,y)\le T(x,z)$ whenever $y\le z$;

(iii) $T(x,y)=T(y,x)$;

(iv) $T(x,T(y,z))=T(T(x,y),z)$.

Define d_T: $[0,1]\times[0,1]\to[0,1]$ by

$$d_T(x,y) = \begin{cases} 1-T(1-x,1-y)-T(x,y)\ , & \text{if } x\neq y\ , \\ 0\ , & \text{if } x=y\ . \end{cases}$$

In [1] it was proved that if T satisfies the Lipschitz condition

$$(*) \qquad T(c,a)-T(b,a)\le c-b$$

for all a, b, c in $[0,1]$ with $b\le c$, then d_T is a metric. If d_T is a metric and T is a continuous t-norm, does T satisfy $(*)$?

Recently Darsow and Frank have found that the answer is negative if T is not continuous or not associative, but for continuous t-norms the general answer is not known.

REFERENCES

1. C. Alsina, On some metrics induced by copulas. In: W. Walter, General Inequalities 4, p.397. Birkhäuser Verlag, Basel - Boston - Stuttgart, 1984.

Claudi Alsina, Departament de Matemàtiques i Estadîstica (ETSAB), Universitat Politècnica de Catalunya, Diagonal 649, 08028 Barcelona, Spain

International Series of
Numerical Mathematics, Vol. 80

THREE PROBLEMS

Wolfgang Eichhorn

Let us call a sequence of functions

$$I^n: \mathbb{R}_+^n \to \mathbb{R}_+^n , \qquad n = 2,3,\ldots ,$$

an <u>inequality measure</u>, if it satisfies <u>strict Schur convexity</u>, i.e.,

(1) $I^n(B\underline{x}) < I^n(\underline{x})$ for all $\underline{x} \in \mathbb{R}_+^n$, $\underline{x} \neq (a,\ldots,a)$ and for all bistochastic matrices B such that $B\underline{x}$ is not a permutation of the components of $\underline{x}$,

$I^n(B\underline{x}) = I^n(\underline{x})$ otherwise, and

(2) $I^n(a,\ldots,a) = 0$ for all $a \in \mathbb{R}_+$ (<u>normalization</u>), and

(3) $I^n(x_1,\ldots,x_n) \leq I^{n+1}(x_1,\ldots,x_n,0)$ for all $\underline{x} \in \mathbb{R}_+^n$ (<u>extension</u>).

Note that (1) is equivalent to I^n being <u>symmetric</u> and satisfying the <u>strict principle of progressive transfers</u>. Note further that (1) implies the <u>mean value property</u> of I^n.

Desiderata for inequality measures are, among others ($\underline{1}$ denotes the vector $(1,1,\ldots,1)$):

(4) quasiconvexity of I^n;

(5) homogeneity of degree zero: $I^n(\lambda\underline{x}) = I^n(\underline{x})$ for all $\underline{x},\lambda \in \mathbb{R}_+$, $\lambda > 0$;

(6) leftists' property: $I^n(\underline{x}+\rho\underline{1}) = I^n(\underline{x})$ for all $\underline{x},\rho \in \mathbb{R}$ such that $\underline{x}+\rho\underline{1} \in \mathbb{R}_+$;

(7) $I^n(\underline{x}) = I^n(\underline{x}+\tau(\mu\underline{x}+(1-\mu)\underline{1}))$ for all $\underline{x},\tau \in \mathbb{R}$ such that $\underline{x}+\tau(\mu\underline{x}+(1-\mu)\underline{1}) \in \mathbb{R}_+$ and for a fixed $\mu \in (0,1)$;

(8) proportionality of type A:

$$I^{n\cdot m}\Big(\underbrace{\frac{x_1}{m},\dots,\frac{x_1}{m}}_{m\text{ times}},\dots,\underbrace{\frac{x_n}{m},\dots,\frac{x_n}{m}}_{m\text{ times}}\Big) = I^n(x_1,\dots,x_n) \quad \text{for all } \underline{x}\in\mathbb{R}^n_+,\ m\in\mathbb{N};$$

(9) proportionality of type B:

$$I^{n\cdot m}(\underbrace{x_1,\dots,x_1}_{m\text{ times}},\dots,\underbrace{x_n,\dots,x_n}_{m\text{ times}}) = I^n(x_1,\dots,x_n) \quad \text{for all } \underline{x}\in\mathbb{R}^n_+,\ m\in\mathbb{N};$$

(10) continuity of I^n;

(11) boundedness of I^n.

Problems:

(1) Determine the consistent subsets of properties (1) to (11).

(2) Determine the independent subsets of properties (1) to (11).

(3) Characterize well-known inequality measures (for instance, the Gini measure or the variance) by subsets of (1) to (11) together with one or more further properties.

Remarks:

(a) Quasiconvexity and symmetry imply Schur convexity. The converse is not true.

(b) If I^n is Schur convex and homogeneous of degree zero, then it is bounded.

(c) Symmetry and properties (2), (5), (6) are consistent for all $n \geq 3$. Properties (1), (2), (5), (6) are inconsistent.

(d) All nonconstant functions I^n satisfying properties (5), (6) are discontinuous at $\underline{x} = (a,\dots,a)$, $a\in\mathbb{R}_+$.

(e) The sets of properties {(1),(2),(3),(5),(10)}, {(1),(2),(3),(6),(10)} and {(1),(2),(3),(7),(10)} are consistent and independent.

Wolfgang Eichhorn, Institut für Wirtschaftstheorie und Operations Research, Universität Karlsruhe, D-7500 Karlsruhe, West Germany

International Series of
Numerical Mathematics, Vol. 80

PROBLEMS ON LANDAU'S INEQUALITY

M.K. Kwong and A. Zettl

Let $1 \le p \le \infty$, $J = \mathbb{R} = (-\infty,\infty)$ or $J = \mathbb{R}^+ = (0,\infty)$, $M = Z = \{\ldots,-2,-1,0,1,2,\ldots\}$ or $M = \mathbb{N} = \{0,1,2,\ldots\}$. Let $W^p(J) = \{y \in L^p(J)$: y' is absolutely continuous on all compact subintervals of J and $y'' \in L^p(J)\}$. It is known that there exist positive numbers K and C such that the following inequalities hold:

(1) $$\|y'\|_p^2 \le K\|y\|_p\|y''\|_p \quad \text{for all } y \text{ in } W^p(J) \quad ,$$

(2) $$\|\Delta x\|_p^2 \le C\|x\|_p\|\Delta^2 x\|_p \quad \text{for all } x \text{ in } \ell^p(M) \quad .$$

Let $K = K(p,J)$ and $C = C(p,M)$ be the smallest, i.e. best, constants in (1) and (2), respectively. We pose the following problems.

<u>Problem 1.</u> It is known that $K(\infty,\mathbb{R}) = C(\infty,Z) = 2$, $K(\infty,\mathbb{R}^+) = C(\infty,\mathbb{N}) = 4$, $K(2,\mathbb{R}) = 1 = C(2,Z)$, $K(2,\mathbb{R}^+) = C(2,\mathbb{N}) = 2$, $K(1,\mathbb{R}) = C(1,Z) = 2$, $K(1,\mathbb{R}^+) = C(1,\mathbb{R}^+) = 5/2$.

Find $K(p,J)$ and $C(p,M)$ for $1 < p < \infty$, $p \neq 2$, $J = \mathbb{R}$ or $J = \mathbb{R}^+$, $M = Z$ or $M = \mathbb{N}$.

<u>Problem 2.</u> Are $C(p,Z) = K(p,\mathbb{R})$, $C(p,\mathbb{N}) = K(p,\mathbb{R}^+)$, $1 \le p \le \infty$?

<u>Problem 3.</u> Are there extremals of $C(p,M)$ for $1 < p < \infty$, $p \neq 2$, $M = Z$ or $M = \mathbb{N}$?

An extremal of $C(p,M)$ is a nonzero sequence x in $\ell^p(M)$ for which equality holds in (2). (It is known that there exist extremals for $C(\infty,Z)$, but not for $C(\infty,\mathbb{N})$, $C(2,Z)$, $C(2,\mathbb{N})$, $C(1,Z)$ and $C(1,\mathbb{N})$. The nonexistence of extremals for $C(\infty,\mathbb{N})$, $C(1,\mathbb{N})$ and $C(1,Z)$ has been established recently by Kwong and Zettl and is not yet published.)

Problem 4. For $\frac{1}{p}+\frac{1}{q}=1$ is $K(p,\mathbb{R})=K(q,\mathbb{R})$ and $C(p,Z)=C(q,Z)$?

Problem 5. Are $K(p,J)$ and $C(p,M)$ decreasing on $[1,2]$ and increasing on $[2,\infty)$ for $J=\mathbb{R}$ or $J=\mathbb{R}^+$ and $M=Z$ or $M=\mathbb{N}$?

Problem 6. Are the extremals for $K(p,\mathbb{R}^+)$, $1<p<\infty$, essentially unique? Oscillatory? By essentially unique we mean that if y is an extremal then all other extremals are of the form $ay(bt)$, where a and b are constants. (For $p=2$ it is known that the extremals are essentially unique and oscillatory.)

Problem 7. For $M=Z$ and $M=\mathbb{N}$, $C(p,M)$ is a continuous function of p, $1\le p\le\infty$.

Problem 8. Find "good" upper bounds for $K(p,\mathbb{R})$ when $1<p<2$ and for $K(p,\mathbb{R}^+)$, $1<p<\infty$.

Problem 9. One extension of Landau's inequality is the following:

$$\|y'\|_\infty^2 \le K(a,J)\|y\|_\infty^{1-a}\|y''|y|^a\|_\infty$$

with $K(a,J)=\frac{2}{1-a}$ when $J=\mathbb{R}$ and $K(a,J)=\frac{4}{1-a}$ when $J=\mathbb{R}^+$, both with $0\le a<1$.

(i) Is the constant $K(a,J)$ best possible for $J=\mathbb{R}$ or $\mathbb{R}^+$ and $0<a<1$ as it is known to be when $a=0$?

(ii) Is there a similar inequality for L^p norms (with some appropriate constant), $1\le p\le\infty$?

Problem 10. Characterize the class of weight functions w for which inequality (1) holds for some finite constant K when the norm $\|\cdot\|_p$ is replaced by the weighted norm $\|\cdot\|_{p,w}$

$$\|f\|_{p,w} = \int_J |f(t)|^p w(t)dt\,, \qquad 1\le p<\infty\,,$$

and $\|f\|_{\infty,w}=$ ess. sup $|f(t)|w(t)$, $t\in J$.

For what class of weight functions w is the best constant K the same as for $w\equiv 1$?

Note added in June of 1986:

1. We have recently shown that $C(p,\mathbb{Z}) > K(p,\mathbb{R})$ for some values of p in the range $3 < p < \infty$.

Man Kam Kwong and Anton Zettl, Department of Mathematical Sciences, Northern Illinois University, DeKalb, IL 60015, USA

International Series of
Numerical Mathematics, Vol. 80

A GENERALIZATION OF YOUNG'S INEQUALITY

Zsolt Páles

A function $\phi: [0,\infty) \to [0,\infty)$ is said to be a Young function if
(i) ϕ is increasing and right continuous on $[0,\infty)$,
(ii) $\lim\limits_{x\to\infty} \phi(x) = \infty$.

The right inverse $\phi^{(-1)}$ of a Young function ϕ is defined by

$$\phi^{(-1)}(y) = \begin{cases} 0 & \text{if } y \in [0,\phi(0)) , \\ \sup \{x \geq 0 \mid \phi(x) \leq y\} & \text{if } y \in [\phi(0),\infty) . \end{cases}$$

It is easy to see that $\phi^{(-1)}$ is also a Young function.

The classical inequality of Young states that

$$xy \leq \int_0^x \phi(t)dt + \int_0^y \phi^{(-1)}(s)ds \tag{1}$$

for any Young function ϕ and any $x,y \geq 0$ (see Losonczi [1]).

The usual geometrical proof of this inequality goes as follows: Let $x,y \geq 0$ be fixed and let

$$H := \{(t,s) \mid 0 \leq t \leq x,\ 0 \leq s \leq y\} ,$$

$$H' := \{(t,s) \mid 0 \leq t \leq x,\ 0 \leq s \leq \phi(t)\} ,$$

$$H'' := \{(t,s) \mid 0 \leq s \leq y,\ 0 \leq t \leq \phi^{(-1)}(s)\} .$$

Then it is easy to see that

$$H \subseteq H' \cup H'' \tag{2}$$

and that each of these sets is Lebesgue mearurable. Thus it follows from (2) that

$$m(H) \leq m(H') + m(H'') , \tag{3}$$

where m denotes the planar Lebesgue measure. On calculating m(H),

$m(H')$, $m(H'')$ we obtain (1).

Our main observation is that (3) holds, not only for planar Lebesgue measure, but for any measure m that is absolutely continuous with respect to Lebesgue measure.

THEOREM. *Let* $A: [0,\infty) \times [0,\infty) \to \mathbb{R}$ *be a* C^2 *function and assume that*

$$(4) \qquad \partial_1 \partial_2 A(s,t) = \frac{\partial}{\partial s}\frac{\partial}{\partial t} A(s,t) \geq 0$$

for all $s,t \geq 0$. *Then*

$$(5) \qquad A(x,y) \leq A(0,0) + \int_0^x \partial_1 A(t,\phi(t))dt + \int_0^y \partial_2 A(\phi^{(-1)}(s),s)ds$$

for any Young function ϕ *and for all* $x,y \geq 0$.

Proof. Suppose $D \subset \mathbb{R}^2$ is measurable. Then the function m, defined by

$$m(D) = \iint_D \partial_1 \partial_2 A(s,t)\, ds\, dt$$

is clearly a measure on $[0,\infty)^2$.

Now let $x,y \geq 0$ be fixed and define the sets H, H' and H" as above. An easy calculation yields:

$$m(H) = A(x,y) - A(x,0) - A(0,y) + A(0,0) \quad ,$$

$$m(H') = \int_0^x \partial_1 A(t,\phi(t))dt - A(x,0) + A(0,0) \; ,$$

$$m(H'') = \int_0^y \partial_2 A(\phi^{(-1)}(s),s)ds - A(0,y) + A(0,0) \; .$$

Applying (3) we obtain (5), which was to be proved.

If we let $A(x,y) = xy$, then (4) is obviously satisfied and (5) reduces to (1).

REFERENCES

1. L. Losonczi, Inequalities of Young-type. Monatsh. Math. 27 (1984), 125-132.

Zsolt Páles, Department of Mathematics, Kossuth Lajos University, Debrecen, 4010 Pf. 12, Hungary

International Series of
Numerical Mathematics, Vol. 80
©1987 Birkhäuser Verlag Basel

ON PACKINGS OF CONGRUENT BALLS IN THE UNIT BALL

Jürg Rätz

Let $(X,\langle\cdot,\cdot\rangle)$ be an infinite-dimensional real or complex inner product space. If $I \neq \emptyset$, $r \in \mathbb{R}_+ := (0,\infty)$, $u_j \in X$ (for all $j \in I$), $u_j \neq u_k$ for $j \neq k$, then the family $(B_r(u_j)\colon j \in I)$ of open r-balls with centers u_j is said to be packable if

(i) $B_r(u_j) \cap B_r(u_k) = \emptyset$ (for $j,k \in I$; $j \neq k$) ,

(ii) $B_r(u_j) \subset B_1(0)$ (for $j \in I$) .

Furthermore we define

$s_n = \sup\{r \in \mathbb{R}_+\colon$ There exist $u_1,\ldots,u_n \in X$ such that $(B_r(u_j)\colon j \in \{1,\ldots,n\})$ is packable$\}$ ($n \in \mathbb{N}$, $n \geq 2$),

$s_\infty = \sup\{r \in \mathbb{R}_+\colon$ There exist $u_j \in X$ (for $j \in I$), where I is infinite, such that $(B_r(u_j)\colon j \in I)$ is packable$\}$.

The following results are known:

(1) $s_n = \dfrac{\sqrt{n}}{\sqrt{2(n-1)} + \sqrt{n}}$ $(n \geq 2)$ (cf. [1], p.78, for an equivalent formulation) ,

(2) $s_\infty = \sqrt{2} - 1$ (cf. [2]) .

It is the purpose of this remark to connect these two results in several respects.

First of all, for card $I = n \in \mathbb{N}$, it does not matter whether we consider the packing problem in X or in its finite-dimensional subspace $M := \operatorname{lin}\{u_1,\ldots,u_n\}$ as the sets $M \cap B_r(u_1),\ldots,M \cap B_r(u_n)$ completely reflect the packing properties of $B_r(u_1),\ldots,B_r(u_n)$.

Secondly, (1) and (2) imply that

(3) $\qquad s_n > s_{n+1}$ $(n \in \mathbb{N},\ n \geq 2)$ and $s_n \to s_\infty \quad (n \to \infty)$.

Finally, (1) and (2) may be based on the same type of auxiliary inequalities:

(1') $\qquad n \in \mathbb{N},\ n \geq 2,\ u_j \in X,\ \|u_j\| \leq a \quad (j = 1,\dots,n)$

$$\Longrightarrow \quad \max\{\mathrm{Re}\, \langle u_j, u_k\rangle :\ j,k \in \{1,\dots,n\},\ j \neq k\} \geq -\frac{a^2}{n-1}\ ,$$

(2') $\qquad I$ infinite, $u_j \in X,\ \|u_j\| \leq a \quad (j \in I)$

$$\Longrightarrow \quad \sup\{\mathrm{Re}\, \langle u_j, u_k\rangle :\ j,k \in I,\ j \neq k\} \geq 0\ .$$

These inequalities are sharp in the following sense: Equality in (1') holds for every regular n-star of vectors in X, i.e., for every set of vertices of a regular (n-1)-simplex centered at 0. Equality in (2') occurs for orthogonal sets $\{u_j :\ j \in I\}$.

REFERENCES

1. T. Bonnesen und W. Fenchel, Theorie der konvexen Körper. Chelsea Publ. Comp., New York, 1948.

2. R.A. Rankin, On packings of spheres in Hilbert space. Proc. Glasgow Math. Assoc. 2 (1955), 145-146.

Jürg Rätz, Universität Bern, Mathematisches Institut, Sidlerstr.5, CH-3012 Bern, Switzerland

INDEX